Computer-aided Design and Diagnosis Methods for Biomedical Applications

Computer-aided Design and Diagnosis Methods for Biomedical Applications

Edited by

Varun Bajaj and G.R. Sinha

CRC Press
Taylor & Francis Group
Boca Raton London New York

CRC Press is an imprint of the
Taylor & Francis Group, an **informa** business

First edition published 2021 by
CRC Press
6000 Broken Sound Parkway NW, Suite 300, Boca Raton, FL 33487-2742

and by
CRC Press
2 Park Square, Milton Park, Abingdon, Oxon, OX14 4RN

Library of Congress Cataloging-in-Publication Data

Names: Bajaj, Varun, editor. | Sinha, G. R., editor.
Title: Computer-aided design and diagnosis methods for biomedical
applications / [edited by] Varun Bajaj and G.R. Sinha.
Description: First edition. | Boca Raton : CRC Press, 2021. | Includes
index.
Identifiers: LCCN 2020049437 (print) | LCCN 2020049438 (ebook) | ISBN
9780367638832 (hardback) | ISBN 9781003121152 (ebook)
Subjects: LCSH: Biomedical materials. | Computer-aided design. | Biomedical
engineering.
Classification: LCC R857.M3 C628 2021 (print) | LCC R857.M3 (ebook) | DDC
610.28--dc23
LC record available at https://lccn.loc.gov/2020049437
LC ebook record available at https://lccn.loc.gov/2020049438

ISBN: 978-0-367-63883-2 (hbk)
ISBN: 978-0-367-63884-9 (pbk)
ISBN: 978-1-003-12115-2 (ebk)

Typeset in Times
by Deanta Global Publishing Services, Chennai, India

Dedicated to my father, the late Mahendra Bajaj, and family members.

Varun Bajaj

Dedicated to my late grandparents, my teachers and the revered Swami Vivekananda

G.R. Sinha

Contents

Preface

Computer-aided Design (CAD) and Diagnosis plays a key role in improving biomedical systems for various applications, such as prosthesis design, blood flow analysis, surgical implant design, and computer-assisted surgery in their creation, modification, analysis, or optimization. It can also use the design, analysis, modeling, and manipulation of the modalities of biomedicine, such as EEG, ECG, EMG, PCG, COP, EOG, MRI, and FMRI, for automatic identification and classification for the diagnosis of any disorder or physiological state. CAD tools and technologies developed for mechanical designs, but they have been successfully applied for geometric modeling, animating, visualizing, and analyzing the behavior of anatomical structures, including human skeletal and vascular systems. This book highlights techniques and methods for improving healthcare systems, including the detection, identification, prediction, analysis, and classification of disease, management of chronic conditions, and the delivery of health services.

The book emphasizes the real challenges for CAD with a variety of applications for analysis, classification, and identification of different states for the improvement of biomedical systems. Each chapter starts with an introduction, before assessing certain principles of applying CAD for the identification and improvement of biomedical systems. In addition, the chapters can be read independently by research scholars, graduate students, faculty members, and R&D engineers who wish to explore research in the fields of computer sciences, electronics, medical sciences, or biomedical engineering. This book also provides information about the identification of diseases, such as Parkinson's disease, epileptic seizures, physical action, and cancer detection.

The chapters are presented as follows: Chapter 1 discusses the identification of emotions using recurrence quantification analysis (RQA) using healthy controls and patients suffering from Parkinson's disease. Chapter 2 investigates the performance of DEn and one of its variant fluctuation based dispersion entropies (FDEn) computed from the wavelet sub-bands of EEG sleep recordings. A random forest classifier is employed for classification with computed entropies as features. The performance of the algorithm is further compared using the DEn and FDEn separately as well as in combination for both entropies. In Chapter 3, a novel technique for the detection of epileptic electroencephalogram (EEG) signals is proposed using sequential visibility graph (VG) motifs. VG is a technique which converts time-domain signals into an undirected and binary graph while retaining their temporal characteristics. However, one limitation of a VG based approach is that the variations present in a time series are captured on a global scale instead of providing insight into the local fluctuations present in a signal. Chapter 4 describes the surface electromyography (sEMG) technique that is used to record and evaluate electrical signals generated by muscles during contractions. The analysis of the EMG signal of a particular yoga pose helps to improve posture, muscle activity, and also suggests the right way of doing the poses. In Chapter 5, the synthetic minority over-sampling technique (SMOTE) is discussed to solve imbalanced data problems along with ensemble machine learning methods to improve the prediction rate of early Parkinson's disease and the detection rate of scan without evidence of dopaminergic deficit (SWEDD) subjects. Chapter 6 highlights computer-aided diagnosis techniques in modern diagnostics systems that could help healthcare experts to understand different aspects of certain tests: biochemical or imaging and in decision-making. Nowadays, in healthcare systems, CAD is considered as a second opinion tool in the interpretation of medical images in cancer diagnostics. Early detection and effective treatment not only help to increase treatment options but also increase the chances of survival of the patient. In Chapter 7, computer vision methods have been widely preferred to detect COVID-19 by using chest X-ray and computed tomography (CT) images. Different widely-known CNNs, such as AlexNet, VGG-16, VGG-19, SqueezeNet, GoogleNet, MobileNet-V2, ResNet-18, ResNet-50, ResNet-101, and Xception, were used to identify COVID-19 infections. Chapter 8 summarizes the basics of MRI sequencing and documents some of the guidelines reported in the literature. It aims to present the most important approaches being used for abnormality detection in brain MRI, especially in recent years. This study will be a valuable reference for researchers dealing with suspicious region diagnosis in brain MRI. Chapter 9 aims to forecast a classification model for forecasting different types of ailments by using Cleveland and Statlog heart dataset. Prediction is implemented using six techniques, the naïve Bayes, k-NN, random forest, logistic regression, SVM, and decision tree, on different datasets along with a comparative analysis of the classification model for better accuracy and results. Chapter 10 relates to strategies that promote access to care, the functionality of doctor appointments, and the development of supporting information systems. The chapter proposes that, in addition to general feedback systems in healthcare, feedback systems focusing specifically on aspects of care intensity should be developed to better reflect the patient's view of successful diagnostics and care. In Chapter 11, nuclear protein Ki67 is discussed as the most common proliferation marker used to measure cell proliferation activity. Recent studies have shown promising correlation results between the Ki67 labeling index with prognosis and predicting the recurrence of a disease. Manual counting based on a printed image is a well-known technique as it delivers results with high accuracy. However, the manual annotations and counting processes of this technique are highly subjective, as they are based on a pathologist's experience. Chapter 12 explores data mining applications, challenges, and future directions in health care. In particular, it discusses data mining and its applications within healthcare in major areas. This hospital-based survey also explores utilities of various data mining techniques such as association rule, clustering, and classification in the healthcare domain. This chapter also defines the cancer site and the morphology pattern among various cancer patients with the help of the above-defined data mining techniques. In Chapter 13, three ensemble meta classifiers "AdaBoostM1, bagging, and random subspace" have been used with a Pearson correlation features selection technique. The reduced error pruning decision tree generates a candidate subsample for pruning and formatting in a leaf node, but reduced error pruning has a drawback in that it cannot manage error complexity and accuracy as an ensemble model. So the reduced error pruning tree is tested in three different environments, AdaBosstM1, bagging, and a random subsample. Chapter 14 presents the proper design and reliability assessment of the COVID-19 diagnosis systems (e.g., proper feature selection, classification, and performance assessment). Also, advanced statistical methods (e.g., multistate and competing risk models) are required to avoid the risk of bias in prognosis systems. Moreover, many studies may be too small and poorly designed to be helpful, merely adding to the COVID-19 noise. Chapter 15 proposes a cost-effective and remote screening method to investigate the possible detection of skin cancer. It highlights a sub-surface visualization methodology known as linear frequency modulated thermal wave imaging to diagnose different stages of skin cancer.

Acknowledgments

Dr Bajaj expresses his heartfelt appreciation to his mother Prabha, wife Anuja, and daughter Avadhi, for their wonderful support and encouragement throughout the completion of this important book on *Computer-aided Diagnosis and Design Methods for Biomedical Applications*. His deepest gratitude goes to his mother-in-law and father-in-law for their constant motivation. This book is the outcome of much sincere effort and would not have been completed without the great support of family. He also give thanks to Prof Sanjeev Jain, Director of PDPM IIITDM Jabalpur, for his support and encouragement.

Dr Sinha expresses his gratitude and gives sincere thanks to his family members, wife Shubhra, daughter Samprati, parents, and teachers.

We would like to thank all our friends, well-wishers, and all those who keep us motivated in doing more and more; better and better. We sincerely thank all contributors for writing relevant theoretical background and real-time applications of CAD used for biomedical applications.

We express our humble thanks to Dr Gagandeep Singh, publisher (Engineering), and all the editorial staff at CRC Press for their generous support, necessary help, appreciation, and quick responses. We also wish to thank CRC Press for giving us this opportunity to contribute to a relevant topic with a reputed publisher. Finally, we want to thank everyone, in one way or another, who helped us with editing this book.

Dr Bajaj would especially like to thank his family, who encouraged him throughout the editing book of this book. This book is heartily dedicated to his father, who took the journey to heaven before the completion of this book.

Last but not least we would also like to thank God for showering us his blessings and strength to do this type of novel and quality work.

Varun Bajaj
G.R. Sinha

Editors' Biographies

Varun Bajaj (Ph.D., MIEEE 16 SMIEEE20) is a faculty member in the discipline of Electronics and Communication Engineering at Indian Institute of Information Technology, Design and Manufacturing (IIITDM) Jabalpur, India, since 2014. He worked as a visiting faculty member at IIITDM Jabalpur from September 2013 to March 2014. He worked as an Assistant Professor at the Department of Electronics and Instrumentation, Shri Vaishnav Institute of Technology and Science, Indore, India, during 2009–2010. He received a B.E. degree in Electronics and Communication Engineering from Rajiv Gandhi Technological University, Bhopal, India in 2006, an M.Tech. degree with honors in Microelectronics and VLSI design from Shri Govindram Seksaria Institute of Technology and Science, Indore, India in 2009. He received his Ph.D. degree in the Discipline of Electrical Engineering at the Indian Institute of Technology Indore, India in 2014.

He is an Associate Editor of *IEEE Sensor Journal* and subject editor-in-chief of *IET Electronics Letters*. He served as a subject editor of *IET Electronics Letters* from November 2018 to June 2020. He is a senior member IEEE, since June 2020, an MIEEE, 2016–2020, and also an active technical reviewer of the leading international journals of IEEE, IET, and Elsevier. He has authored more than 100 research papers in various reputed international journals/conferences like *IEEE Transactions*, Elsevier, Springer, and IOP. He has edited *Modelling and Analysis of Active Biopotential Signals in Healthcare*—Volume 1 and 2, published by IOP books. Currently, he is editing a book for CRC Press (Taylor & Francis Group). The his publications have been cited around 1998 times, he has a h index of 20, and an i10 index of 44 (Google Scholar September 2020). He has guided six (three completed and three in-process) Ph.D. scholars and 6 M.Tech. scholars. He has been a recipient of various reputed national and international awards. His research interests include biomedical signal processing, image processing, time-frequency analysis, and computer-aided medical diagnosis.

G.R. Sinha is an Adjunct Professor at the International Institute of Information Technology Bangalore (IIITB) and currently deputed as Professor at the Myanmar Institute of Information Technology (MIIT) Mandalay Myanmar. He obtained his B.E. (Electronics Engineering) and M.Tech. (Computer Technology) with a Gold Medal from the National Institute of Technology Raipur, India. He received his Ph.D. in Electronics and Telecommunication Engineering from Chhattisgarh Swami Vivekanand Technical University (CSVTU) Bhilai, India. He was a Visiting Professor (Honorary) at the Sri Lanka Technological Campus Colombo for one year 2019–2020. He has published 254 research papers, book chapters, and books at the international and national level, which includes *Biometrics* published by Wiley India, a subsidiary of John Wiley; *Medical Image Processing* published by Prentice Hall of India; and five edited books with IOP, Elsevier, and Springer. He is an active reviewer and editorial member of more than 12 reputed international journals of the IEEE, IOP, Springer, and Elsevier. He has 21 years of teaching and research experience. He has been Dean of Faculty and Executive Council Member of the CSVTU and is currently a member of the Senate of MIIT. Dr Sinha has been delivering ACM lectures across the world as an ACM distinguished speaker in the field of DSP since 2017. Some of his more important assignments include being an expert member for the Vocational Training Programme at Tata Institute of Social Sciences (TISS) for two years (2017–2019); Chhattisgarh Representative of the IEEE MP Sub-Section Executive Council (2016–2019); distinguished speaker in the field of digital image processing at the Computer Society of India (2015). He is the recipient of many awards and recognitions like the TCS Award 2014 for Outstanding Contributions to the Campus Commune of the TCS, the Rajaram Bapu Patil ISTE National Award 2013 for Promising Teacher in Technical Education by the ISTE New Delhi, Emerging Chhattisgarh Award 2013, Engineer of the Year Award 2011, Young Engineer Award 2008, Young Scientist Award 2005, IEI Expert Engineer Award 2007, ISCA Young Scientist Award 2006 Nomination, and the Deshbandhu Merit Scholarship for five years. He served as Distinguished IEEE Lecturer on the IEEE India council, Bombay section. He is a senior member of the IEEE, fellow of the Institute of Engineers India and fellow of the IETE India.

He has delivered more than 50 keynote/invited talks and chaired many technical sessions at international conferences across the world. His special session on "Deep Learning in Biometrics" was included in the IEEE International Conference on Image Processing 2017. He is also a member of many national professional bodies like the ISTE, CSI, ISCA, and IEI. He is a member of various committees of the university and has been vice president of the Computer Society of India for the Bhilai Chapter for two consecutive years. He is a consultant of various skill development initiatives of the NSDC, Govt. of India. He is a regular referee for the Project Grants under DST-EMR scheme and several other schemes of the Govt. of India. He has received some important consultancy support, such as grants and travel support. Dr Sinha has supervised eight Ph.D. scholars, 15 M.Tech. scholars, and is currently supervising one Ph.D. scholar. His research interests include biometrics, cognitive science, medical image processing, computer vision, outcome-based education (OBE) and ICT tools for developing employability skills.

Contributors

Emina Aličković
Department of Electrical Engineering
Linkoping University
Linkoping, Sweden

Waleed Alshuaib
Department of Physiology
Faculty of Medicine
Kuwait University
Kuwait

Vanita Arora
School of Electronics
Indian Institute of Information
    Technology, Una
Himachal Pradesh, India

Varun Bajaj
Department of Electronics and
    Communication
PDPM Indian Institute of Information
    Technology
Design and Manufacturing
Jabalpur, India

Harald Binder
Institute of Medical Biometry and
    Statistics
Faculty of Medicine and Medical
    Center
University of Freiburg
Freiburg, Germany

Rohit Bose
Department of Bio-Engineering
University of Pittsburgh
Pittsburgh, Pennsylvania

Ali K. Bourisly
Department of Physiology
Faculty of Medicine
Kuwait University
Kuwait

Soumya Chatterjee
Department of Electrical Engineering
Techno India University
Kolkata, India

Maja von Cube
Institute of Medical Biometry and
    Statistics
Faculty of Medicine and Medical
    Center
University of Freiburg
Freiburg, Germany

Pankaj Dadheech
Associate Professor
Department of Computer Science &
    Engineering
Swami Keshvanand Institute of
    Technology, Management &
    Gramothan
Rajasthan, India

S. R. Dogiwal
Associate Professor
Department of Information Technology
Swami Keshvanand Institute of
    Technology, Management &
    Gramothan
Rajasthan, India

Geetika Dua
Electronics and Communication
    Engineering
Thapar Institute of Engineering and
    Technology
Punjab, India

Fahmi Akmal Dzulkifli
Faculty of Electronic Engineering
    Technology
Universiti Malaysia Perlis (UniMAP)
Perlis, Malaysia

Enrique Herrera-Viedma
Andalusian Research Institute in
    Data Science and Computational
    Intelligence
University of Granada
Granada, Spain

Hasnan Jaafar
Department of Pathology
School of Medical Sciences
Health Campus Universiti Sains
    Malaysia
Malaysia

Vipin Jain
Associate Professor
Department of Information Technology
Swami Keshvanand Institute of
    Technology, Management &
    Gramothan
Rajasthan, India

Mislav Jordanic
Biomedical Engineering Research
    Centre (CREB)
Automatic Control Department (ESAII)
Universitat Politècnica de Catalunya-
    Barcelona Tech (UPC)
Barcelona, Spain

Sadaf Khademi
Biomedical Engineering Department
Engineering Faculty
University of Isfahan
Isfahan, Iran

Johra Khan
Department of Medical Laboratory
    Sciences
College of Applied Medical Sciences
Majmaah University
Majmaah, Kingdom of Saudi Arabia

Smith K. Khare
Department of Electronics and
    Communication
PDPM Indian Institute of Information
    Technology
Design and Manufacturing
Jabalpur, India

Ayca Kirimtat
Faculty of Informatics and
    Management
Center for Basic and Applied Research
University of Hradec Kralove
Hradec Kralove, Czech Republic

Ondrej Krejcar
Faculty of Informatics and Management
Center for Basic and Applied Research
University of Hradec Kralove
Hradec Kralove, Czech Republic

Kamil Kuca
Faculty of Informatics and Management
Center for Basic and Applied Research
University of Hradec Kralove
Hradec Kralove, Czech Republic

Ankit Kumar
Assistant Professor
Department of Computer Science &
    Engineering
Swami Keshvanand Institute of
    Technology, Management &
    Gramothan
Rajasthan, India

Sandeep Kumar
Associate Professor
Department of Computer Science &
    Engineering
Amity University, Jaipur
Rajasthan, India

Rajani Kumari
Assistant Professor
Department of Computer Science &
    Engineering
JECRC University, Jaipur
Rajasthan, India

Miguel Ángel Mañanas
Biomedical Engineering Research
    Centre (CREB)
Automatic Control Department (ESAII)
Universitat Politècnica de Catalunya-
    Barcelona Tech (UPC)
Barcelona, Spain
and
Biomedical Research Networking
    Center in Bioengineering,
    Biomaterials, and Nanomedicine
    (CIBER-BBN)
Spain

Marjan Mansourian
Epidemiology and Biostatistics
    Department
Health School
Isfahan University of Medical Sciences
Isfahan, Iran
and
Biomedical Engineering Research
    Centre (CREB)
Automatic Control Department (ESAII)
Universitat Politècnica de Catalunya-
    Barcelona Tech (UPC)
Barcelona, Spain

Hamid Reza Marateb
Biomedical Engineering Research
    Centre (CREB)
Automatic Control Department (ESAII)
Universitat Politècnica de Catalunya-
    Barcelona Tech (UPC)
Barcelona, Spain

and
Biomedical Engineering Department
Engineering Faculty
University of Isfahan
Isfahan, Iran

Mohd Yusoff Mashor
Faculty of Electronic Engineering
    Technology
Universiti Malaysia Perlis (UniMAP)
Perlis, Malaysia

Arka Mitra
Electronics and Electrical
    Communication Engineering
IIT Kharagpur
Kharagpur, India

Sudip Modak
Department of Electrical Engineering
Techno India University
Kolkata, India

Ravibabu Mulaveesala
InfraRed Imaging Laboratory (IRIL)
Department of Electrical Engineering
Indian Institute of Technology Ropar
Punjab, India

M. Murugappan
Department of Electronics and
    Communication Engineering
Intelligent Signal Processing Research
    Lab
Kuwait College of Science and
    Technology
Doha, Kuwait

Neelamshobha Nirala
Department of Biomedical Engineering
National Institute of Technology Raipur
Raipur, India

Fatih Ozyurt
Department of Software Engineering
College of Engineering
Firat University
Elazig, Turkey

Saurabh Pal
Department of Computer Applications
VBS Purvanchal University
Uttar Pradesh, India

Linesh Raja
Assistant Professor
Department of Computer Applications
Manipal University Jaipur
Rajasthan, India

Yousef Rasmi
Department of Biochemistry
Faculty of Medicine
Urmia University of Medical Sciences
Urmia, Iran
and
Cellular and Molecular Research Center
Urmia University of Medical Sciences
Urmia, Iran

Marjo Rissanen
Prime Multimedia Ltd.

Sayanjit Singha Roy
Department of Electrical Engineering
Techno India University
Kolkata, India

Ajit Kumar Sahoo
Department of Electronics and
    Communication Engineering
National Institute of Technology,
    Rourkela
Rourkela, India

Padmini Sahu
Department of Biomedical Engineering
National Institute of Technology Raipur
Raipur, India

Sitanshu Sekhar Sahu
Department of Electronics and
    Communication Engineering
Birla Institute of Technology, Mesra
    Ranchi
Jharkhand, India

Kaniska Samanta
Department of Electrical Engineering
Techno India University
Kolkata, India

Ali Selamat
Faculty of Informatics and Management
Center for Basic and Applied Research
University of Hradec Kralove
Hradec Kralove, Czech Republic
and
School of Computing
Faculty of Engineering
Universiti Teknologi Malaysia UTM
Johor, Malaysia
and
Malaysia-Japan International Institute
    of Technology (MJIIT)
Universiti Teknologi Malaysia Jalan
    Sultan Yahya Petra
Kuala Lumpur, Malaysia

Maryam Ahmad Sharifuddin
Department of Pathology
School of Medical Sciences
Health Campus Universiti Sains Malaysia
Malaysia

Rajeev Sharma
School of Electronics Engineering
VIT-AP University
Andhra Pradesh, India

Rishi Raj Sharma
Department of Electronics Engineering
Defense Institute of Advanced
    Technology
Maharatra, India

Bikesh Kumar Singh
Department of Biomedical Engineering
National Institute of Technology Raipur
Raipur, India

Vijander Sing
Assistant Professor
Department of Computer Science &
    Engineering
Manipal University Jaipur
Rajasthan, India

G.R. Sinha
Myanmar Institute of Information
    Technology (MIIT)
Mandalay, Myanmar

Abdulhamit Subasi
Institute of Biomedicine
University of Turku
Turku, Finland

Turker Tuncer
Department of Digital Forensics
    Engineering
College of Technology
Firat University
Elazig, Turkey

Abhay Upadhyay
Department of Electronics and
    Communication Engineering
Institute of Engineering and
    Technology
Bundelkhand University
Utter Pradesh, India

Martin Wolkewitz
Institute of Medical Biometry and
    Statistics
Faculty of Medicine and Medical
    Center
University of Freiburg
Freiburg, Germany

Dhyan Chandra Yadav
Department of Computer Applications
VBS Purvanchal University
Uttar Pradesh, India

Anis Yazidi
Department of Computer Science
Oslo Metropolitan University
Oslo, Norway

1 Electroencephalogram Signals Based Emotion Classification in Parkinson's Disease Using Recurrence Quantification Analysis and Non-Linear Classifiers

*M Murugappan, Smith K. Khare, Waleed Alshuaib,
Ali K Bourisly, Varun Bajaj, and G.R. Sinha*

CONTENTS

1.1   INTRODUCTION

Parkinson's disease (PD) is an acute neurological disease and it is considered a movement disorder. The cause of PD is yet to be discovered; studies suggest that more than 10 million people living around the world suffer from PD [1]. Neurologists and

1

psychologists use clinical assessment tools to diagnose PD and its different stages. These tools utilize a set of questionnaires to assess the level or stage of PD in a clinical environment. The probability of contracting PD increases with age, and it affects males 1.5 times more than females [2,3]. The symptoms are characterized by fatigue, depression, speech problems, anxiety, dementia, motor deficits, and cognitive impairments. The research shows that dysfunction in social cognition appears before motor deficits in PD [4]. The research also suggests that more than 50% of people diagnosed with PD show emotional disfunction [5–7]. There is growing evidence of the cognitive and social impairments associated with the disease, particularly in emotion processing. Moreover, patients suffering from PD are unable to use their facial expressions to express emotions. The research on the assessment of emotions has gained much attention in the last decade. Neuroscientists and psychologists claim to have a good understanding of the emotions of PD patients and their impairments. The impairments include cognitive, expressive, and subjective. Several hypotheses were reported in the literature on emotional impairment and understanding. Thus, medication is needed to improve social behavior, and to detect the emotions, of the patients with PD. The detection of emotions can be accomplished with facial expressions, speech, gestures, and bio-signals. However, the detection of emotions from facial expressions introduces the probability of a false prediction. Moreover, facial expressions, audio, and speech can be altered deliberately and as such the effectiveness of these models is limited [8–10]. Bio-signals, such as electroencephalograms (EEG), electrocardiograms (ECG), electromyograms (EMG), and electrooculograms (EOG) proved to be a promising choice for the detection of emotions. But methods employing ECG, EMG, and EOG are error-prone and are limited in their performance [11, 12]. EEG signals are the most widely used bio-signals in the analysis of various physiological and pathological disorders, like motor imagery, seizures, drowsiness, sleep, schizophrenia, and focal disorders [13–23]. Moreover, EEG signals have got an added advantage due to their non-radioactive, non-invasive, and non-intrusive characteristics.

Researchers have proposed several methods to identify emotions using EEG signals. EEGs are low-frequency signals that are useful for recording the brain's electrical activities, which are analyzed at <60 Hz. In general, the filtering approach is the most popular in EEG signal processing, which is used to extract different frequency bands of information from EEGs for a variety of applications. In [24], researchers filtered the EEG signals into different frequency bands, and t-test analysis was performed by discriminating between different tasks. Analysis of alpha (8–13 Hz), beta (13–30 Hz), delta (0.1–4 Hz), theta (4.1–8 Hz), and gamma (31–60 Hz) bands was carried out in different studies. The power spectral density and energy of these rhythms were studied to isolate emotions using the analysis of variance (ANOVA) statistical test [25–27]. Various statistical moments were extracted from the filtered EEG signals in [28] and they were classified with a decision tree (DT) classifier. In [29], EEG signals were filtered to obtain rhythms. Later, the power spectral densities of different rhythms were studied and classified using a support vector machine (SVM) and k-nearest neighbor (k-NN). Different entropies were explored to differentiate between emotions in [30]. Statistical analysis in the form of a boxplot and

ANOVA test was used to select features and then classify them with the k-NN and probabilistic neural network (PNN) classifiers. In [31, 32], rhythms were extracted from the EEG signals of normal controls, PD right-side-affected, and PD left-side-affected subjects using filtering. The features of power spectral density and higher-order statistics elicited from the rhythms were given to the classifiers. Six types of emotions were classified using k-NN and a SVM.

The utility of recurrence quantification analysis (RQA) was explored to isolate different emotions in patients with Parkinson's disease and a normal control. Features extracted using RQA from alpha, beta, and gamma rhythms were classified with different kernels of an extreme learning machine [33]. The signals obtained by filtering the EEG were used to extract higher-order statistical features. These features were then classified using different machine learning algorithms, such as DT, fuzzy k-NN (Fk-NN), k-NN, naïve Bayes (NB), PNN, and SVM [34]. A combination of filtering, cross-correlation, genetic algorithm, and artificial neural networks was employed for feature extraction, selection, and classification [35]. In [36], partial directed coherence was used to extract features which were classified with a machine learning algorithm. The frequency-domain features extracted using a fast Fourier transform were classified with an NB classifier [37]. Features evaluated using local binary patterns were classified with MB and logistic regression classification techniques [38]. Linear and self-similarity features based on inter-channel similarity, correlation coefficients, and linear predictive coefficients were separated using a SVM [39]. In [40], features extracted using single value decomposition were classified using k-NN. Features were extracted from the sub-bands of the empirical wavelet transform and the empirical wavelet packet transform in [41]. The statistically significant features were classified using k-NN, PNN, and ELM. Features based on a spectral, wavelet, and non-linear dynamic analysis were analyzed. Independent component analysis and a SVM were used to reduce the feature dimensionality and classification [42].

Entropy boundary minimization, S-transform, and Bayesian neural networks were employed in [43]. Coherence, phase synchronization, and correlation synchronization were used to isolate emotions in [44, 45]. These features were classified with a SVM. The behavioral changes and analysis of delta responses were studied using an ANOVA in [46]. The utility of advanced versions of the wavelet transforms, namely the tunable Q wavelet transform, multi-wavelet analysis, and flexible analytic wavelet transform, were used for feature extraction. The features elicited from the sub-bands were classified with a k-NN, least square support vector machine, and ELM [47, 48, 49, 50]. In [51], the two-stage filtering of EEG signals was explored. Noisy intrinsic mode functions and modes were removed by filtering. Later, features were extracted and classified with the least square support vector machine. The utility of time-frequency analysis and deep convolutional networks were explored in [52, 53, 54]. The EEG signals were transformed into time-frequency representations using the smoothed pseudo-Wigner-Ville distribution and short-time Fourier transform, and they were analyzed using convolutional neural networks.

In this chapter, an effective model for the accurate classification of emotions from normal control and patients with Parkinson's disease is presented. The EEG signals were separated into alpha, beta, and gamma rhythms using filtering. Later, recurrence quantification analysis was applied to each of the rhythms, and several features were extracted. These features were tested for their discriminability using an ANOVA and boxplot analysis. The selected features were then classified with different machine learning algorithms. The remainder of this chapter is organized as follows: Section 1.2 describes the methodology, the results are covered in Section 1.3, and conclusions are presented in Section 1.4.

1.2 METHODOLOGY

The proposed emotion recognition method based on EEG signals consists of a data-set, filtering, recurrence quantification analysis, feature extraction, statistical analysis, and classification techniques. The steps involved in the proposed method are illustrated in Figure 1.1.

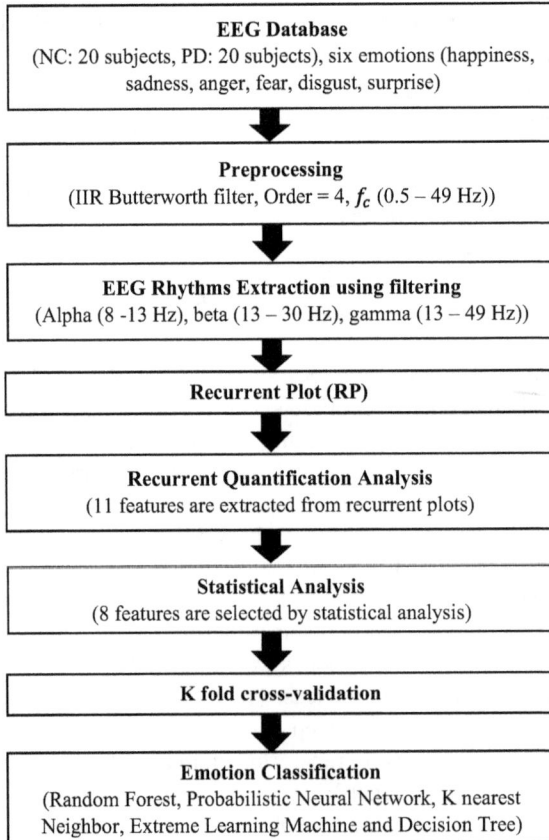

FIGURE 1.1 Flowchart of the proposed emotion recognition.

1.2.1 DATASET

We used a Parkinson's disease dataset recorded at the UKM medical hospital in Kuala Lumpur, Malaysia. The dataset contained the EEG recordings of 20 Parkinson's disease patients who were right-handed (10 males and 10 females) and 20 right-handed healthy controls (9 males and 11 females). All the subjects were age, gender, and education matched, and the average age of the subjects was 58.7 years. The average duration of formal education of the healthy controls and patients with PD was 11.05 ± 3.34 and 10.45 ± 4.86 years, respectively. The average duration of the disease was 5.75 ± 3.52 years. An Emotiv Wireless EEG EPOC Headset operating at a 2.4 GHz band with 14 channels was used to record EEG signals at a sampling rate of 128 samples/sec. All the electrodes were placed according to the international 10–20 electrode system and used to collect the EEG signals of six different emotions, namely happiness (H), sadness (S), surprise (Su), fear (F), anger (A), and disgust (D). The details of the dataset are available in [31, 32, 33, 42, 44, 45]. In this work, 6-sec segments of EEG samples were considered for processing the data for emotion classification. For each subject, six audio-visual clips were shown to the subjects, and this was repeated six times, as shown in Figure 1.2 [31, 33]. The devise arrangement for the EEG recording is shown in Figure 1.3.

1.2.2 PRE-PROCESSING

EEG signals are contaminated with eye movements and power line noise. These artifacts and noises were identified and removed using a fourth-order IIR Butterworth filter with a frequency range of 0.5–49 Hz. Later, the frequency bands were separated into three rhythms, namely alpha (8–13 Hz), beta (13–30 Hz), and gamma (30–49 Hz).

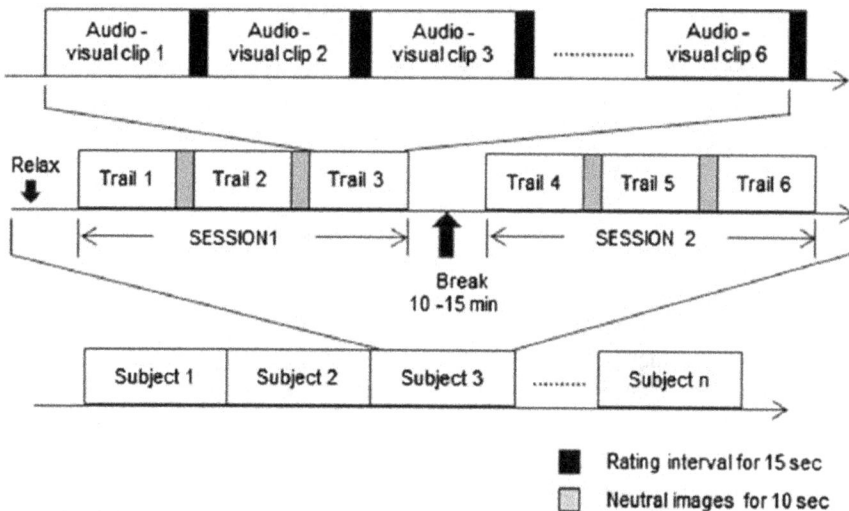

FIGURE 1.2 Schematic representation of experimental protocol [32, 33].

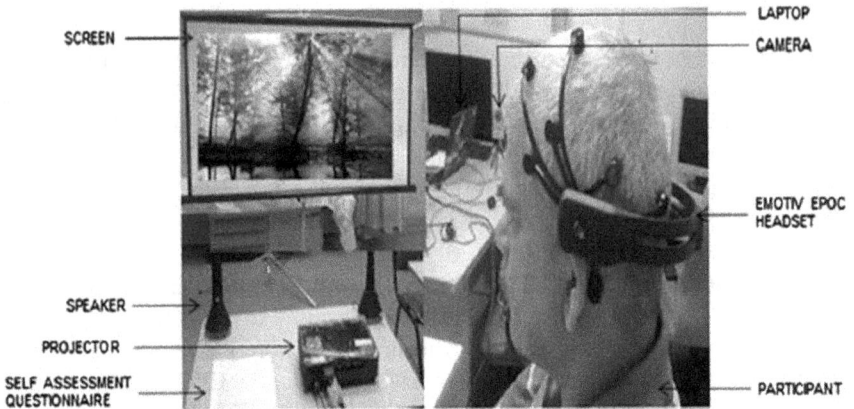

FIGURE 1.3 Experimental setup for emotion assessment using multimodal stimuli.

1.2.3 RECURRENCE QUANTIFICATION ANALYSIS

A recurrence plot (RP) is a two-dimensional graphical representation used to display the recurrence behavior of any given system. A RP is used to identify the periodicities that are not noticeable in the time domain, and to measure the non-stationarity of the time-series signal. Whenever the distance between two states (i and j) falls below the threshold (ε) value then recurrence occurs. Let a_i and a_j be the ith and j^{th} points, respectively, in the orbit of an m dimensional space. Whenever the two points are close to each other, a dot is placed on the location (i,j). This RP plot is symmetric along the axis $i=j$ because if a_i close to a_j, then a_j close to a_i. Finally, the array of dots form a square matrix of dimension $N \times N$. They can be viewed as dots of black and white with a solid diagonal line. The black color dot refers to the recurrence in a system. RQA is used to measure the number of recurrences in a RP and the duration of recurrences in a dynamical system [55, 56]. RQA is used to measure the complexity and non-linearity present in the input signal through a set of measures, namely the recurrence rate (RR), laminarity (LAM), mean diagonal line length (MDL), maximum vertical line length (MV), entropy (ENT), percent determination (%DET), Maximal diagonal line length, trapping time (TT), recurrence time of the first type, recurrence time of the second type, and recurrence probability density entropy (RPDE).

In general, RQA measures are exploited by the physiognomies of the system in the phase space, which gives a clear demonstration of how the phase space of the two systems varies or changes due to the coupling and the recurrence plot analysis. The recurrence plot visualizes the recurrence in the dynamical system and gives a strong idea of the statistical deeds of the orbits. The system trajectory revisits the same state or reoccurrence in time. The recurrence matrix is given by:

$$R(i, j) = H(\varepsilon - \|x(i) - x(j)\|), x(i) \in \mathbb{R}^m, \quad i, j = 1, \ldots, N, \tag{1.1}$$

where, N is the total number of samples in a given state; $x(i)$ = considerable state; ε = threshold distance; and H is the Heaviside function

1.2.4 FEATURES

To get the most representative signals from the samples, the statistical parameters were evaluated. These statistical parameters are called features that represent the distinguishable characteristics of any entity being analyzed. In this method, 11 features were evaluated to test their discrimination abilities. These features were RR, Laminarity (LAM), mean diagonal line length (MDL), maximum vertical line length (MV), entropy (ENT), percent determination (%DET), maximal diagonal line length, TT, recurrence time of the first type, recurrence time of the second type, and RPDE. The details of these features are available in [57, 58]. The mathematical formulation of these features is explained in Table 1.1.

The elementary parameters deliberated for the RQA analysis were the embedding dimension (d), delay (τ), and radius (threshold distance). Embedding dimension states must match to count as recurrence. It was possible to use the dimensions with categorical data, and the window size was firm for theoretical and practical concerns. Embedding dimensions were chosen by considering their false nearest neighbors. In the case of delay, the temporal ordering was preserved. A time delay was considered by taking the first local minimum of the mutual information function. The radius is the distance between the units required in order to count as a recurrent point. The parameters were selected based on the heuristic approach, and the selected values were $d = 1$, and $\tau = 8$ that gave the highest mean emotion detection rate in normal control (NC) and PD. The RQA measures were computed in three different EEG frequency bands, namely alpha (8–13 Hz), beta (13–30 Hz), and γ (30–49 Hz). Among the 11 features, 8 features gave a mean classification rate above 50% in the three frequency ranges considered in this work. The final set of eight features are shown in Table 1.2.

1.2.5 CLASSIFICATION TECHNIQUES

The feature matrix was fed into different machine learning algorithms. In this chapter, the classification of emotions was carried out using a SVM, k-NN, random forest, DT, PNN, and ELM. The details of these classifiers are available in [60, 61, 62]. A SVM works by drawing a hyperplane in a non-linear domain such that it draws the decision boundaries of the classifiers very smoothly. The k-NN assigns the rank to the nearest samples among the selected neighbors and assigns a proper class to them. A decision tree classifies the data using a supervised learning principle of a number of splits called a tree branch and assigns a proper class to the data. PNN assigns a probability to the available data points in the matrix. A class is assigned to the sample with the highest probability of occurrence. A random forest is yet another powerful machine learning algorithm that provides a decision by constructing characteristics during training, and the output is assigned during testing to the class that is matched during training. ELM is a single hidden layer feed-forward

TABLE 1.1
List of Features Considered in RQA for Emotion Classification in HC and PD

No.	Measure	Definition
1	Recurrence Rate (RR)	The percentage of recurrence points in a recurrence plot (RP). It resembles the recurrence point density of RP. $RR = \dfrac{1}{N^2} \sum_{i,j=1}^{N} R_{i,j}$ $R_{i,j}$ refers to the representation of RP. N is the total number of recurrence point in a $R_{i,j}$ phase.
2	Percent Determinism (%D)	The percentage of recurrence points which form diagonal lines. $DET = \dfrac{\sum_{\ell=\ell_{min}}^{N} \ell P(\ell)}{\sum_{\ell=1}^{N} \ell P(\ell)} \times 100$ $P(l)$ is the histogram of the lengths, l of the diagonal lines. Determinism is also predictability measure that indicates the proportion of diagonal structures of recurrence points.
3	Average Diagonal Line Length (MEANLINE)	The average length of the diagonal lines. $L_{avg} = \dfrac{\sum_{\ell=\ell_{min}}^{N} \ell P(\ell)}{\sum_{\ell=\ell_{min}}^{N} P(\ell)}$ The average time is reflected in L_{avg} as having two parts of trajectory locations near. This is also interpreted as the average prediction interval.
4	Maximal Diagonal Line Length (MD)	The length of the longest diagonal line. $L_{max} = \max\left(\{\ell_i ; i = 1,\ldots,N_\ell\}\right)$
5	Entropy of The Diagonal Line Length (ENTR)	The Shannon entropy of the probability distribution of the diagonal line lengths, $p(\ell) = \dfrac{P(\ell)}{\sum_{\ell=\ell_{min}}^{N} P(\ell)}$ $ENTR = -\sum_{\ell=\ell_{min}}^{N} p(\ell) \ln p(\ell)$
6	Laminarity (LAM)	The percentage of recurrence points which form vertical lines. $LAM = \dfrac{\sum_{v=v_{min}}^{N} v P(v)}{\sum_{v=1}^{N} v P(v)}$ $P(v)$ is the histogram of lengths, v of the vertical lines. LAM infers the occurrence of laminar states and does not describe the length of the laminar phases.
7	Trapping Time (TT)	The average length of the vertical lines. $L_{avg} = \dfrac{\sum_{v=v_{min}}^{N} v P(v)}{\sum_{v=v_{min}}^{N} P(v)}$ The average time can be found using TT, in which the structure remains in a certain state or trapped.

(Continued)

TABLE 1.1 (CONTINUED)

List of Features Considered in RQA for Emotion Classification in HC and PD

8	Maximal Vertical Line Length (MV)	The length of the longest vertical line. $V_{max} = \max\left(\{v_i; i = 1, \ldots, N_v\}\right)$
9	Recurrence Time of First Type (T_1)	$T_1 = \left\|\{i, j : \vec{x}_i, \vec{x}_j \in R_i\}\right\|$
10	Recurrence Time of Second Type (T_2)	$T_1 = \left\|\{i, j : \vec{x}_i, \vec{x}_j \in R_i; \vec{x}_{j-1} \notin R_i\}\right\|$
11	Recurrence Probability Density Entropy (RPDE)	$RPDE = \dfrac{-\sum_{i=1}^{T_{max}} p(i)\ln p(i)}{\ln T_{max}}$ where T_{max} is the maximum recurrence time found in the embedded state space, with discrete-valued density $p(i), i = 1, 2, \ldots, M$

TABLE 1.2

Final List of Features Considered in RQA for Emotion Classification in NC and PD

Feature Name	RQA Features
F1	Recurrence Rate (RR)
F2	Percent Determination (%D)
F3	Average Diagonal Line Length (MEANLINE)
F4	Maximum Line Length (ML)
F5	Entropy (ENT)
F6	Laminarity (LAM)
F7	Trapping Time (TT)
F8	Maximal Vertical Line Length (MV)

network that makes the decision with a number of hidden layers and number of hidden neurons.

1.3 RESULTS

This methodology aimed to develop an automatic emotion recognition system uisng normal controls and patients with Parkinson's disease. We used a dataset from the UKM medical hospital in Kuala Lumpur, Malaysia. All the EEG signals were pre-processed to eliminate artifacts and power line noises. Three frequency bands, namely alpha, beta, and gamma, were elicited from the filtered EEG signals. Recurrence quantification analysis was applied to each of these frequency bands separately. The recurrence plots obtained for each emotion for alpha and beta bands are shown in Figures 1.4–1.11. As shown in these figures, each emotion for normal

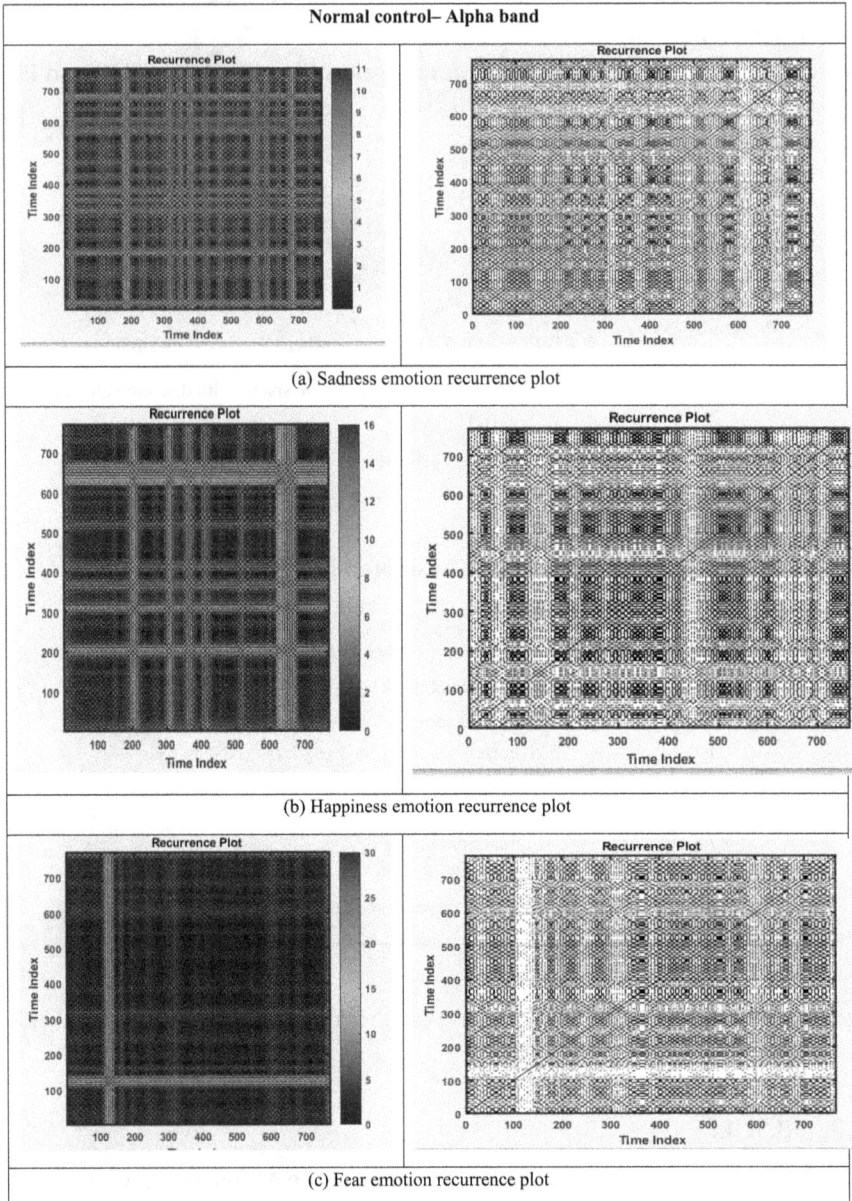

(a) Sadness emotion recurrence plot

(b) Happiness emotion recurrence plot

(c) Fear emotion recurrence plot

FIGURE 1.4 Recurrence plot of the alpha band for sadness, happiness and fear emotion of normal control.

(d) Disgust emotion recurrence plot

(e) Surprise emotion recurrence plot

(f) Anger emotion recurrence plot

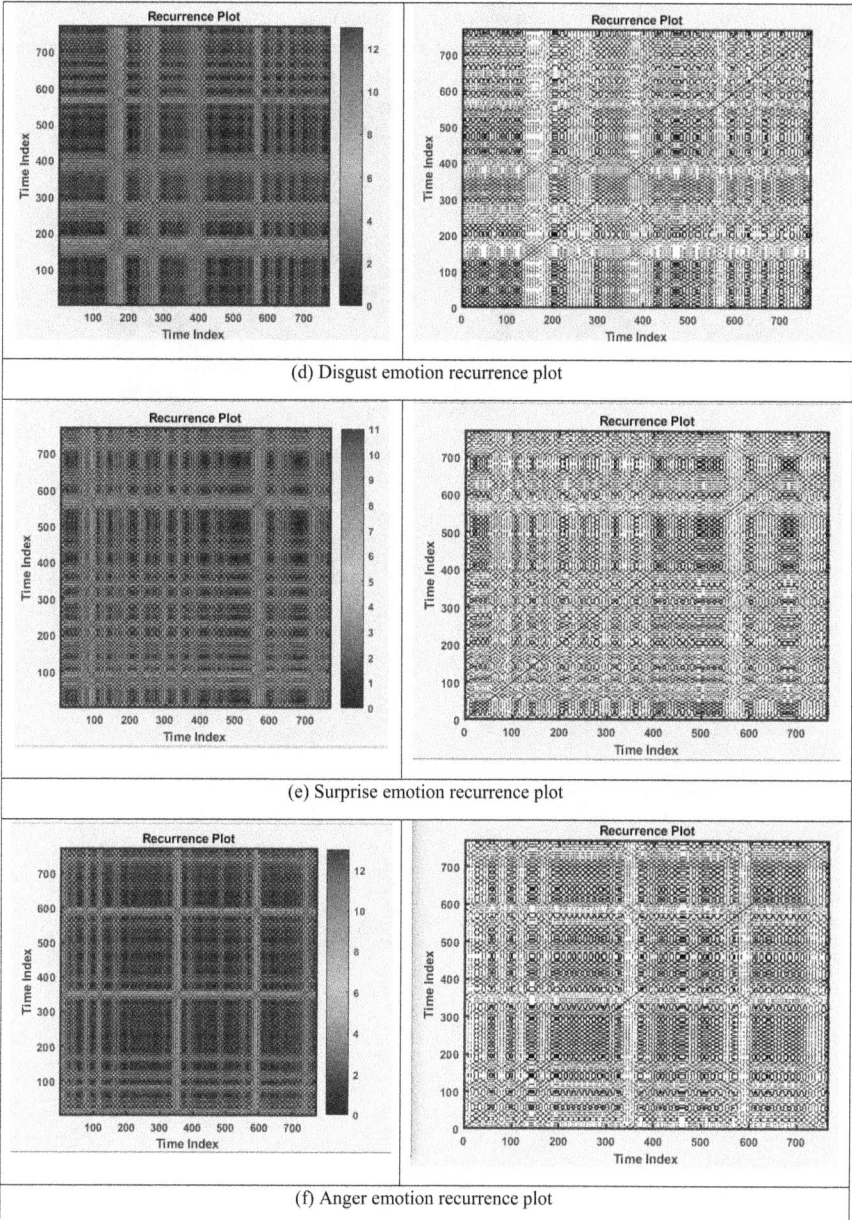

FIGURE 1.5   Recurrence plot of alpha bands for disgust, Surprise and Anger emotion of normal control.

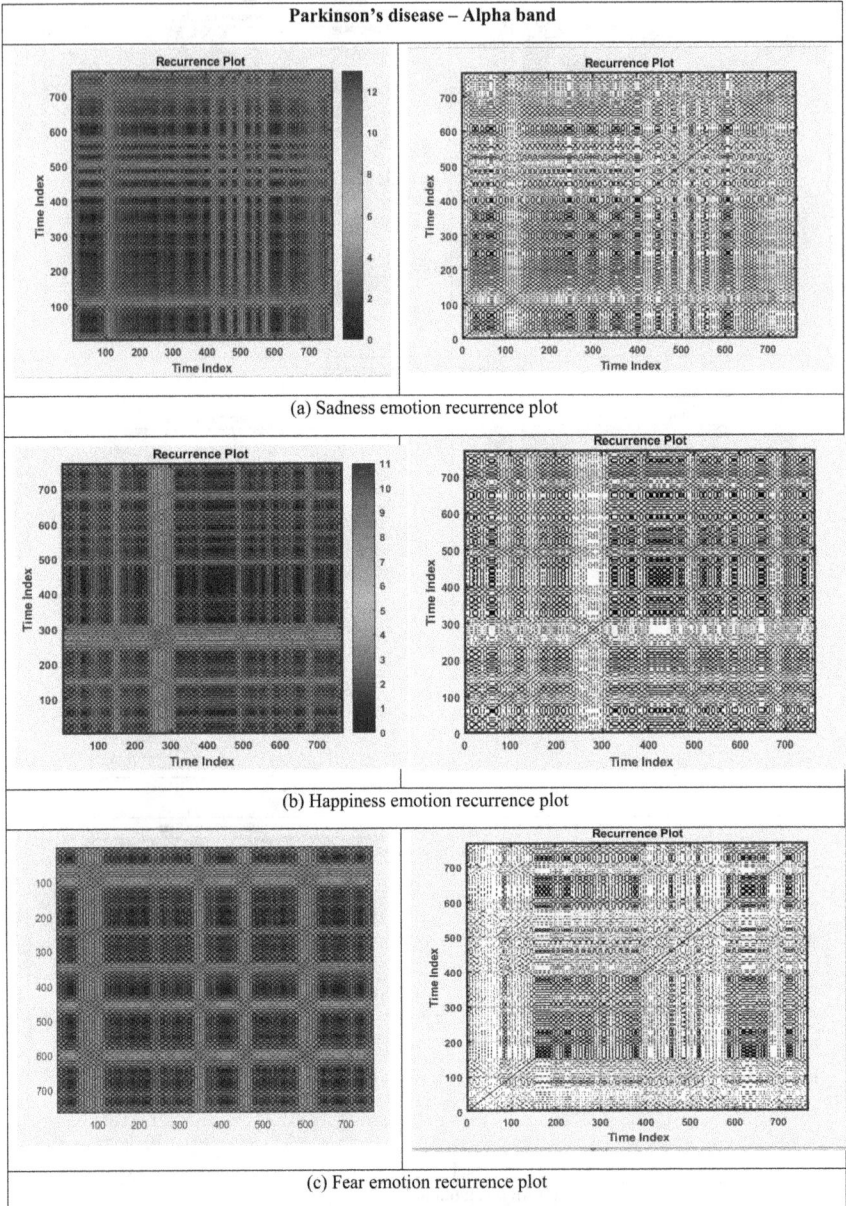

FIGURE 1.6   Recurrence plot of alpha for sadness, happiness and fear emotion of PD.

(d) Disgust emotion recurrence plot

(e) Surprise emotion recurrence plot

(f) Anger emotion recurrence plot

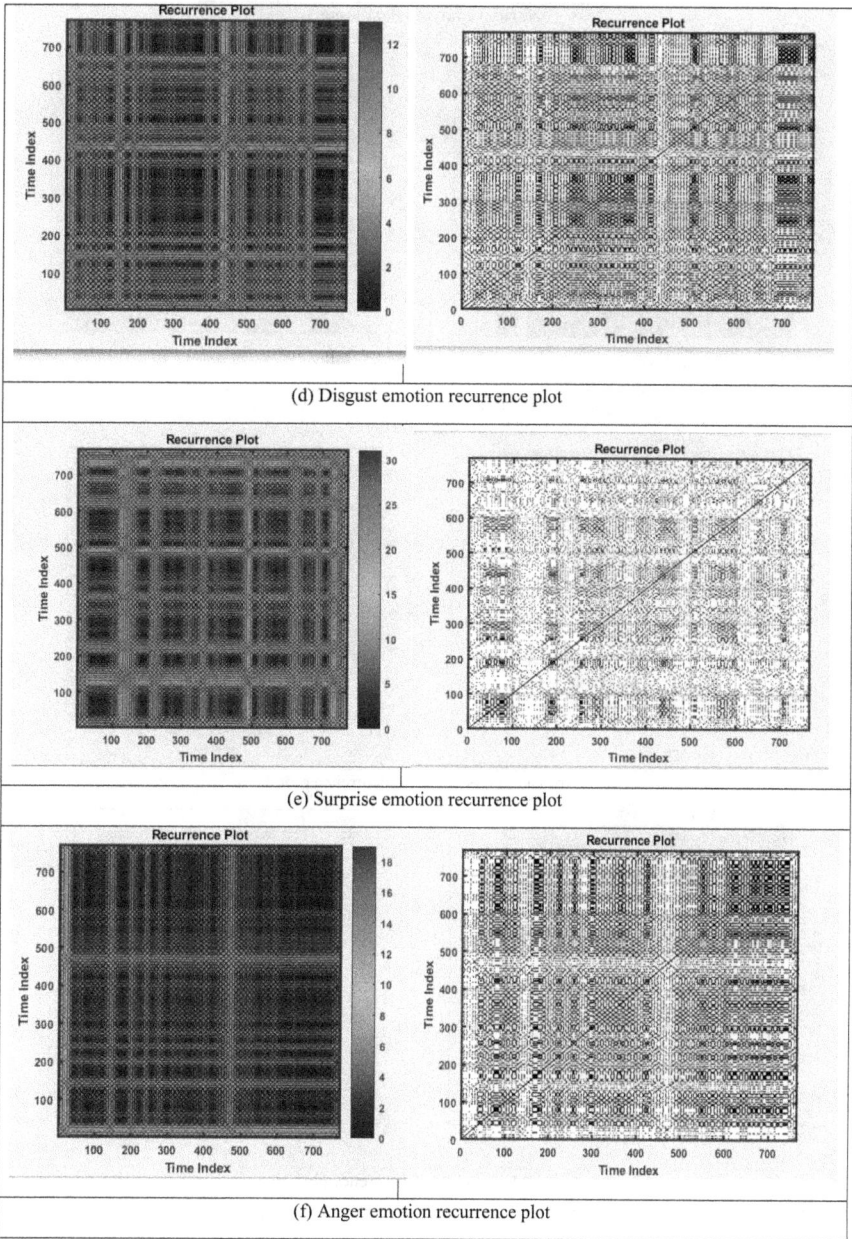

FIGURE 1.7   Recurrence plot of the alpha band for disgust, surprise and anger emotion of PD.

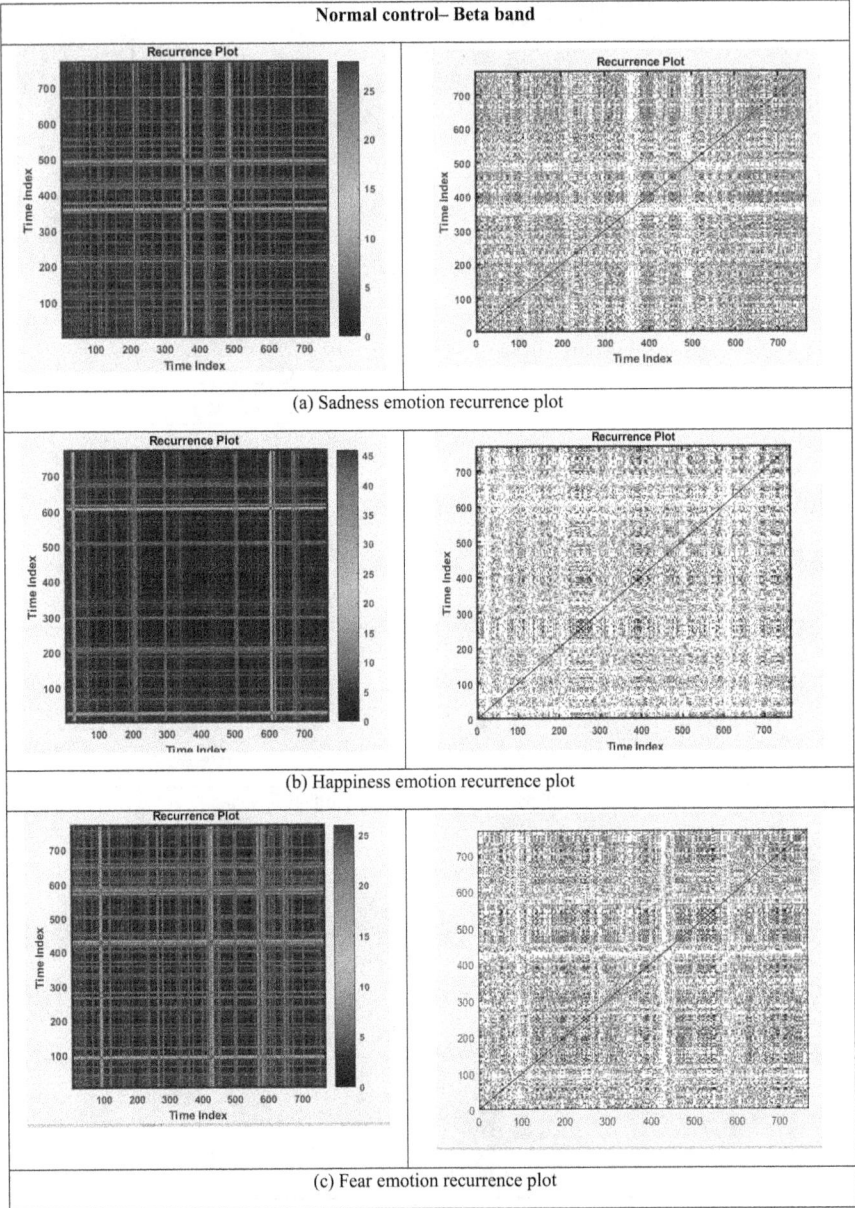

FIGURE 1.8 Recurrence plot of the beta band for sadness, happiness and fear emotion of normal control.

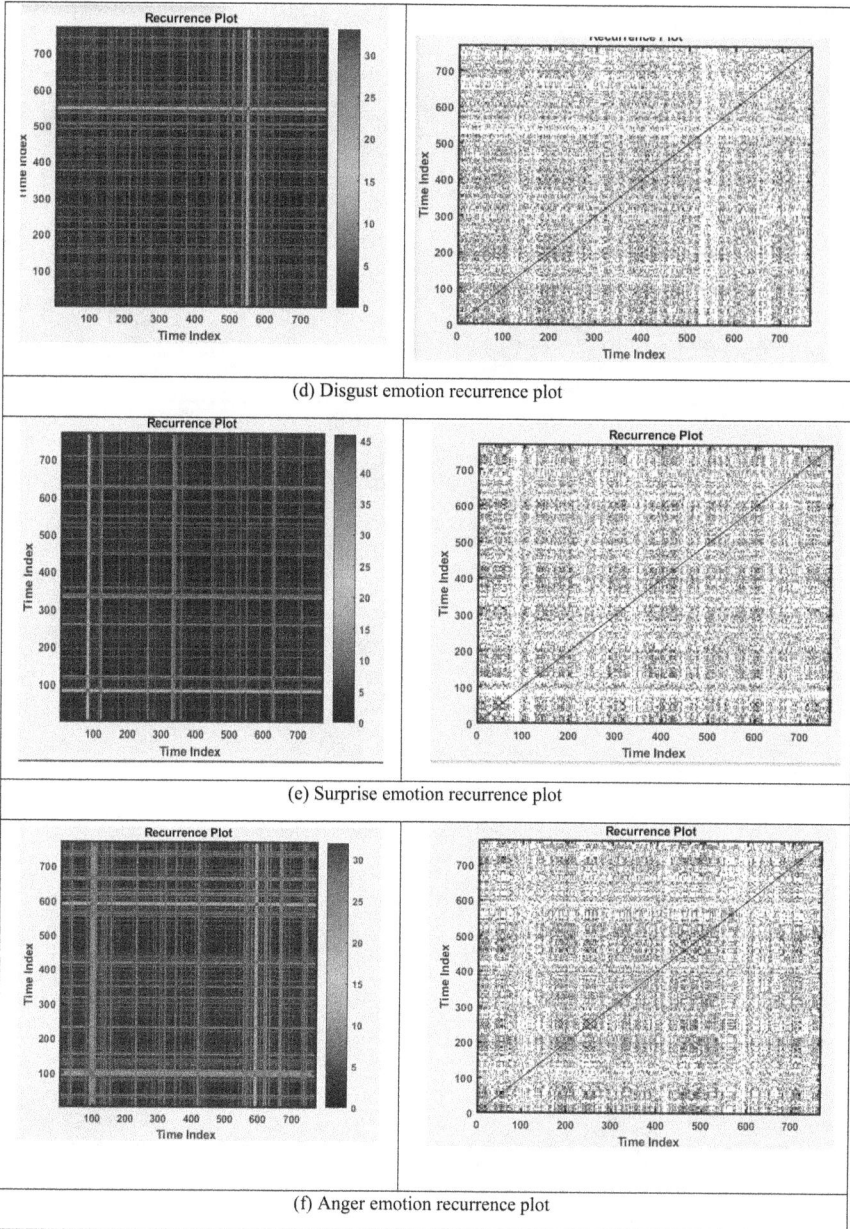

(d) Disgust emotion recurrence plot

(e) Surprise emotion recurrence plot

(f) Anger emotion recurrence plot

FIGURE 1.9 Recurrence plot of the beta band for disgust, surprise and anger emotion of normal control.

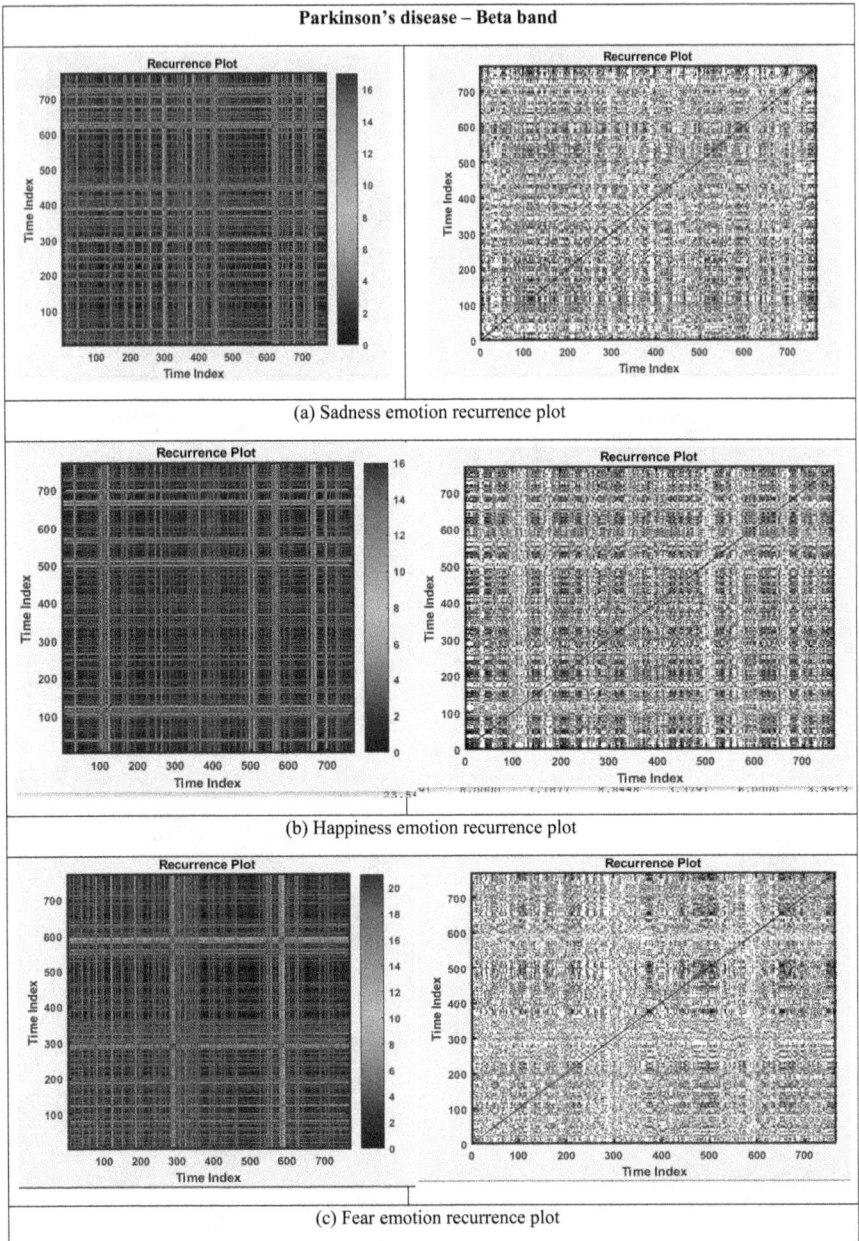

(a) Sadness emotion recurrence plot

(b) Happiness emotion recurrence plot

(c) Fear emotion recurrence plot

FIGURE 1.10   Recurrence plot of the beta band for sadness, happiness and fear emotion of PD.

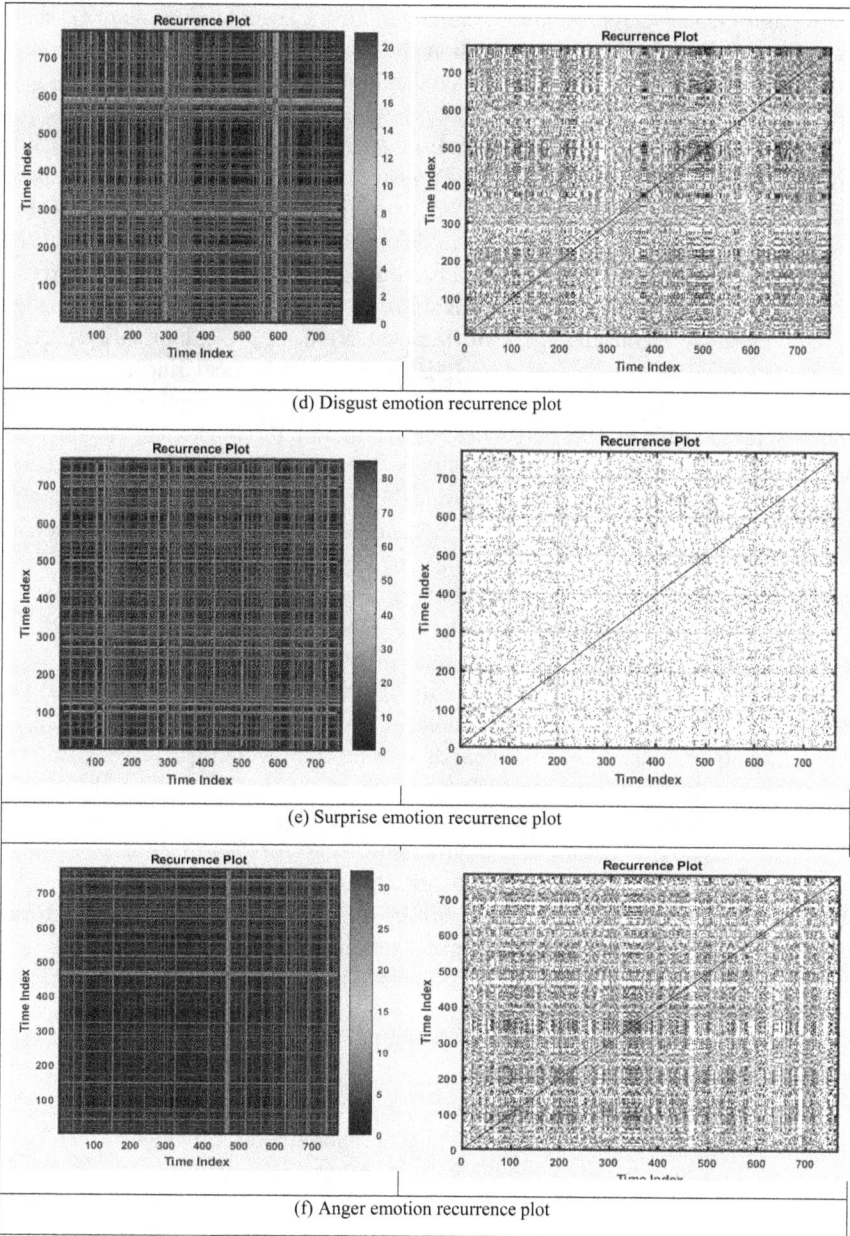

(d) Disgust emotion recurrence plot

(e) Surprise emotion recurrence plot

(f) Anger emotion recurrence plot

FIGURE 1.11 Recurrence plot of the beta band for disgust, surprise and anger emotion of PD.

control and PD shows representative characteristics. Eleven features were extracted from these plots to find the discrimination ability of each emotion in normal control and patients with PD. However, to reduce the dimensionality and to obtain the most distinguishable features, a statistical analysis was performed.

Two types of statistical analysis were used to select the feature set, namely a box-plot analysis and the ANOVA test. Eleven features, namely RR, laminarity (LAM), mean diagonal line length (MEANLINE), maximum vertical line length (MV), entropy (ENT), percent determination (%D), maximal diagonal line length, TT, recurrence time of the first type, recurrence time of the second type, and recurrence probability density entropy, were evaluated from the recurrence plot. Among these 11 features, 8 were considered for further evaluation, i.e., RR, laminarity (LAM), mean diagonal line length (MEANLINE), maximum vertical line length (MV), entropy (ENT), percent determination (%D), maximum line length (ML), and TT. The boxplot analysis of RR, percentage determination (%D), average diagonal line length (MEANLINE), maximum line length (MV), entropy (ENT), and laminarity (LAM) for six emotions are shown in Figure 1.12. Figure 1.13 shows the boxplot for TT, maximal vertical line length (MV), RR and percent determination (%D), percent determination (%D) and maximal vertical line length (MV), RR and maximal vertical line length (MV), and RR, percent determination (%D) and maximal vertical line length (MV), respectively. To confirm the separability of these features, the ANOVA test was performed over features of the alpha, beta, and gamma bands. Tables 1.3.1– 1.6 show the values of F and the probability of chi. A value of chi less than 0.05 was considered to be significant and separable. As evident from Tables 1.3–1.5, all the features of alpha, beta, and gamma bands showed significant separable characteristics. Table 1.6 shows the value of the probability of chi for the combined features of the alpha, beta, and gamma bands. The results of Tables 1.3–1.6 suggest it would be beneficial to use these features and their combinations for classification.

The classification accuracy achieved for alpha band RQA features using different machine learning algorithms is shown in Table 1.7. A SVM with a RBF kernel proved to be best over a linear and sigmoid kernel with an accuracy of 92%, 89.6%, 88.4%, 87.7%, 85.6%, and 88.1% for NC, and for PD it was 90.1%, 85.4%, 86.3%, 84.6%, 83.9%, and 87.7% for S, H, F, D, Su, and A emotions, respectively, using a RR feature. The average accuracy was 88.6% and 86.3% for NC and PD, respectively. A Manhattan kernel of k-NN proved to be the best among the other kernels of k-NN. The average accuracy obtained with a Manhattan kernel was 87.1% and 84.7% for NC and PD, respectively, using a RR feature. A random forest provided an 87.6% and 85.4% accurate separation of emotions using RR and RR and %D features for NC and PD. A RR feature was the best for PNN with an accuracy of 86.9% and 84.9% for NC and PD. A Tanh kernel of ELM was superior over others with an accuracy of 87.2% and 85.1% for NC and PD using a RR feature. Overall, the best alpha band RQA features were provided by a SVM-RBF using RR features.

The accuracy of the RQA features of the beta band is shown in Table 1.8. A RR feature and a SVM-RBF kernel provided the highest average accuracy of 95.7% and 94.1% for NC and PC, respectively. The individual accuracy provided by S, H, F, D, Su, and A emotions for NC was 97.6%, 97.2%, 95.4%, 95.1%, 93.9%, and

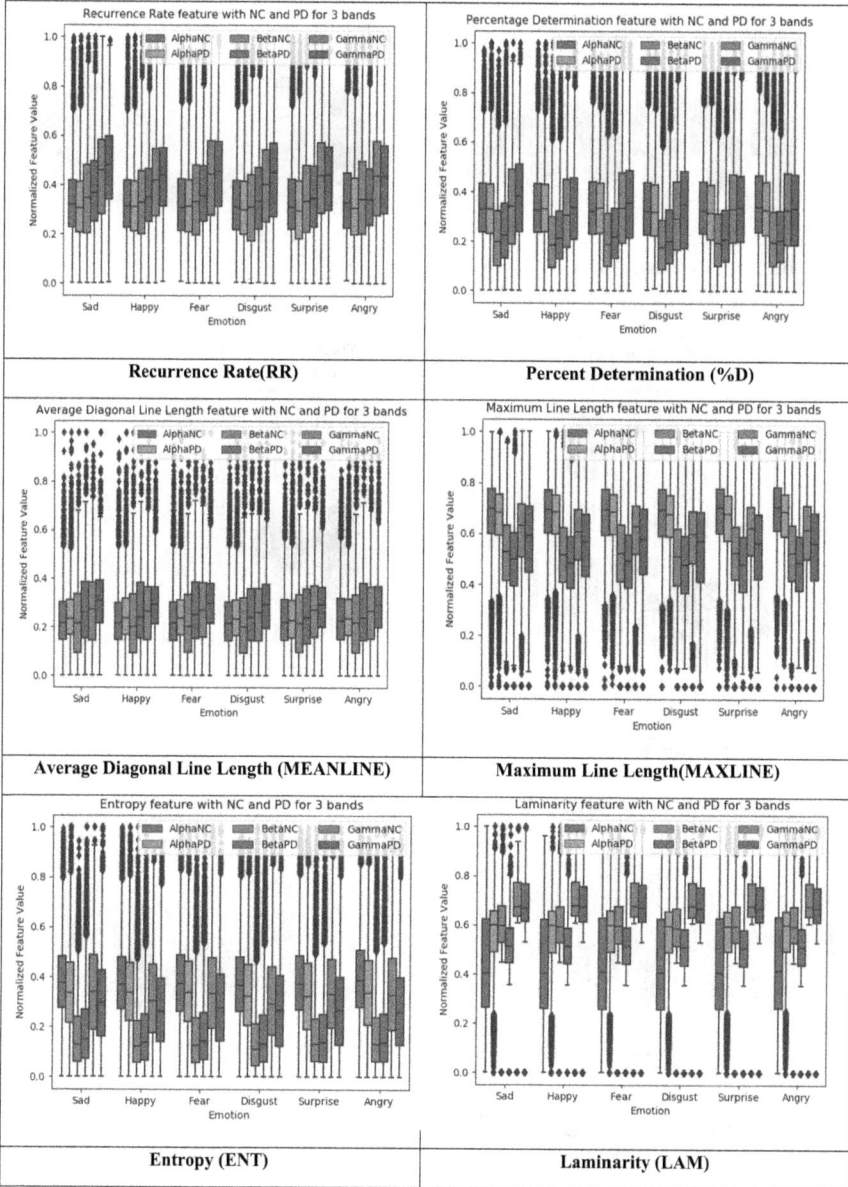

FIGURE 1.12 Boxplot analysis of RQA features in three frequency bands (alpha, beta, and gamma).

95.2%, while for PD it was 96.3%, 94.1%, 93.2%, 93.7 %, 93.1%, and 93.9%. The average accuracy obtained with a DT, random forest, and PNN for NC was 89.2%, 94.9%, and 94.5% with RR and %D features, while for PD, the accuracy was 86.5%, 92.7%, and 92.7% using a RR feature. The highest average accuracy obtained with a

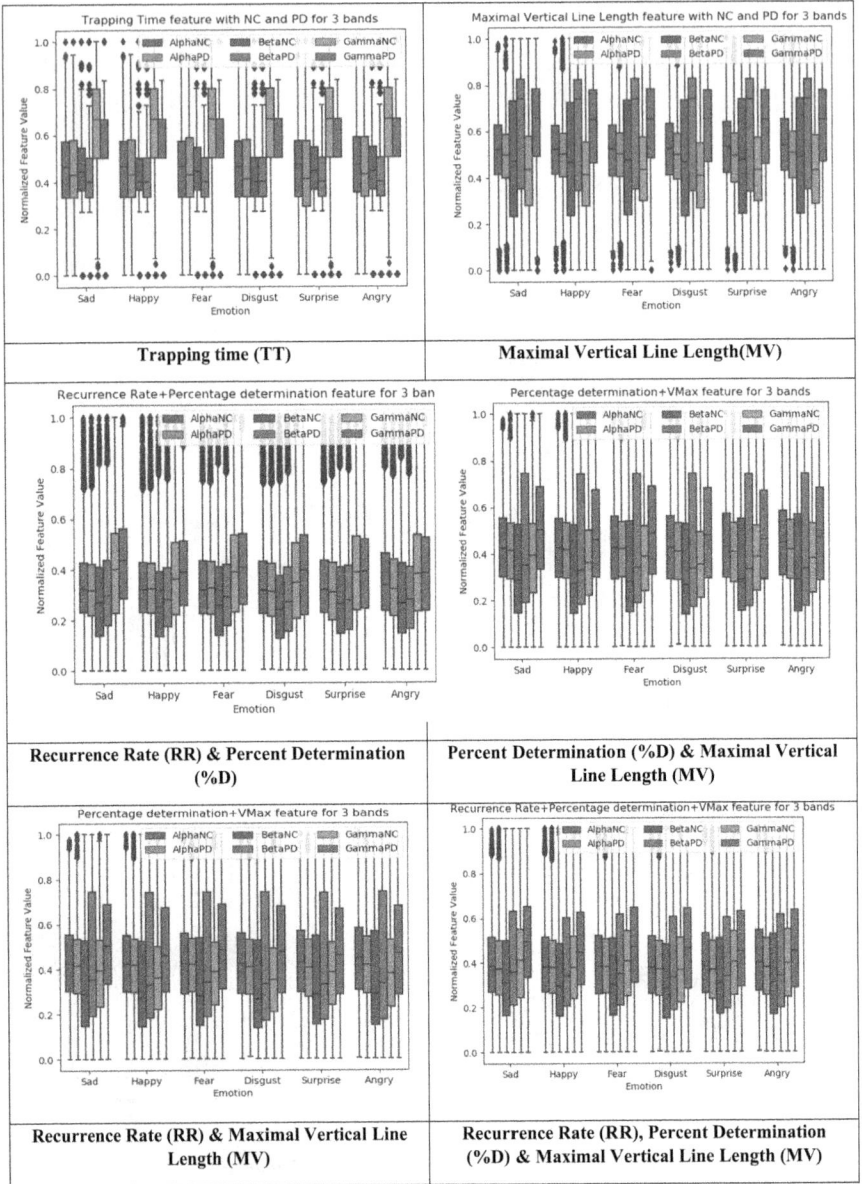

FIGURE 1.13   Boxplot analysis of RQA features in three frequency bands (alpha, beta, and gamma).

Manhattan kernel of k-NN and a Gaussian kernel of ELM was 94.3% and 94.6% for NC while for PD it was 92.3% and 92.4% using RR features.

The accuracy of the RQA features in the beta band is shown in Table 1.9. A RR and %D feature and a SVM-RBF kernel provided the highest average accuracy of

TABLE 1.3
Results of the ANOVA Test on Alpha Band RQA Features

Feature Name	RQA Features	Alpha Band			
		NC		PD	
		F-Value	p-Value	F-Value	p-Value
F1	Recurrence Rate (RR)	24.49	9.82×10^{-25}	8.68	3.01×10^{-08}
F2	Percent Determination (%D)	25.09	2.25×10^{-25}	10.5	4.24×10^{-10}
F3	Average Diagonal Line Length (MEANLINE)	17.77	1.19×10^{-17}	8.49	4.77×10^{-08}
F4	Maximum Line Length (ML)	30.99	1.25×10^{-31}	22.84	5.46×10^{-23}
F5	Entropy (ENT)	25.08	2.34×10^{-25}	11.81	1.92×10^{-11}
F6	Laminarity (LAM)	15.56	2.45×10^{-15}	4.48	4.399×10^{-4}
F7	Trapping Time (TT)	14.88	1.25×10^{-14}	10.92	1.58×10^{-10}
F8	Maximal Vertical Line Length (MV)	27.73	3.65×10^{-28}	17.52	2.15×10^{-17}

TABLE 1.4
Results of the ANOVA Test on Beta Band RQA Features

Feature Name	RQA Features	Beta Band			
		NC		PD	
		F-Value	p-Value	F-Value	p-Value
F1	Recurrence Rate (RR)	40.56	7.93×10^{-42}	26.97	2.32×10^{-27}
F2	Percent Determination (%D)	24.83	4.29×10^{-25}	18.93	7.10×10^{-19}
F3	Average Diagonal Line Length (MEANLINE)	15.42	3.40×10^{-15}	29.1	1.27×10^{-29}
F4	Maximum Line Length (ML)	21.59	1.13×10^{-21}	45.33	6.45×10^{-47}
F5	Entropy (ENT)	24.28	1.63×10^{-24}	13.56	2.95×10^{-13}
F6	Laminarity (LAM)	90.55	2.30×10^{-95}	37.55	1.31×10^{-38}
F7	Trapping Time (TT)	66.42	1.70×10^{-69}	28.33	8.32×10^{-29}
F8	Maximal Vertical Line Length (MV)	13.62	2.55×10^{-13}	39.29	1.83×10^{-40}

96.1% and 94.3% for NC and PC. The individual accuracy provided by S, H, F, D, Su, and A emotions for NC was 97.7%, 97%, 95.8%, 95.5%, 93.4%, and 97.7%, while for PD it was 96.6%, 93.2%, 92.7%, 93.1%, 92.7%, and 96.6%. The average accuracy obtained with a DT, random forest, and PNN for NC was 87.4%, 95.1%, and 95.1% with RR and %D features, while for PD the accuracy was 85.7%, 92.8%, and 93.1% using RR and %D features. The highest average accuracy obtained with an Euclidean kernel of k-NN and a Tanh kernel of ELM was 94.9% and 94.1% for NC while for PD it was 92.9% and 92.4% using RR features.

TABLE 1.5

Results of the ANOVA Test on Gamma Band RQA Features

Feature Name	RQA Features	Gamma Band			
		NC		PD	
		F-Value	p-Value	F-Value	p-Value
F1	Recurrence Rate (RR)	55.1	2.32×10^{-57}	74.72	2.19×10^{-78}
F2	Percent Determination (%D)	54.89	3.82×10^{-57}	65.88	6.48×10^{-69}
F3	Average Diagonal Line Length (MEANLINE)	32.99	9.26×10^{-34}	56.08	2.03×10^{-58}
F4	Maximum Line Length (ML)	53.83	5.27×10^{-56}	76.74	1.51×10^{-80}
F5	Entropy (ENT)	54.89	3.82×10^{-57}	47.54	2.81×10^{-49}
F6	Laminarity (LAM)	26.82	3.34×10^{-27}	36.15	4.06×10^{-37}
F7	Trapping Time (TT)	3.92	1.462×10^{-3}	39.37	1.47×10^{-40}
F8	Maximal Vertical Line Length (MV)	53.3	1.94×10^{-55}	67.06	3.54×10^{-70}

As seen from Tables 1.7–1.9, the best-performing feature was RR for alpha and beta while for the gamma band it was RR and %D in combination with the a RBF kernel of a SVM. To understand the system further, sensitivity and specificity were evaluated, as shown in Table 1.10. The performance parameters, i.e., accuracy, sensitivity, and specificity, obtained for the alpha band for NC and PD was 88.58 ± 1.11%, and 86.33 ± 1.18%, 93.2%, and 91.8%, 65.8%, and 59%. For the beta band accuracy was 95.74 ± 0.54% and 94.06 ± 0.79% for NC and PD, sensitivity was 97.4% and 96.4%, while specificity was 87.2% and 82.2% for NC and PD. Finally, the accuracy obtained for NC and PD was 96.07 ± 0.46% and 94.32 ± 1.13%, sensitivity was 97.4% and 96.2%, and specificity was 87.2% and 81.2% using the gamma band. The best separation was given by the gamma band followed by the beta and alpha band.

1.4   CONCLUSION

This chapter aimed to classify the emotions of patients who have Parkinson's disease. An analysis of alpha, beta, and gamma frequency bands was carried out using recurrence quantification analysis and machine learning algorithms. An analysis of the emotions sadness, happiness, fear, disgust, surprise, and anger using EEG signals was carried out. Recurrence rate and percentage determination features provided the best separation among other feature sets. A radial basis function kernel with a support vector machine provided the best separation among all other classifiers. A detailed insight into the emotions was best given by gamma bands followed by beta and alpha bands. This evolutionary method can be employed in the real-time emotion identification of patients with Parkinson's disease with a performance of success of about 95%.

TABLE 1.6

Results of ANOVA Test on the Combination of RQA Features Over Three Frequency Bands

Frequency band	Features combination	Name of the features	NC		PD	
			F-Value	p-Value	F-Value	p-Value
Alpha	**F1F2**	**Recurrence Rate (RR) & Percent Determination (%D)**	**47.05**	**8.51 x 10^{-49}**	**18.06**	**5.83 x 10^{-18}**
	F1F8	RR & Maximal Vertical Line Length(MV)	11.01	1.29×10^{-10}	4.2	8.05×10^{-4}
	F2 F8	Percent Determination(%D) and Maximal Vertical Line Length(MV)	11.38	5.24×10^{-11}	5.22	8.37×10^{-05}
	F1F2F8	RR, Percent Determination(%D) & Maximal Vertical Line Length(MV)	26.17	1.57×10^{-26}	10.58	3.46×10^{-10}
Beta	**F1F2**	**RR & Percent Determination (%D)**	**37.88**	**5.46 x 10^{-39}**	**20.13**	**3.87 x 10^{-20}**
	F1F8	RR & Maximal Vertical Line Length (MV)	17.41	2.77×10^{-17}	10.84	1.88×10^{-10}
	F2 F8	Percent Determination(%D) & Maximal Vertical Line Length(MV)	22.63	8.83×10^{-23}	20.42	1.92×10^{-20}
	F1F2F8	RR, Percent Determination(%D) & Maximal Vertical Line Length (MV)	29.75	2.50×10^{-30}	17.46	2.44×10^{-17}
Gamma	**F1F2**	**RR & Percent Determination (%D)**	**56.55**	**5.70 x 10^{-59}**	**68.01**	**2.84 x 10^{-71}**
	F1F8	RR & Maximal Vertical Line Length (MV)	17.11	5.82×10^{-17}	24.13	2.31×10^{-24}
	F2 F8	Percent Determination(%D) & Maximal Vertical Line Length(MV)	34.66	1.47×10^{-35}	47.54	2.58×10^{-49}
	F1F2F8	RR, Percent Determination(%D) & Maximal Vertical Line Length (MV)	37.19	2.92×10^{-38}	47.62	2.05×10^{-49}

TABLE 1.7

Results of Alpha Band RQA Features That Give Maximum Classification Rate (in %)

Classifier	Type	Network Parameters	Feature (High ACC)	Individual Class Accuracy						
				S	H	F	D	Su	A	ACC
SVM-RBF	**NC**	C = 4, γ = 16	**RR**	**92**	**89.6**	**88.4**	**87.7**	**85.6**	**88.1**	**88.6**
	PD	C = 4, γ = 16	**RR**	**90.1**	**85.4**	**86.3**	**84.6**	**83.9**	**87.7**	**86.3**
SVM-Linear	NC	C = 1024	ML + MV + RR	78.3	81.7	77.6	77.2	77.3	80.3	77.2
	PD	C = 1024	ML + MV + RR + %D	74.9	77.4	78	75.7	77.8	77.7	76.9
SVM-Sigmoid	NC	C = 128, γ = 0.0078	RR	72.5	76.4	77.7	81.1	71.7	76	77.6
	PD	C = 64, γ = 0.03125	RR	70.4	78.1	78.2	76.2	79.3	77.2	76.6
Decision Tree	NC	Default values	RR	85.3	81.8	81.9	81.0	78.0	82.1	81.7
	PD	Default values	RR	82.9	80.6	79.9	78.9	79.2	81.2	80.4
KNN Chebyshev	NC	NN = 8	RR	90.1	86.4	85.4	84.9	82.4	85.1	85.7
	PD	NN = 8	RR	88.2	83.8	83.6	82.0	81.0	84.4	83.8
KNN Euclidean	NC	NN = 10	RR	91.0	87.4	86.9	85.7	83.7	86.3	86.8
	PD	NN = 9	RR	89.3	84.3	84.1	82.2	83.1	85.6	84.8
KNN Manhattan	NC	NN = 8	RR	91.0	88.2	87.8	85.6	83.7	86.1	87.1
	PD	NN = 8	RR	89.2	84.3	84.1	82.5	82.5	85.6	84.7
KNN Minkowski	NC	NN = 15	RR	90.5	87.4	86.6	84.9	83.8	86.1	86.6
	PD	NN = 12	RR	89.3	84.4	83.5	82.5	82.7	85.3	84.6
Random Forest	NC	NE = 210	RR + %D	90.5	88.0	88.3	87.4	84.6	87.0	87.6
	PD	NE = 390	RR	88.8	85.5	85.3	83.8	83.1	85.7	85.4

(Continued)

TABLE 1.7 (CONTINUED)

Results of Alpha Band RQA Features That Give Maximum Classification Rate (in %)

Classifier	Type	Network Parameters	Feature (High ACC)	Individual Class Accuracy						
				S	H	F	D	Su	A	ACC
PNN	NC	Sigma = 0.1	RR	91.1	87.9	87.1	85.4	84.0	86.0	86.9
	PD	Sigma = 0.09	RR	90.0	84.4	83.9	82.4	82.8	85.7	84.9
ELM RBF	NC	NHN = 960, RBFW = 0.02	RR	88.7	87.1	86.5	86.5	84.6	87.2	86.8
	PD	NHN = 1360, RBFW = 0.007	RR	88.8	84.1	84.7	83.6	83.3	85.3	85
ELM Tanh	NC	NHN = 1020	RR	89.9	87.5	86.6	86.4	85.6	87.1	87.2
	PD	NHN = 940	RR	89.2	84.7	85.0	83.4	83.5	85.0	85.1
ELM Sigmoid	NC	NHN = 920	RR	90.0	87.6	86.9	86.2	84.8	87.1	87.1
	PD	NHN = 920	RR	88.9	84.8	84.9	83.2	83.1	85.4	85
ELM Gaussian	NC	NHN = 920	RR	89.8	87.7	86.9	86.2	84.7	86.7	87
	PD	NHN = 960	RR	89.2	84.6	85.1	83.9	83.3	85.1	85.2
ELM Hardlim	NC	NHN = 1360	RR	86.4	83.9	83.0	83.3	81.7	84.4	83.8
	PD	NHN = 1000	RR	83.8	81.3	81.9	80.1	80.8	82.1	81.7

S: Sadness; H: Happiness; D: Disgust; A: Anger; Su: Surprise; F: Fear; NC: Normal Control; PD: Parkinson's disease; RBF: Radial Basis Function; ML: Maximum Line Length; MV: Maximal Vertical Line Length (V_{max}); %D: Percentage of determination; NHN: No of Hidden Neurons; RBFW: RBF Width; NE: No of Estimators; NN: No of Neighbors

TABLE 1.8

Results of Beta Band RQA Features That Give Maximum Classification Rate (in %)

Classifier	Type	Network Parameters	Feature (High ACC)	Individual Class Accuracy						
				S	H	F	D	Su	A	ACC
SVM-RBF	**NC**	**C = 1, γ = 1**	**RR**	**97.6**	**97.2**	**95.4**	**95.1**	**93.9**	**95.2**	**95.7**
	PD	**C = 4, γ = 32**	**RR**	**96.3**	**94.1**	**93.2**	**93.7**	**93.1**	**93.9**	**94.1**
SVM-Linear	NC	C = 256	ML + MV + RR + %D	77.6	81.4	78.3	77.0	76.8	80.8	79.2
	PD	C = 16	RR + %D	77.9	75.8	79.7	77.9	79.3	79.4	78.3
SVM-Sigmoid	NC	C = 256, γ = 0.015	RR + %D + MV	76.7	75.3	76.7	79.6	76.8	80.7	77.6
	PD	C = 1024, γ = 0.003	RR + %D + MV	78.5	76.1	79.8	76.5	79.6	78.5	78.1
Decision Tree	NC	Default values	RR	91.9	91.2	88.9	87.8	86.0	89.1	89.2
	PD	Default values	RR	89.9	86.1	85.3	86.3	85.4	86.0	86.5
KNN Chebyshev	NC	NN = 6	RR	95.3	94.8	92.2	91.7	90.3	91.8	92.7
	PD	NN = 7	RR	94.4	91.3	90.0	90.6	90.4	91.9	91.4
KNN Euclidean	NC	NN = 4	RR	96.4	96.0	93.4	94.0	92.1	93.4	94.2
	PD	NN = 6	RR	94.8	91.7	90.9	91.7	91.0	92.6	92.1
KNN Manhattan	NC	NN = 6	RR	96.6	95.5	93.8	93.6	92.3	93.7	94.3
	PD	NN = 9	RR	95.0	91.7	90.3	92.3	91.3	93.0	92.3
KNN Minkowski	NC	NN = 6	RR+%D	99.6	94.9	93.6	93.1	91.1	93.0	94.2
	PD	NN = 8	RR	95.3	91.8	90.8	91.4	91.0	92.8	92.2

(Continued)

TABLE 1.8 (CONTINUED)
Results of Beta Band RQA Features That Give Maximum Classification Rate (in %)

Classifier	Type	Network Parameters	Feature (High ACC)	S	H	F	D	Su	A	ACC
						Individual Class Accuracy				
Random Forest	NC	NE = 390	RR	97.1	96.2	94.4	94.0	92.8	95.0	94.9
	PD	NE = 390	RR	95.6	92.1	91.4	92.7	91.6	92.5	92.7
PNN	NC	Sigma = 0.1	RR + %D	96.7	95.6	94.3	94.0	92.1	94.1	94.5
	PD	Sigma = 0.08	RR	95.5	92.3	90.9	92.5	91.7	93.2	92.7
ELM RBF	NC	NHN = 1280, RBFW = 0.01	RR	96.6	95.6	93.9	93.3	92.0	93.6	94.2
	PD	NHN = 1400, RBFW = 0.008	RR	95.3	91.9	91.3	91.5	90.7	92.3	92.2
ELM Tanh	NC	NHN = 1360	RR	96.6	95.7	94.3	93.7	92.4	93.8	94.4
	PD	NHN = 1200	RR	95.3	92.1	91.5	92.4	91.2	92.7	92.5
ELM Sigmoid	NC	NHN = 1140	RR	96.7	95.8	93.8	93.6	92.7	94.1	94.5
	PD	NHN = 1040	RR	95.6	91.7	91.1	91.6	91.2	92.2	92.2
ELM Gaussian	NC	NHN = 1140	RR	96.7	96.1	94.3	93.8	92.6	94.0	94.6
	PD	NHN = 1260	RR	95.4	92.0	91.1	92.3	91.3	92.5	92.4
ELM Hardlim	NC	NHN = 1460	RR	93.1	92.2	90.1	89.3	87.3	90.1	90.4
	PD	NHN = 1480	RR	91.0	87.7	87.6	88.4	87.6	88.5	88.5

S: Sadness; H: Happiness; D: Disgust; A: Anger; Su: Surprise; F: Fear; NC: Normal Control; PD: Parkinson's disease; RBF: Radial Basis Function; ML: Maximum Line Length; MV: Maximal Vertical Line Length (V_{max}); %D: Percentage of determination; NHN: No of Hidden Neurons; RBFW: RBF Width; NE: No of Estimators; NN: No of Neighbors

TABLE 1.9
Results of Gamma Band RQA Features That Give Maximum Classification Rate (In %)

Classifier	Type	Network Parameters	Feature (High ACC)	Individual Class Accuracy							ACC
				S	H	F	D	Su	A		
SVM-RBF	**NC**	**C = 2, γ = 8**	**RR + %D**	**97.7**	**97.0**	**95.8**	**95.5**	**93.4**	**97.7**		**96.1**
	PD	**C = 4, γ = 8**	**RR + %D**	**96.6**	**93.2**	**92.7**	**93.1**	**92.7**	**96.6**		**94.3**
SVM-Linear	NC	C = 8192	ML + MV + RR	78.3	81.7	77.6	77.2	77.3	78.3		78.7
	PD	C = 64	ML + MV + RR + %D	78.9	76.8	76.8	79.5	77.3	78.9		77.9
SVM-Sigmoid	NC	C = 256, γ = 0.007	RR + %D + MV	76.9	73.5	76.2	78.3	76.0	76.9		77.8
	PD	C = 32, γ = 0.015	RR + %D	78.0	75.7	79.1	77.2	79.0	78.0		76.3
Decision Tree	NC	Default values	RR + %D	90.0	89.6	86.3	86.8	85.0	90.0		87.4
	PD	Default values	RR	88.8	85.6	85.2	85.4	84.5	88.8		85.7
KNN Chebyshev	NC	NN = 6	RR + %D	96.7	96.2	94.4	93.7	93.4	96.7		94.8
	PD	NN = 4	RR	95.0	92.0	91.0	92.0	91.0	95.0		92
KNN Euclidean	NC	NN = 6	RR	96.7	96.2	94.4	94.1	93.2	96.7		94.9
	PD	NN = 6	RR	95.5	93.0	91.7	92.8	91.8	95.5		92.9
KNN Manhattan	NC	NN = 6	RR	96.5	96.2	94.1	94.2	93.2	96.5		94.8
	PD	NN = 5	RR + %D	95.6	92.9	91.6	92.4	91.3	95.6		92.7
KNN Minkowski	NC	NN = 6	RR	96.9	96.0	94.1	93.6	92.4	96.9		94.5
	PD	NN = 4	RR	95.5	92.2	92.0	92.7	91.7	95.5		92.7
Random Forest	NC	NE = 210, max_features=log2	RR + %D	96.7	96.2	94.5	94.5	93.6	96.7		95.1
	PD	NE = 160, max_features = auto	RR	95.1	92.5	92.5	92.9	91.8	95.1		92.8

(*Continued*)

TABLE 1.9 (CONTINUED)
Results of Gamma Band RQA Features That Give Maximum Classification Rate (In %)

Classifier	Type	Network Parameters	Feature (High ACC)	Individual Class Accuracy						
				S	H	F	D	Su	A	ACC
PNN	NC	Sigma = 0.09	RR	97.2	96.7	94.7	94.2	93.1	97.2	95.1
	PD	Sigma = 0.09	RR	95.3	93.0	92.1	93.1	92.2	95.3	93.1
ELM RBF	NC	NHN = 1240, RBFW = 0.03	RR	96.6	95.6	93.9	93.3	92.0	96.6	94.1
	PD	NHN = 1400, RBFW = 0.007	RR	95.3	91.9	91.3	91.5	90.7	95.3	92
ELM Tanh	NC	NHN = 1200	RR	96.6	95.7	94.3	93.7	92.4	96.6	94.1
	PD	NHN = 1420	RR	95.3	92.1	91.5	92.4	91.2	95.3	92.4
ELM Sigmoid	NC	NHN = 1040	RR	96.7	95.8	93.8	93.6	92.7	96.7	94.2
	PD	NHN = 1040	RR	95.6	91.7	91.1	91.6	91.2	95.6	92.1
ELM Gaussian	NC	NHN = 1000	RR	96.7	96.1	94.3	93.8	92.6	96.7	94.1
	PD	NHN = 1160	RR	95.4	92.0	91.1	92.3	91.3	95.4	92.3
ELM Hardlim	NC	NHN = 1420	RR	93.1	92.2	90.1	89.3	87.3	93.1	90.3
	PD	NHN = 1400	RR + %D	89.7	86.5	86.7	87.4	86.1	89.7	87

S: Sadness; H: Happiness; D: Disgust; A: Anger; Su: Surprise; F: Fear; NC: Normal Control; PD: Parkinson's disease; RBF: Radial Basis Function; ML: Maximum Line Length; MV: Maximal Vertical Line Length (V_{max}); %D: Percentage of determination; NHN: No of Hidden Neurons; RBFW: RBF Width; NE: No of Estimators; NN: No of Neighbors

TABLE 1.10
Summary of Maximum Emotion Classification Rate (In %) Using Three Frequency Bands Using RQA Features

BAND	FEATURE	CLASSIFIER	ACC ± STD		Sensitivity		Specificity	
			NC	PD	NC	PD	NC	PD
Alpha	RR	SVM-RBF	88.58 ± 1.11	86.33 ± 1.18	93.2	91.8	65.8	59
Beta	RR	SVM-RBF	95.74 ± 0.54	94.06 ± 0.79	97.4	96.4	87.2	82.2
Gamma	**RR + %D**	**SVM-RBF**	**96.07 ± 0.46**	**94.32 ± 1.13**	**97.4**	**96.2**	**87.2**	**81.2**

ACKNOWLEDGMENT

This work is financially supported by the Kuwait Foundation for the Advancement of Sciences (KFAS), Kuwait through exploratory research grant scheme. Grand Number: PR18-13MM-08.

REFERENCES

1. C. Marras, J. Beck, J. Bower, E. Roberts, B. Ritz, G. Ross, R. Abbott, R. Savica, S. Van Den Eeden, A. Willis, and C. Tanner. "Prevalence of Parkinson's disease across North America." npj Parkinson's Disease, vol. 4, 2018.
2. E. Mohr, J. Juncos, C. Cox, I. Litvan, P. Fedio, and T. Chase. "Selective deficits in cognition and memory in high-functioning Parkinson's patients." *Journal of neurology, neurosurgery, and psychiatry*, vol. 53, pp. 603–606, 1990.
3. B. Pillon, B. Dubois, and Y. Agid. "Testing cognition may contribute to the diagnosis of movement disorders." *Neurology*, vol. 46, no. 2, pp. 329–334, 1996 [Online]. Available: https://n.neurology.org/content/46/2/329
4. N. Yoshimura and M. Kawamura. "Impairment of social cognition in Parkinson's disease." *No to shinkei Brain and nerve*, vol. 57, no. 2, pp. 107–113, 2005 [Online]. Available: http://europepmc.org/abstract/MED/15856756
5. J. Péron, T. Dondaine, F. Jeune, D. Grandjean, and M. Vérin. "Emotional processing in Parkinson's disease: A systematic review." *Movement disorders: official journal of the Movement Disorder Society*, vol. 27, pp. 186–199, 2012.
6. H. M. Gray and L. Tickle-Degnen. "A meta-analysis of performance on emotion recognition tasks in Parkinson's disease." *Neuropsychology*, vol. 24, pp. 176–191, 2010.
7. N. Yoshimura and M. Kawamura. "Non-motor symptoms in Parkinson's disease." *European journal of neurology*, vol. 15, no. 01 Supplement 1, pp. 14–20, 2008 [Online]. doi: 10.1111/j.1468-1331.2008.02056.x
8. R. Cousins, A. Pettigrew, O. Ferrie, and J. R. Hanley. "Understanding the role of configural processing in face emotion recognition in Parkinson's disease." Journal of Neuropsychology [Online]. Available: https://onlinelibrary.wiley.com/doi/abs/10.1111/jnp.12210
9. G. Mattavelli, E. Barvas, C. Longo, F. Zappini, D. Ottaviani, M. C. Malaguti, M. Pellegrini, and C. Papagno. "Facial expressions recognition and discrimination in Parkinson's disease." Journal of Neuropsychology [Online]. Available: https://onlinelibrary.wiley.com/doi/abs/10.1111/jnp.12209

10. J. Bek, E. Poliakoff, and K. Lander. "Measuring emotion recognition by people with Parkinson's disease using eye-tracking with dynamic facial expressions." *Journal of Neuroscience Methods*, vol. 331, p. 108524, 2020 [Online]. Available: http://www.scie ncedirect.com/science/article/pii/S0165027019303814

11. Joao Perdiz, Gabriel Pires, and Urbano Nunes. Emotional state detection based on EMG and EOG biosignals: A short survey. IEEE 5th Portuguese Meeting on Bioengineering (ENBENG), pp. 1–4, 2017.

12. X. Guo. Study of emotion recognition based on electrocardiogram and RBF neural network. In *Procedia Engineering*, vol. 15, pp. 2408–2412, 2011. doi: 10.1016/j. proeng.2011.08.452

13. V. Bajaj, Y. Guo, and A. Sengur et al. A hybrid method based on time–frequency images for classification of alcohol and control EEG signals. *Neural Computing & Applications*, vol. 28, pp. 3717–3723, 2017. doi: 10.1007/s00521-016-2276-x

14. V. Bajaj and R. B. Pachori. "Classification of seizure and nonseizure EEG signals using empirical mode decomposition." in *IEEE Transactions on Information Technology in Biomedicine*, vol. 16, no. 6, pp. 1135–1142, 2012. doi: 10.1109/TITB.2011.2181403

15. S. K. Khare and V. Bajaj. "Optimized tunable Q wavelet transform based drowsiness detection from electroencephalogram signals." *Innovation and research in Biomedical Engineering*, pp. 1–1, 2020.

16. S. K. Khare, V. Bajaj, and G. R. Sinha. "Automatic drowsiness detection based on variational nonlinear chirp mode decomposition using electroencephalogram signals." *Modelling and Analysis of Active Biopotential Signals in Healthcare*, vol. 1, pp. 5.1–5.25, 2020.

17. S. K. Khare, V. Bajaj, S. Siuly and G. R. Sinha. "Classification of schizophrenia patients through empirical wavelet transformation using electroencephalogram signals." *Modelling and Analysis of Active Biopotential Signals in Healthcare*, vol. 1, pp. 1.1–5.26, 2020.

18. S. L. Ullo, S. K. Khare, V. Bajaj, and G. R. Sinha. "Hybrid computerized method for environmental sound classification." *IEEE Access*, vol. 8, pp. 124 055–124 065, 2020.

19. S. Chaudhary, S. Taran, V. Bajaj, and A. Sengur. Convolutional neural network based approach towards motor imagery tasks EEG signals classification. *IEEE Sensors Journal*, vol. 19, no. 12, pp. 4494–4500, 2019. doi: 10.1109/jsen.2019.2899645

20. S. Taran, V. Bajaj, Motor imagery tasks-based EEG signals classification using tunable-Q wavelet transform. *Neural Comput & Applic* **31**, 6925–6932 (2019). https://doi.org/10 .1007/s00521-018-3531-0

21. V. Bajaj, S. Taran, S. K. Khare, and A. Sengur. "Feature extraction method for classification of alertness and drowsiness states EEG signals." *Applied Acoustics*, vol. 163, p. 107224, 2020.

22. A. R. Hassan, S. Siuly, and Y. Zhang. "Epileptic seizure detection in EEG signals using tunable-Q factor wavelet transform and bootstrap aggregating." *Computer Methods and Programs in Biomedicine*, vol. 137, pp. 247–259, 2016.

23. S. K. Khare and V. Bajaj. "Constrained based tunable Q wavelet transform for efficient decomposition of EEG signals." *Applied Acoustics*, vol. 163, p. 107234, 2020.

24. J. I. Serrano, M. D. del Castillo, V. Cort's, N. Mendes, A. Arroyo, J. Andreo, E. Rocon, M. del Valle, J. Herreros, and J. P. Romero. "EEG microstates change in response to increase in dopaminergic stimulation in typical Parkinson's disease patients." *Frontiers in Neuroscience*, vol. 12, p. 714, 2018 [Online]. Available: https://www.frontiersin.org/ article/10.3389/fnins.2018.00714

25. R. Yuvaraj, M. Murugappan, N. M. Ibrahim, M. I. Omar, K. Sundaraj, K. Mohamad, R. Palaniappan, E. Mesquita, and M. Satiyan. "On the analysis of EEG power, frequency and asymmetry in Parkinson's disease during emotion processing." *Behavioral and Brain Functions*, vol. 10, no. 1, pp. 1-19 2014.

26. R. Yuvaraj, M. Murugappan, and R. Palaniappan. "The effect of lateralization of motor onset and emotional recognition in PD patients using EEG." *Brain Topography*, vol. 30, 2016.

27. R. Yuvaraj, M. Murugappan, Y. Htut, N. Mohamed Ibrahim, K. Sundaraj, K. Mohamad, and M. Satiyan. "Emotion processing in Parkinson's disease: An EEG spectral power study," *The International Journal of neuroscience*, vol. 124, 2013.

28. L. Pepa, M. Capecci, and M. G. Ceravolo. "Smartwatch based emotion recognition in Parkinson's disease." in 2019 IEEE 23rd International Symposium on Consumer Technologies (ISCT), pp. 23–24, 2019.

29. E. Naghsh, M. Sabahi, and S. Beheshti. "Spatial analysis of EEG signals for Parkinson's disease stage detection." *Signal Image and Video Processing*, vol. 14, pp. 397–405, 2019.

30. K. N. Rejith and K. Subramaniam. "Classification of emotional states in Parkinson's disease patients using machine learning algorithms." *Biomedical and Pharmacology Journal*, vol. 11, pp. 333–341, 2018.

31. R. Yuvaraj, M. Murugappan, N. M. Ibrahim, K. Sundaraj, M. I. Omar, K. Mohamad, and R. Palaniappan. "Detection of emotions in Parkinson's disease using higher order spectral features from brain's electrical activity." *Biomedical Signal Processing and Control*, vol. 14, pp. 108–116, 2014 [Online]. Available: http://www.sciencedirect.com/science/article/pii/S1746809414001116

32. R. Yuvaraj, M. Murugappan, N. M. Ibrahim, K. Sundaraj, M. I. Omar, K. Mohamad, R. Palaniappan, and M. Satiyan. "Inter-hemispheric EEG coherence analysis in Parkinson's disease: assessing brain activity during emotion processing." *Journal of Neural Transmission* (Vienna, Austria: 1996), vol. 122, no. 2, pp. 237–252, 2015 [Online]. doi: 10.1007/s00702-014-1249-4

33. M. Murugappan, W. B. Alshuaib, A. Bourisly, S. Sruthi, W. Khairunizam, B. Shalini, and W. Yean. "Emotion classification in Parkinson's disease EEG using RQA and ELM." in 2020 *16th IEEE International Colloquium on Signal Processing Its Applications* (CSPA), pp. 290–295, 2020.

34. R. Yuvaraj, U Rajendra Acharya, Y. Hagiwara, A novel Parkinson's Disease Diagnosis Index using higher-order spectra features in EEG signals. *Neural Comput & Applic* **30**, 1225–1235 (2018). https://doi.org/10.1007/s00521-016-2756-z

35. G. Silva, M. Alves, R. Cunha, B. Bispo, and P. Rodrigues. "Parkinson disease early detection using EEG channels cross-correlation." *International Journal of Applied Engineering Research*, vol. 15, pp. 197–203, 2020.

36. A. P. S. de Oliveira, M. A. de Santana, M. K. S. Andrade, J. C. Gomes, M. C. A. Rodrigues, and W. P. dos Santos. "Early diagnosis of Parkinson's disease using EEG, machine learning and partial directed coherence." *Research on Biomedical Engineering*, pp. 397–405, 2020.

37. S. J. Priya, A. J. Rani, and S. Su Ma. "Diagnosis of Parkinson's disease using fast Fourier transform." in 2020 5th International Conference on Devices, *Circuits and Systems (ICDCS)*, pp. 198–202, 2020.

38. Omer Ertugrul, Yilmaz Kaya, Ramazan Tekin, and M. Almali. "Detection of Parkinson's disease by shifted one dimensional local binary patterns from gait." *Expert Systems with Applications*, vol. 56, pp. 156–163, 2016 [Online]. Available: http://www.sciencedirect.com/science/article/pii/S0957417416301105

39. A. Bhurane, S. Dhok, M. Sharma, R. Yuvaraj, M. Murugappan, and U. R. Acharya. "Diagnosis of Parkinson's disease from electroencephalography signals using linear and self-similarity features." *Expert Systems*, p. e12472. doi: 10.1111/exsy.12472 [Online]. Available: https://onlinelibrary.wiley.com/doi/abs/10.1111/exsy.12472

40. T. Tuncer, S. Dogan, and U. R. Acharya. "Automated detection of Parkinson's disease using minimum average maximum tree and singular value decomposition method with

vowels." *Biocybernetics and Biomedical Engineering*, vol. 40, no. 1, pp. 211–220, 2020 [Online]. Available: http://www.sciencedirect.com/science/article/pii/S0208521619 300853

41. O. Qi Wei, H. M, S. Basah, H. Lee, and V. Vijean. "Empirical wavelet transform based features for classification of Parkinson's disease severity." *Journal of Medical Systems*, vol. 42, 2018.

42. R. Yuvaraj, M. Murugappan, N. M. Ibrahim, K. Sundaraj, M. I. Omar, K. Mohamad, and R. Palaniappan. "Optimal set of EEG features for emotional state classification and trajectory visualization in Parkinson's disease." *International Journal of Psychophysiology*, vol. 94, no. 3, pp. 482–495, 2014 [Online]. Available: http://www .sciencedirect.com/science/article/pii/S0167876014001913

43. Q. T. Ly. "Detection of freezing of gait and gait initiation failure in people with Parkinson's disease using electroencephalogram signals." pp. 1–156 2017., http://hdl .handle.net/10453/127981

44. R. Yuvaraj, M. Murugappan, U. R. Acharya, H. Adeli, N. M. Ibrahim, and E. Mesquita. "Brain functional connectivity patterns for emotional state classification in Parkinson's disease patients without dementia." *Behavioural Brain Research*, vol. 298, no. Pt B, pp. 248–260, 2016 [Online]. doi: 10.1016/j.bbr.2015.10.036

45. R. Yuvaraj and M. Murugappan. "Hemispheric asymmetry non-linear analysis of EEG during emotional responses from idiopathic Parkinson's disease patients." *Cognitive Neurodynamics*, vol. 10, 2016.

46. B. Güntekin, L. Hano˘glu, D. Güner, N. H. YÄ´slmaz, F. Çadirci, N. Mantar, T. Aktürk, D. D. Emek-Sava¸s, F. F. Özer, G. Yener, and E. Basar. "Cognitive impairment in Parkinson's disease is reflected with gradual decrease of EEG delta responses during auditory discrimination." *Frontiers in Psychology*, vol. 9, p. 170, 2018 [Online]. Available: https://www.frontiersin.org/article/10.3389/fpsyg.2018.00170

47. V. Bajaj, A. Hari Krishna, B. S. Aravapalli, K. Priyanka, and S. Taran. "Emotion classification using EEG signals based on tunable Q wavelet transform." *IET Science, Measurement & Technology*, vol. 13, 2018.

48. V. Bajaj, S. Taran, and A. Sengur. "Emotion classification using flexible analytic wavelet transform for electroencephalogram signals." *Health Information Science and Systems*, vol. 6, no. 1, p. 12, 2018.

49. V. Bajaj and R. B. Pachori. "Human emotion classification from EEG signals using multi-wavelet transform." In *2014* International Conference on Medical Biometrics, Shenzhen, pp. 125–130, 2014. doi: 10.1109/ICMB.2014.29

50. S. K. Khare, V. Bajaj and G. R. Sinha, Adaptive Tunable Q Wavelet Transform-Based Emotion Identification, in *IEEE Transactions on Instrumentation and Measurement*, vol. 69, no. 12, pp. 9609–9617, Dec. 2020, doi: 10.1109/TIM.2020.3006611

51. S. Taran and V. Bajaj. "Emotion recognition from single-channel EEG signals using a two-stage correlation and instantaneous frequency-based filtering method." *Computer Methods and Programs in Biomedicine*, vol. 173, pp. 157–165, 2019.

52. O. S. Sushkova, A. A. Morozov, and A. V. Gabova. "Investigation of specificity of Parkinson's disease features obtained using the method of cerebral cortex electrical activity analysis based on wave trains." in 2017 13th International Conference on Signal-Image Technology Internet-Based Systems (SITIS), 2017, pp. 168–172.

53. S. K. Khare and V. Bajaj. "Time-frequency representation and convolutional neural network based emotion recognition." *IEEE Transactions on Neural Networks and Learning Systems*, in Press pp. 1–1, 2020.

54. H. Gunduz, Deep Learning-Based Parkinson's Disease Classification Using Vocal Feature Sets, in *IEEE Access*, vol. 7, pp. 115540–115551, 2019, doi: 10.1109/ ACCESS.2019.2936564

55. Webber, Charles & Marwan, Norbert. *Recurrence Quantification Analysis—Theory and Best Practices.* Springer, 2015.
56. Joseph Zbilut, Nitza Thomasson, and Charles Webber. Recurrence quantification analysis as a tool for nonlinear exploration of nonstationary cardiac signals. *Medical engineering & physics*, vol. 24, pp. 53–60, 2002. doi: 10.1016/S1350-4533(01)00112-6
57. U. R. Acharya, S. V. Sree, S. Chattopadhyay, W. Yu, and P. C. Ang. Application of recurrence quantification analysis for the automated identification of epileptic EEG signals. *The International Journal of Neural Systems*, vol. 21, no. 3, pp. 199–211, 2011. doi: 10.1142/S0129065711002808
58. E. P. Ng, Teik-Cheng Lim, Subhagata Chattopadhyay, and Muralidhar Bairy. Automated identification of epileptic and alcoholic EEG signals using recurrence quantification analysis. *Journal of Mechanics in Medicine and Biology*, vol. 12, 2012. doi: 10.1142/S0219519412400283
59. G. -B. Huang, Q. -Y. Zhu, and C. -K. Siew. "Extreme learning machine: Theory and applications." *Neurocomputing*, vol. 70, no. 1, pp. 489–501, 2006.
60. S. Kotsiantis. "Supervised machine learning: A review of classification techniques." *Informatica (Slovenia)*, vol. 31, pp. 249–268, 2007.
61. A. Sarica, A. Cerasa, and A. Quattrone. "Random forest algorithm for the classification of neuroimaging data in Alzheimer's disease: A systematic review." *Frontiers in Aging Neuroscience*, vol. 9, p. 329, 2017 [Online]. Available: https://www.frontiersin.org/article/10.3389/fnagi.2017.00329
62. S. Siuly, S. K. Khare, V. Bajaj, H. Wang and Y. Zhang, "A Computerized Method for Automatic Detection of Schizophrenia Using EEG Signals," in *IEEE Transactions on Neural Systems and Rehabilitation Engineering*, vol. 28, no. 11, pp. 2390–2400, Nov. 2020, doi: 10.1109/TNSRE.2020.3022715

2 Sleep Stage Classification Using DWT and Dispersion Entropy Applied on EEG Signals

Rajeev Sharma, Sitanshu Sekhar Sahu,
Abhay Upadhyay, Rishi Raj Sharma,
and Ajit Kumar Sahoo

CONTENTS

2.1   INTRODUCTION

Sleep is an essential part of human life. To gain deep insight into sleep disorders requires sleep analysis [1]. A healthy brain reflects several distinct features during sleep which are categorized as sleep stages [2]. The manual marking of sleep stages using sleep electroencephalogram (EEG) recordings is a tedious task. The visual observation of sleep recordings helps to identify any abnormalities related to sleep, such as sleep apnea, narcolepsy, and insomnia. Although visual observation is useful for the diagnosis of sleep abnormalities, it is a time-consuming process. Therefore, an accurate and automatic scoring of EEG records is important for sleep specialists because they usually perform scoring through visual observation. There are existing standards that are used for manual identification of the sleep stages using EEG signals and some other physiological signals. One of the standards for manual sleep scoring is by Rechtschaffen and Kales, which is referred to as the R&K standard [3].

According to the R&K system an entire sleep duration can be categorized into several stages: wake (W), four stages from stage 1 to stage 4 (S1, S2, S3, S4), and rapid eye movement (REM). Other activities related to the movement of the body parts are referred to as movement time (M). Although the system proposed by Rechtschaffen and Kales is useful, it is also prone to subjective interpretation [4, 5]. Recently, the American Academy of Sleep Medicine (AASM) modified the R&K system of sleep scoring and proposed new guidelines for scoring rules in the AASM standard [6]. In the AASM standard, the sleep stages S3 and S4 are considered as a single stage. Hence, the AASM standard only has W, S1 S2, S3, and REM sleep stages. Using computer-aided algorithms for automatic identification of these stages can result in a more efficient diagnosis of sleep abnormalities. Such techniques which employ EEG recordings for the automatic detection of different sleep stages can reduce the diagnosis time for sleep disorders.

In the literature, many techniques have been introduced in recent years for performing automatic sleep scoring using polysomnogram (PSG) recordings. During PSG recording several types of physiological signals are collected, such as EEG, electrocardiogram (ECG), electromyogram (EMG), electrooculogram (EOG), and nasal airflow. In [7], segmentation and self-organization techniques were used for automatic sleep stage detection using PSG recordings. Segmented PSG recordings and their subsequently derived features, such as dominant rhythm, amplitude, frequency-weighted energy, and an index related to alpha-slow-wave, etc. were used in the sleep analysis. They used a k-means clustering classifier for the classification of the sleep stages. Similarly, in another study [8], an adaptive neuro-fuzzy classifier was employed for sleep scoring using PSG recording. In another study [9], PSG recordings from 20 healthy subjects were used to compute 74 measures for automatic sleep scoring. The classification was performed using the quadratic discriminant analysis classifier.

The methods mentioned above are based on the analysis of PSG recordings related to sleep. Although the recording of multiple physiological signals helps to accurately identify sleep stages, it requires a large number of sensors to be placed on the subject's body. Further, performing a PSG sleep recording is a time-consuming process; generally, it continues for longer than the sleep duration of the subject. Moreover, the placement of the electrodes related to the different types of signal make it difficult for the subject to sleep properly during the recording. Therefore, techniques that only use a single physiological signal, such as an EEG, are of special interest in the development of automatic sleep scoring systems. Hence, several methodologies have been developed which detect different sleep stages by only using EEG signals. In [10], nonlinear dynamic features were extracted from EEG signals. These features were extracted using a higher-order spectra (HOS) analysis and a recurrence quantification analysis (RQA) of the EEG signals. The nonlinearity of the EEG signal was quantified using HOS- and RQA-based features. The HOS measured the deviation of the EEG signals from the Gaussianity; the RQA-based features also quantified the irregularity of the EEG signals. The discrimination ability of these features was measured using the F-values and p-values obtained as a result of the ANOVA test. In [11], multiscale entropy was used for extracting features from EEG recordings. An autoregressive model for the analysis of EEG sleep signals was also used. They performed a six-class classification to show the performance of the developed technique, obtaining results of 76.9% sensitivity. Energy-based parameters were analyzed in [12] for developing a sleep scoring technique. The five stages of the classification were performed using EEG signals. They obtained an 87.2% classification accuracy as a result of a five-class classification using a recurrent neural network classifier. In another study [13], a technique was developed for the detection of the REM stage. The developed method used a spectral edge frequency of 8–16 Hz, the absolute and relative power of the EEG signals. They detected REM with an 83% sensitivity. In another study [14], a method was developed using a stacked sparse autoencoder, and classifier performance was validated using a 20-fold cross-validation technique. They handled the class imbalance problem using a random sampling method. EEG sleep signals were analyzed using time-frequency representation in a study [15] where the Wigner–Ville distribution was used for computing a time-frequency image; the features were extracted from the segmented part of the time-frequency image. The segmentation of the time-frequency image was performed with consideration of the frequency range of EEG rhythms. The histogram of the segmented image was used for feature extraction, and a support vector machine (SVM) was used for classification. The analysis of the EEG signals was performed using the Hilbert–Huang transform, and bandwidth parameters were used to classify the sleep and wake stages [16]. The bandwidth features were amplitude modulation bandwidth and frequency modulation bandwidth. A fuzzy clustering algorithm used the parameters as input features to detect the sleep and wake stages. The events of sleep apnea, a disorder related to sleep, were identified using information extracted from EEG signals [17]. For extracting the information, the Hermite coefficients optimized using a particle swarm optimization were used. The classification was performed using a SVM to detect apnea events. Another method of detecting sleep apnea was developed based on Hermite coefficients optimized using artificial bee colony algorithms [18]. The features were extracted from the Hermite coefficients and used as inputs for a least-squares SVM classifier and extreme learning machine. Similarly, in another study [19], a horizontal visibility algorithm was used for the analysis of EEG sleep recordings. The authors used the mean degree and the distribution degree as features and a SVM as a classifier. They studied different multi-class classification problems, combining the different sleep stages. An ensemble empirical mode decomposition with adaptive noise, a modified form of empirical mode decomposition (EMD), was used to decompose EEG signals [20]. They extracted statistical moments from the intrinsic mode functions (IMFs) and used them with bootstrap aggregating techniques for classification of the sleep stages. In [21], another signal decomposition technique, known as the wavelet transform with tunable Q factor, also called TQWT, was used to analyze EEG signals. They computed spectral features from the subband obtained as a result of the TQWT of the EEG sleep epochs. Spectral features and a random forest classifier were used for different multi-class classification problems. In another work [22], the subbands related to five rhythms were extracted from EEG signals using filter-banks. The energy of the subbands was computed, and various classifiers were trained using the energy of the subbands. The highest classification frequency was obtained using the random forest classifier.

Several techniques were developed using neural network-based classifiers for the detection of the sleep stages. In one method [23], EMD was employed to denoise EEG signals, and a detrended fluctuation analysis (DFA) extracted the scale characteristics of the signals. An artificial neural network (ANN) was then used for classification of the sleep stages. In [24], a deep neural network (DNN) was designed for the automatic detection of sleep stages using single-channel EEG signals. They studied the effect of data size, the regularization techniques, and the choice of architecture. The different architectures of a recurrent DNN were evaluated. It was reported that a network designed using convolution layers and long short-term memory layers achieved a significant agreement with a human annotator. A deep learning model was developed which used PSG recordings for the classification of sleep stages [25]. The EEG and EOG signals present in the PSG recordings were used to develop a one-dimensional convolution neural network (1DCNN). The 3000 samples of the epoch served as the input to the CNN. The sleep stage classification was performed for two to six classes of sleep stage.

Apart from EEG signals, other modalities such as EOG and ECG signals are also used for developing sleep stage detection techniques. In a recent study [26], an automatic sleep scoring technique was presented based on the analysis of a RR time-series analysis. Similarly, another study [27] presented a technique for detecting sleep stages using EOG signals. For the analysis, signals were decomposed using the discrete wavelet transform (DWT). The Daubechies-4 (DB4) was used as a mother wavelet for a four-level DWT-based decomposition. The statistical features related to spectral entropy, moment-based parameters, refined-composite versions of the dispersion entropy, and coefficients computed using the autoregressive model were computed. The features with the highest discrimination ability were selected by performing a one-way analysis of the variance and boxplots to reduce the model complexity.

In [28], a summary was presented of the recently developed classification techniques for sleep stage classification using EEG signals. They also proposed an efficient, fast, and easy technique and implemented it using hardware. The EEG signals were decomposed into 10-sec epochs instead of 20- or 30-sec epochs. The subbands of the filtered EEG epochs, corresponding to the delta, theta, alpha, and beta frequency, were used to compute the features. The new energy and distant based statistical features computed from the subband signals were fed into different classifiers, such as a decision tree, support vector machine, neural network, naïve Bayes, etc. The best results were reported when the decision tree classifier was used.

It was observed from the literature review that dealing with the non-stationary nature of the EEG signal while extracting the sleep-stage properties was very important for accurate detection of the sleep stages. In general, EEG signals are decomposed using various techniques, such as EMD, TQWT, etc. On decomposition, the EEG signal can be represented as a sum of simpler components. The components are used to extract various linear and nonlinear parameters to be used as the features for classification. Recently, dispersion entropy (DEn) [29] and fluctuation-based dispersion entropy (FDEn) [30] were proposed for quantifying the uncertainty of a signal. DEn overcomes certain shortcomings of sample entropy (SpEn) [31] and permutation entropy (PEn) [32], as discussed in [29,30]. DEn measures the distribution of symbolic dynamics which are also called patterns. DEn measurements can track simultaneous changes in amplitude and frequency [30]. Whenever fluctuations in the signals are important, FDEn can be considered as a relevant measure. It is important to note that sleep stages exhibit different patterns in amplitude and certain frequency ranges associated with EEG rhythms. Hence, in this chapter, we study the performance of DEn and FDEn for sleep stage classification. An algorithm for the classification of sleep stages using EEG recordings based on the DWT, DEn, and FDEn is presented. The DWT is used to extract subband signals which are associated with a certain range of frequencies in EEG signals. Hence, the subsequent computation of DEn and FDEn from subband signals can lead to effective feature extraction. Therefore, DEn and FDEn are computed separately from the subband signals and used as features. Further, a random forest is used as a classifier to evaluate classification performance. A comparison of classification accuracy is presented where DEn and FDEn are used as features, both separately as well as in combination. The proposed method was applied to different types of multi-class classifications associated with sleep stages.

The remaining part of the chapter is organized as follows: In Section 2.2, the different concepts used in the proposed method are discussed, such as the DWT, DEn, FDEn, and the random forest classifier. Results and discussion are elaborated in Section 2.3. The conclusion of the chapter is presented in Section 2.4.

2.2 METHODOLOGY

The development of a sleep stage classification system includes several steps, such as analysis of the signals, feature extraction algorithm, and classifier. Therefore, in this section descriptions of the database used, signal decomposition using the DWT, DEn, and FDEn, and the classifiers are presented.

2.2.1 SLEEP-EDF DATABASE

The sleep-EDF database [33] contains several PSG sleep recordings that cover a whole night's duration. The database can be obtained from the Physionet website [34]. It can be accessed without any restriction, and it is freely available to download. The recordings contain the signals of several physiologies, such as EEG, EOG, respiration, body temperature, Chin EMG, and markers related to sleep events. The sleep stages have been manually scored by trained experts according to the R&K system of sleep scoring. The recordings have two types of file extension indicated by "SC" and "ST". Both types of files have EEG signals recorded with two channels: FPz-Cz and Pz-Oz [33]. In the presented study we used four files labeled with "SC" namely SC4002E0, SC4012E0, SC4102E0, SC4112E0, and four files labeled with "ST" namely ST7022J0, ST7052J0, ST7121J0, and ST7132J0. The SC recordings were of healthy subjects, and the ST recordings were obtained from subjects exhibiting difficulty with falling asleep [33]. The EEG signals present in the PSG recordings are sampled at 100 Hz. The literature review suggests that some of the studies used the Pz-Oz channel, and some of them used the FPz-Cz channel. We used EEG signals recorded from both channels in this study.

The annotations provided in the dataset were considered as the gold standard for comparing the performance of the proposed method. The annotations were obtained from the hypnogram file available in the dataset. Each 30-sec epoch of the EEG signal was provided with annotation of a sleep stage from S1–S4, W, R, M, or U (unknown state). We did not use epochs related to the U annotation. The multi-class classification problem was studied for several cases, considering the different sets of classes combined.

We used a six-class classification, considering classes S1 to S4, W, and REM as suggested by the R&K standard, which has also been used in previous studies [11,19,20]. The newly proposed AASM standard suggests the use of a five-class classification, obtained by combining the S3 and S4 stages into a single class. Therefore, we considered a five-class classification as the most important type of classification. The sleep stages S1 and S2 can be combined as they are known as shallow sleep [20,35]. Further, the combination of S1 and S2 as a single class gives rise to a four-class classification. The S1, S2, S3, and S4 stages fall into non-REM categories. The classification between W, REM, and non-REM sleep stages can be helpful for the diagnosis of REM-sleep-related disorders [36,37]. Sometimes it is of interest to separate EEG signals related to sleep and awake conditions; therefore, we also studied a two-class classification. Figure 2.1 depicts an example of the 30-sec epochs of the different sleep stages of 3000 samples. Table 2.1 includes the total number of epochs which were used in this study.

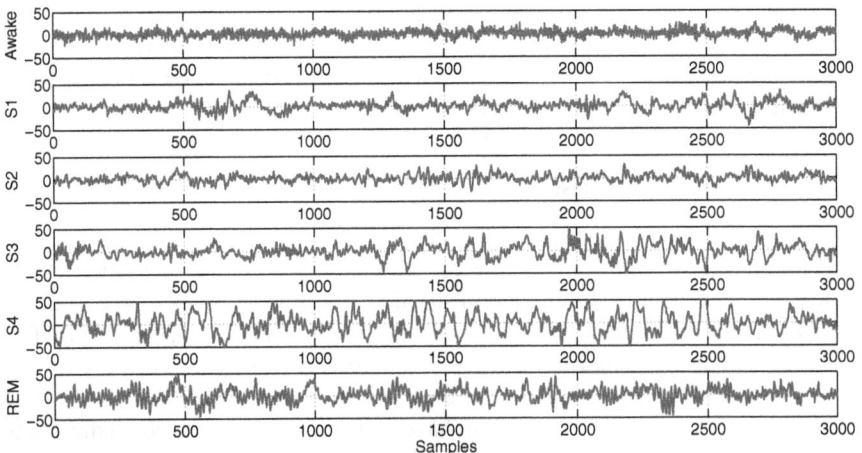

FIGURE 2.1 The epochs of different sleep stages: Awake—Awake condition; S1—Sleep Stage 1; S2—Sleep Stage 2; S3—Sleep Stage 3; S4—Sleep Stage 4; REM—Rapid Eye Movement.

TABLE 2.1

Number of Epochs of Different Sleep Stages Used in the Study

Sleep stage	Number of epochs used
Awake (W)	8006
S1	604
S2	3621
S3	672
S4	627
REM	1620
Total	**15,150**

2.2.2 DISCRETE WAVELET TRANSFORM

There are several methods for analyzing non-stationary signals like biomedical, earthquake, financial, and mechanical signals, etc. The wavelet transform is considered a powerful method for decomposing non-stationary signals into different scales, and provides a time-scale representation [38]. The wavelet transform is used for resolving the fixed time-frequency resolution problem of the short-time Fourier transform (STFT) [39]. The wavelet transform consists of two properties, namely space and frequency localization and multi-resolution analysis. Theoretically, the wavelet transform is broadly classified as the continuous wavelet transform (CWT) and the discrete wavelet transform (DWT). The CWT of signal $y(t)$ is computed as follows [40]:

$$Y_{CWT}(\tau,s) = \int_{-\infty}^{\infty} y(t)\psi_{\tau,s}^{*}(t)dt \qquad (2.1)$$

where, $\tau \in R$, $s \in R^{+}$, and $\psi(t)$ represent the translation parameter, scale parameter, and mother wavelet. In Equation (2.1), ψ^{*} indicates the conjugate operation of ψ. The $\psi_{\tau,s}(t)$ is expressed in terms of τ and s as follows [40]:

$$\psi_{\tau,s}(t) = \frac{\psi\left((t-\tau)/s\right)}{\sqrt{s}} \qquad (2.2)$$

The discrete version of the CWT is known as the DWT, which is explored for practical purposes [4]. Analyzing the signals using the DWT with a successive resolution is represented by the following expression [41]:

$$Y(t) = \sum_{p,q} c_{p,q}\psi_{p,q}(t) \qquad (2.3)$$

The translation and scale parameters of the mother wavelet for the DWT are as follows [4]:

$$\psi_{p,q}(t) = 2^{-p/2}\psi\left(2^{-p}t - q\right) \tag{2.4}$$

where p and q are the scaling and translation parameters. The coefficients of the DWT $c_{p,q}$ are computed as follows [41]:

$$c_{p,q} = \left\langle \psi_{p,q}(t),\, y(t) \right\rangle \tag{2.5}$$

Equation (2.5) shows that the coefficients of the DWT can be obtained from the inner product $\psi_{p,q}(t)$ and $y(t)$. In the case of the DWT, the mother wavelet has no closed-form, but it is generated using Equation (2.4) with dyadic, orthogonal, and non-redundant mapping in the time-frequency space [42].

In this work, a four-level DWT decomposition of the EEG signals was performed. The obtained subbands were reconstructed in the time domain for each EEG signal. The approximate and detailed components are shown in Figure 2.2. An epoch of a REM sleep stage was extracted from the dataset and subjected to a four-level wavelet decomposition. For the REM epoch obtained, the detailed components at each level, and the fourth-level approximate components, are depicted in Figure 2.2.

2.2.3 DISPERSION ENTROPY AND FLUCTUATION-BASED DISPERSION ENTROPY

Dispersion entropy (DEn) [29] is an entropy proposed recently to quantify the patterns in a time series. DEn can be used to characterize a time series and as a measure

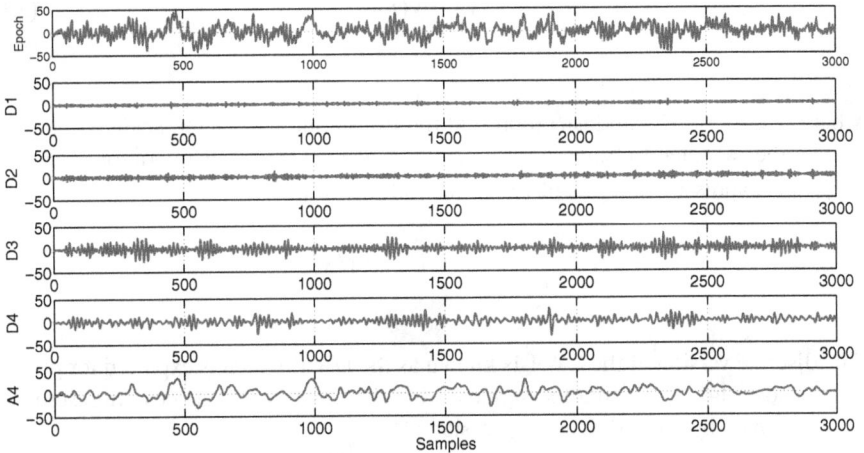

FIGURE 2.2 An epoch of 30 seconds and its wavelet components: A4—Approximate component level 4, D4—Detail component level 4, D3—Detail component level 3, D2— Detail component level 2, D1—Detail component level 1.

of uncertainty. It performs the mapping of dispersion patterns of the given time series; the mapping can be linear or nonlinear [29,30]. The concept of DEn was developed for use in computing permutation entropy (PEn) and sample entropy (SpEn) [30]. The details of the computation of DEn for time series $u = (u_1, u_1, \ldots, u_N)$ are explained in the following steps [29]:

1. In the first step mapping of the given array u is performed onto a set of predecided classes, say $s_1, s_2, \ldots, s_c$. Using a cumulative distribution function (NCDF), the first time series is mapped onto $v = (v_1, v_2, \ldots, v_N)$ from 0 to 1. Next, the integer number is assigned to each value of v from 1 to c using certain mapping formulas.

2. The elements of the embedded time series $v_j^{m,c}$ with embedding dimension m and delay τ are mapped onto a dispersion pattern. The dispersion pattern can be represented by the symbol $\pi_{l_1 l_2 \ldots l_m}$. Each embedded time series has m elements.

3. The relative frequency which indicates the probabilistic occurrence of pattern $\pi_{l_1 l_2 \ldots l_m}$ can be computed as:

$$P\left(\pi_{l_1 l_2 \ldots l_m}\right) = \frac{\#\{j \mid j \leq N - (m-1)\tau, v_j^{m,c} \text{ has type } \pi_{l_1 l_2 \ldots l_m}}{N - (m-1)\tau}$$

Which indicates the probability that dispersion pattern $\pi_{l_1 l_2 \ldots l_m}$ is mapped onto the embedded time series $v_j^{m,c}$.

4. Lastly, DEn can be computed as:

$$DEn(x, m, c, \tau) = \sum_{\pi=1}^{c^m} P\left(\pi_{l_1 l_2 \ldots l_m}\right) \ln\left(P\left(\pi_{l_1 l_2 \ldots l_m}\right)\right)$$

DEn quantifies the extent to which the dataset is spread out. The highest value of DEn corresponds to a situation where the probability of occurrence of all the patterns is equal. For further details and examples of DEn refer to references [29,30].

In this study, we also used a modified version of DEn known as fluctuation-based DEn (FDEn), as proposed in [30]. In the computation of FDEn, the difference between two adjacent dispersion patterns can be quantified. The difference in measurement between DEn and FDEn is in the way the potential patterns are computed. Further details can be obtained from [30], which also details possible fluctuations in the patterns with examples. In this study, we used a normalized version of DEn and FDEn.

2.2.4 RANDOM FOREST CLASSIFIERS

A random forest classifier consists of a group of tree classifiers [43]. Each tree classifier is generated by collecting a sample of a random vector from the input vector [43]. Let us assume the input vector is (v), and the class prediction vector is:

$$C_{RF}^M = \text{Majority vote} \left\{ \widehat{C_k}(v) \right\}_1^M ,$$

where $\widehat{C_k}(v)$ is the prediction vector of the vth random forest tree. The random forest classifier is used to increase the diversity of trees using a bootstrap aggregating technique [43]. In the random forest classifier, trees are used as base classifiers as follows [44]:

$$\left\{ h(v, \Theta_k) \right\},$$

where v and $\{\Theta_k\}$ are the input and independent, identically distributed, vectors. Due to the above-mentioned approach, it can be concluded that some data may be repeatedly used for training. The random forest classifier can then detect the small variations in input vectors [45].

Generally, the random forest classifier algorithm is summarized as follows [43]:

1. Assuming that Q number vector sets are considered as training data, then for every sampling process Q number of vector sets are required for training samples in a random manner. In this type of selection of the input vector sets, it may be possible that all the training data are not used at once; on the other hand, some data are used for training more than once.
2. Assuming that the p number of features is randomly selected from the total input P features. Then the best split on these p-dimensional features is used to split the node.
3. Each tree keeps growing until all the training samples are completely separated without pruning.

Any correlation between the trees is used to compute the error rate and strength of the random forest classifier [43]. The random forest classification accuracy and sensitivity parameters consist of the trade-off between the error rate and strength factors. The different combinations of trees control the error rate and strength factors of the random forest classifier.

2.3   RESULTS AND DISCUSSION

In this work, we used an sleep-EDF database available online, as discussed in Section 2.2.1 The sleep-EDF database [33] contains several PSG sleep recordings of a whole night's duration. In the given database, several EEG sleep recordings were present which were recorded from channels Pz-Oz and Fpz-Cz. The number of epochs used in this work for each sleep stage is shown in Table 2.1. The annotations provided in the dataset were used for evaluating the performance of the proposed method. In the present study, a ten-fold cross-validation was performed for evaluation of the model created during the classification. The multi-class classification problems was studied with several different classifications problems created by combining different sets of classes. We used a six-class classification which considered classes S1 to S4, W, and REM, as suggested by the R&K standard, which has also been used in previous studies [12,19,20]. Each of the sleep stages mentioned forms a separate class in a six-class classification. According to the AASM system for sleep scoring, five sleep stages are recognized. Therefore, a five-class classification was considered as the most important classification among the different types of multi-class classification experiments.

The EEG signals of both channels were decomposed using the DWT method. The DB4 wavelet [46] was used as a mother wavelet, as it was found suitable for the decomposition of the EEG signals due to its smoothening characteristic [47]. The rhythms are the frequency bands of the EEG spectrum [48] and have an important relation with sleep characteristics. Since the sampling frequency of the EEG signals is 100 samples/sec, the approximate component at the fourth level typically represents frequency content related to the delta rhythms. Therefore, we used a four-level wavelet decomposition in this study. DEn and FDEn were computed from the reconstructed subband components of the EEG signals. DEn values and FDEn values were used to make the feature vector for classification. For computing DEn and FDEn, the values of embedding dimension m and time delay τ were fixed as 3 and 1. We used both channels of the EEG signal available in the PSG recordings, and DEn and FDEn features were extracted from the wavelet subband of both EEG signals. The feature vectors of both the entropies were formed separately and their performance was compared in terms of the classification accuracy obtained using the random forest classifier. The random forest classifier is a tree-based classifier, which had shown good performance in our previous study of sleep stage classification [49]. The random forest classifier was used for assessing the performance of the features obtained from the EEG sleep signals.

Table 2.2 presents the accuracy of the classification obtained when the DEn and FDEn parameters were applied over the decomposed components of the EEG signals. The accuracy values are tabulated for each multi-class classification, as discussed in Section 2.2.1. It can be observed that accuracies for DEn and FDEn features of the six-class classification were 85.41% and 86.34%, respectively. The same

TABLE 2.2
The Classification Accuracy Obtained as a Result of the Classification with DEn Features and FDEn Features Using a Random Forest Classifier with 100 Trees

Multi-class classification case	Classification Accuracy (%) with	
	DEn features	FDEn features
Six-class	85.41	86.34
Five-class	86.73	87.68
Four-class	87.84	88.74
Three-class	92.78	93.63
Two-class	96.27	97.09

accuracies obtained for five-class classification were 86.73% and 87.68%, respectively. Similarly, for a four-class classification, the obtained accuracies were 87.84% and 88.74%, respectively. In a three-class classification, DEn feature classification accuracy was 92.78%, and FDEn classification accuracy was 93.09%. In the classification of REM and NREM, the classification accuracies were 96.27% and 97.09%, respectively.

Based on close observation of the classification accuracy values for the different cases, it was evident that FDEn features result in a slightly better performance of sleep stage classification as compared to DEn features. The experiments showed that, as features for classification, the use of FDEn was more effective as compared to the use of DEn. In other words, FDEn captured the signatures for discriminating against the different sleep stages more effectively. Therefore, FDEn can be considered more suitable for implementing in a sleep stage classification system. However, it was considered of interest to use both of these entropies to make a combined feature set for the classification system. Therefore, we further performed the experiments by combining both DEn and FDEn. The resultant feature vector was 15,150×20 in size, and the results are presented in Table 2.3.

It can be observed from Table 2.3 that the results obtained when combining DEn and FDEn improved the classification accuracies for each multi-class classification. When comparing the results presented in Table 2.2 and Table 2.3 it shows that the use of both DEn and FDEn in combination is more effective for sleep stage classification. Therefore, further experiments were performed using both entropies simultaneously.

In previous experiments, the number of trees used in the random forest classifier was fixed at 100. In further experiments, the accuracies were computed by varying the value of the number of trees (Tr) while keeping other values the same. The experiments were performed for all multi-class classifications. The number of trees vs classification accuracy (Acc) for different combinations of classes is shown in Figure 2.3 to Figure 2.7. The values of Tr were varied with the values of Acc. We took different values of Tr ranging from 50 to 400 with a difference of 50 trees and plotted the obtained accuracies. Figure 2.3 shows the plot Acc vs Tr for the classification of

---

TABLE 2.3

The Classification Accuracy Obtained as a Result of the Classification of Combining DEn and FDEn Features Using a Random Forest Classifier with 100 Trees

Multi-class classification case	Classification Accuracy (%) obtained using both DEn and FDEn as features
Six-class	86.77
Five-class	88.20
Four-class	89.12
Three-class	93.94
Two-class	97.31

six classes. The best value of *Tr* corresponds to the highest value of *Acc*. The highest value of *Acc* was 86.90%, and the corresponding value of *Tr* was 250. Similarly, for a five-class classification *Acc* vs *Tr* is shown in Figure 2.4. The highest value of *Acc* obtained, in this case, was 88.20%, and the value of *Tr* was 100. In a similar manner, *Acc* is plotted against *Tr* in Figure 2.5 for a four-class classification, in which the highest *Acc* obtained was 89.37%, with a corresponding value of *Tr* of 300. Another *Acc* vs *Tr* plot for a three-class classification is plotted in Figure 2.6.

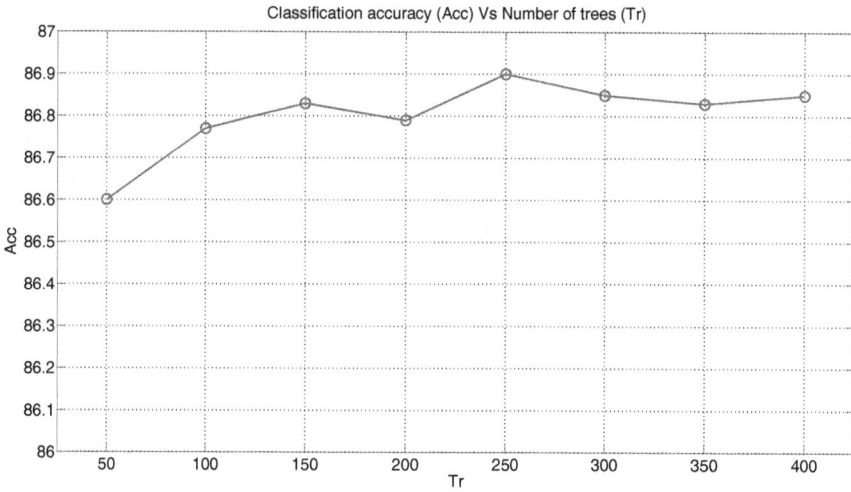

FIGURE 2.3 Plot of classification accuracy vs the number of trees used in a random forest classifier for a six-class classification.

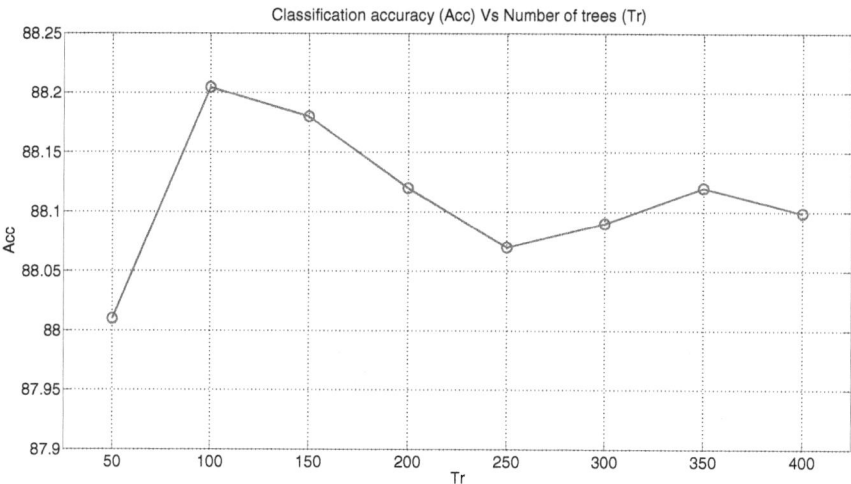

FIGURE 2.4 Plot of classification accuracy vs the number of trees used in a random forest classifier for a five-class classification.

It can be observed that the highest value of *Acc* was 94.07%, and the corresponding *Tr* value was 350. Similarly, a plot is shown in Figure 2.7, which presents the variation in *Acc* with respect to *Tr*. It was observed that on using *Tr* as 300, 350, and 400 the highest value of accuracy (97.37%) was obtained. Although, multiple values of *Tr* resulted in the same classification accuracy, the minimum value of *Tr*, which was 300, and recommended for classification. The variation in classification accuracies

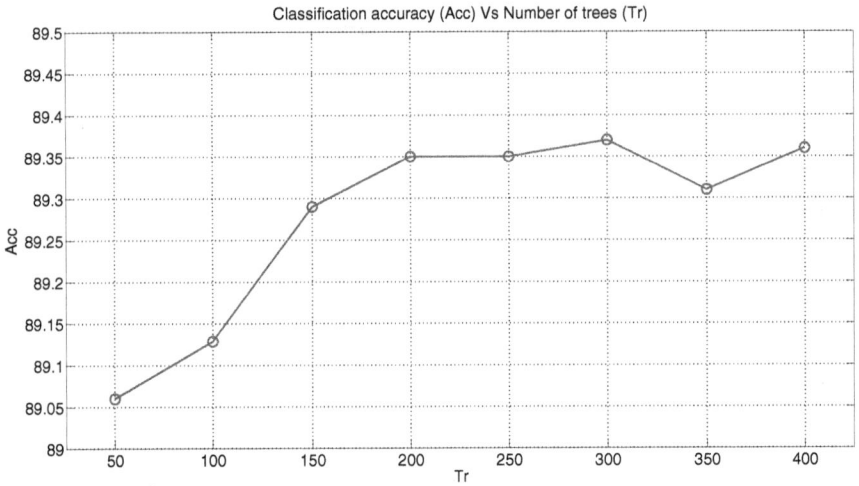

FIGURE 2.5   Plot of classification accuracy vs the number of trees used in a random forest classifier for a four-class classification.

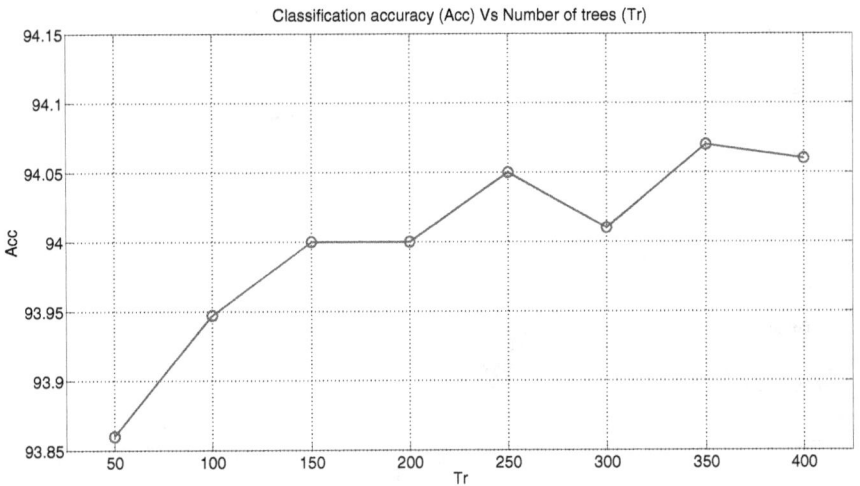

FIGURE 2.6   Plot of classification accuracy vs the number of trees used in a random forest classifier for a three-class classification.

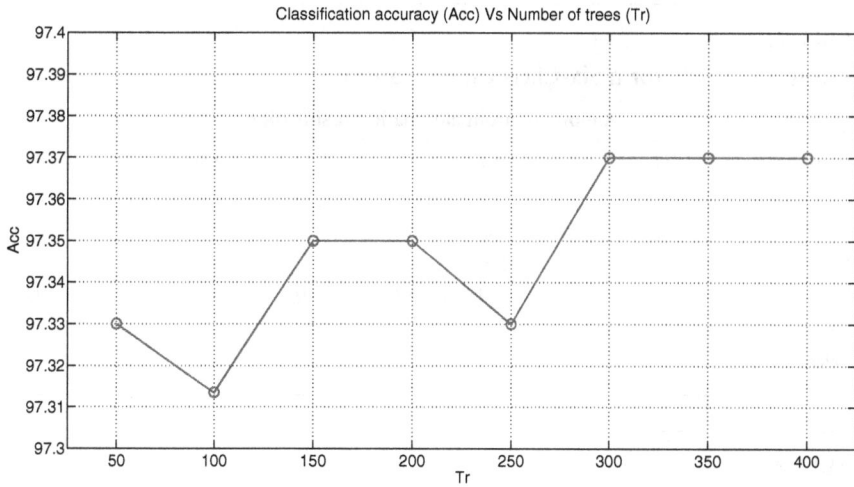

Classification accuracy (Acc) Vs Number of trees (Tr)

FIGURE 2.7 Plot of classification accuracy vs the number of trees used in a random forest classifier for a two-class classification.

TABLE 2.4

Classification Accuracies (*Acc*) on a Varying Number of Trees (*Tr*) for Each Multi-Class Classification

Multi-class classification	Number of trees used in the random forest classifiers							
	50	100	150	200	250	300	350	400
Six-class	86.60	86.77	86.83	86.79	86.90	86.85	86.83	86.85
Five-class	88.01	88.20	88.18	88.12	88.07	88.09	88.12	88.09
Four-class	89.06	89.12	89.29	89.35	89.35	89.37	89.31	89.36
Three-class	93.86	93.94	94.00	94.00	94.05	94.01	94.07	94.06
Two-class	97.33	97.31	97.35	97.35	97.33	97.37	97.37	97.37

with respect to variation in *Tr* is also presented in Table 2.4. From Table 2.4, the exact accuracy values can be obtained for each multi-class classification.

In order to evaluate the performance of a multi-class classification, *Acc* is not enough, and the values of sensitivity (SEN) and specificity (SPE) are also important for different multi-class classifications. Therefore, a confusion matrix was presented corresponding to the highest accuracy of each multi-class classification, along with the values of SEN and SPE for each class. Table 2.5 to Table 2.9 depicts the confusion matrix, SEN, and SPE values for six-class to two-class classification, respectively. It can be observed from the SEN values given in Table 2.5 to Table 2.6 that the S1 stage was misclassified as REM in the six-class and five-class classifications. Hence, the SEN of S1 was less when compared to other classes. Similarly, the S3 was classified as S2 resulting in the smallest SEN value, as can be observed in Table 2.5. However, S2, REM, and Awake were identified effectively with significant accuracy.

TABLE 2.5

Confusion Matrix for a Six-Class Classification

Expert annotation	Annotation from automatic classification						SEN	SPE
	S1	S2	S3	S4	REM	Awake		
S1	168	93	2	0	219	122	27.8	99.2
S2	32	3270	50	72	119	78	90.3	92.5
S3	3	427	116	101	4	21	17.3	99.4
S4	1	188	39	387	0	12	61.7	98.8
REM	42	150	1	5	1341	81	82.8	96.9
Awake	32	12	0	0	78	7884	98.5	95.6

TABLE 2.6

Confusion Matrix for a Five-Class Classification

Expert annotation	Annotation from automatic classification					SEN	SPE
	S1	S2	S3/S4	REM	Awake		
S1	182	94	2	208	118	30.1	99.2
S2	33	3169	229	112	78	87.5	93.7
S3/S4	3	467	795	7	27	61.2	98.2
REM	45	153	10	1337	75	82.5	97.0
Awake	37	12	2	75	7880	98.4	95.8

TABLE 2.7

Confusion Matrix for a Four-Class Classification

Expert annotation	Annotation from automatic classification				SEN	SPE
	S1/ S2	S3/S4	REM	Awake		
S1/ S2	3583	224	247	171	80.8	92.6
S3/S4	471	801	2	25	61.7	98.3
REM	257	8	1291	64	79.7	97.7
Awake	75	1	65	7865	98.2	96.4

The technique presented in this work can be used for detecting sleep stages in a computer-assisted sleep stage identification system. The results presented here indicate the effectiveness of DEn and FDEn features for sleep stage classification. The developed technique is based on two channels of EEG signals. However, the performance of the studied entropy parameters can be evaluated for the individual channels of EEG signals.

The nature of the EEG signal is non-stationary; therefore, in the first step of the developed algorithm, the EEG signal was first decomposed into simpler components

TABLE 2.8
Confusion Matrix for a Three-Class Classification

Expert annotation	Annotation from automatic classification				
	S1/ S2/ S3/S4	REM	Awake	SEN	SPE
S1/ S2 /S3/S4	5131	230	163	92.9	96.1
REM	289	1267	64	78.2	97.8
Awake	85	67	7854	98.1	96.8

TABLE 2.9
Confusion Matrix for a Two-Class Classification

Expert annotation	Annotation from automatic classification			
	S1/ S2/ S3/S4/ REM	Awake	SEN	SPE
S1/ S2 /S3/S4/ REM	6964	180	97.5	97.30
Awake	217	7789	97.3	97.50

using the DWT. This component can be helpful in effectively capturing the properties of various sleep stages. The various sleep stages show the different characteristics associated with the different frequencies [50]. For example, the S1 stage exhibits a low voltage in mixed frequencies and the highest voltage at 2–7 Hz. Similarly, the S3 stage can be characterized by 2 Hz or slower oscillations. The DWT filters out certain frequency bands and provides subband signals associated with a specific range of frequencies. However, the subband frequencies depend on the sampling frequencies. Hence, extracting features from the subband components can be effective.

In one study [51], changes in multiscale signal complexity were explored using EEG signals. The power-law scaling exponents were computed from EEG frequency spectra to quantify complexity during the human sleep cycle. They used multiscale DEn and observed the dependency of entropy during the sleep cycle on the time scale. Their observations were presented in terms of a graph of multiscale DEn at fine and coarse time scales on power spectral density (PSD) slopes and statistical analysis. In [27], DEn was computed from EOG signals, in the DWT domain, for sleep stage classification. They used a version of DEn, namely refined-composite multiscale DEn (RCMDEn); also, spectral entropy, moment-based measures, and autoregressive model coefficients were computed along with the RCMDEn. They used a random forest support vector machine and a random under-sampling boost for classification. The developed algorithm resulted in significantly improved performance over the compared methods.

Similarly, another study [26] used dispersion entropy, approximate entropy, sample entropy, statistical features, time-frequency-based entropy measures, and autocorrelation-based features for the analysis of a time-frequency image of heart rate signals. They developed an algorithm that could determine if a patient of obstructive sleep apnea/hypopnea syndrome was sleeping or awake. The above discussion indicates that DEn and its variants can effectively discriminate between the various sleep stages using different modalities. Therefore, in the proposed method, DEn and FDEn computed from EEG signals were explored for developing an algorithm for the development of an automatic sleep stage detection system.

Sleep physiology is studied using PSG recordings. In PSG, a number of channels are present, and they are used to identify the patterns of different sleep stages. In general, annotations provided using PSG recordings are considered as the gold standard; therefore, in this work, the annotations provided by experts were used to develop and evaluate the proposed method. However, the placement of different electrodes corresponding to EEG, EMG, and ECG, etc. can cause challenges in subject preparation and the pasting of electrodes on the skin of the subject. Therefore, developing a sleep identification system based on a single modality like an EEG signal can be useful and convenient for the subject during the recording process.

We used data from the eight recordings available in the EDF sleep database. The sleep stage annotations were assigned to each 30-sec epoch taken from different subjects. In this work, the EEG epochs taken from different subjects were combined without including any prior information about the patients to form different classes. Therefore, it would be interesting to study the effect of subject variability on the performance of the proposed algorithm. Hence, in the future, more recordings taken from more patients could be studied to evaluate the performance of the proposed methodology.

2.4 CONCLUSION

In this chapter, DEn and FDEn entropy measures were studied for automated sleep stage classification. The DWT was used to decompose EEG epochs of a 30-sec duration so that each epoch could be represented by simpler subband signals associated with certain frequencies. The DEn and FDEn measures were computed from the subband signals so that the patterns in the associated frequencies could be captured effectively. The classification was performed by considering each of the DEn and FDEn features separately, one at a time. The presented results were compared in terms of classification accuracy. It was observed that classification accuracy was higher when FDEn was used for feature computation. We also presented the results of a method when both DEn and FDEn were combined to extract features; the overall classification performance improved when combining DEn and FDEn.

The presented method was applied to classify sleep stages associated with different multi-class classifications. It was evident from the results that the presented method identified various classes with significant classification accuracy values across all multi-class classifications. The EEG signals extracted from both channels were employed for evaluating the performance of the presented method. The sleep scoring system based on single-channel data could be implemented with less complexity and therefore be more suitable for implementation in wearable devices for sleep scoring. Hence, in the future, the suitability of each channel can be evaluated using the presented algorithm.

ACKNOWLEDGMENTS

This work has been carried out in Signal Processing Lab, Department of Electronics and Communication Engineering, Birla Institute of Technology, Mesra, Ranchi, India. The study has been partially funded by MHRD-NPIU, Govt. of India grant # 1-5737336180.

REFERENCES

1. Hsu Y-L, Yang Y-T, Wang J-S and Hsu C-Y. 2013. Automatic sleep stage recurrent neural classifier using energy features of EEG signals. *Neurocomputing*, **104**, 105–14.
2. Šušmáková K and Krakovská A. 2008. Discrimination ability of individual measures used in sleep stages classification. *Artif. Intell. Med.*, **44**, 261–77.
3. Allan Hobson J. 1969. A manual of standardized terminology, techniques and scoring system for sleep stages of human subjects. *Electroencephalogr. Clin. Neurophysiol.*, **26**, 644.
4. Danker-Hopfe H. et.al 2001. Interrater reliability between scorer from eight European sleep labs in subjects with different sleep disorders. *J. Sleep Res.*13, 63–69.
5. Penzel T, Behler PG, Von Buttlar M, Conradt R, Meier M, Möller A and Danker-Hopfe H. 2003. Reliablität der visuellen schlafauswertung nach Rechtschaffen und Kales von acht aufzeichnungen durch neun schlaflabore. *Somnologie*, **7**, 49–58.
6. Iber C, Ancoli-Israel S, Chesson AL and Quan SF. 2007. The American Academy of Sleep Medicine (AASM) Manual for the Scoring of Sleep and Associated Events: Rules, Terminology and Technical Specifications, Vol. 1. Westchester, IL: American Academy of Sleep Medicine,
7. Agarwal R and Gotman J. 2001. Computer-assisted sleep staging *IEEE Trans. Biomed. Eng.*, **48**, 1412–23.
8. Held CM, Heiss JE, Estévez PA, Perez CA, Garrido M, Algarín C and Peirano P. 2006. Extracting fuzzy rules from polysomnographic recordings for infant sleep classification. *IEEE Trans. Biomed. Eng.*, **53**, 1954–62.
9. Krakovská A and Mezeiová K. 2011. Automatic sleep scoring: A search for an optimal combination of measures. *Artif. Intell. Med.*, **53**, 25–33.
10. Acharya UR, Bhat S, Faust O, Adeli H, Chua ECP, Lim WJE and Koh JEW. 2015. Nonlinear dynamics measures for automated EEG-based sleep stage detection *Eur. Neurol.*, **74**, 268–87.
11. Liang SF, Kuo CE, Hu YH, Pan YH and Wang YH. 2012. Automatic stage scoring of single-channel sleep EEG by using multiscale entropy and autoregressive models. *IEEE Trans. Instrum. Meas.*, **61**, 1649–57.
12. Hsu YL, Yang YT, Wang JS and Hsu CY. 2013. Automatic sleep stage recurrent neural classifier using energy features of EEG signals. *Neurocomputing*, **104**, 105–14.
13. Imtiaz SA and Rodriguez-Villegas E. 2014. A low computational cost algorithm for REM sleep detection using single channel EEG. *Ann. Biomed. Eng.*, **42**, 2344–59.
14. Tsinalis O, Matthews PM and Guo Y. 2016. Automatic sleep stage scoring using time-frequency analysis and stacked sparse autoencoders. *Ann. Biomed. Eng.*, **44**, 1587–97.
15. Bajaj V and Pachori RB. 2013. Automatic classification of sleep stages based on the time-frequency image of EEG signals. *Comput. Methods Programs Biomed.*, **112**, 320–8.

16. Rai K, Bajaj V and Kumar A. 2015. Hilbert-Huang transform based classification of sleep and wake EEG signals using fuzzy c-means algorithm. 2015. International Conference on Communication and Signal Processing, ICCSP 2015 (Institute of Electrical and Electronics Engineers Inc.), pp. 460–4.
17. Taran S, Bajaj V and Sharma D. 2017. Robust Hermite decomposition algorithm for classification of sleep apnea EEG signals. *Electron. Lett.*, **53**, 1182–4.
18. Taran S and Bajaj V. 2020. Sleep apnea detection using artificial bee colony optimize hermite basis functions for EEG signals. *IEEE Trans. Instrum. Meas.*, **69**, 608–16.
19. Zhu G, Li Y and Wen PP. 2014. Analysis and classification of sleep stages based on difference visibility graphs from a single-channel EEG signal. *IEEE J. Biomed. Heal. Informatics*, **18**, 1813–21.
20. Hassan AR and Bhuiyan MIH. 2016. Computer-aided sleep staging using complete ensemble empirical mode decomposition with adaptive noise and bootstrap aggregating. *Biomed. Signal Process. Control*, **24**, 1–10.
21. Hassan AR and Bhuiyan MIH. 2016. A decision support system for automatic sleep staging from EEG signals using tunable Q-factor wavelet transform and spectral features. *J. Neurosci. Methods*, **271**, 107–18.
22. Giannakeas N. 2018. EEG-based automatic sleep stage classification. *Biomed. J. Sci. Tech. Res.*, **7**, 6032–7.
23. Zhang Y, Xing J, Guo C, Yang C and Liu X. 2018. Sleep stage classification based on EEG signal by using EMD and DFA algorithm. *ACM International Conference Proceeding Series* (Association for Computing Machinery), pp. 156–60.
24. Bresch E, Großekathöfer U and Garcia-Molina G. 2018. Recurrent deep neural networks for real-time sleep stage classification from single channel EEG. *Front. Comput. Neurosci.*, **12**, 1–12.
25. Yildirim O, Baloglu UB and Acharya UR. 2019. A deep learning model for automated sleep stages classification using PSG signals. *Int. J. Environ. Res. Public Health*, **16**(4):599, 1-21.
26. Tripathy RK and Acharya UR. 2018. Use of features from RR-time series and EEG signals for automated classification of sleep stages in deep neural network framework. *Biocybern. Biomed. Eng.*, **38**, 890–902.
27. Rahman MM, Bhuiyan MIH and Hassan AR. 2018. Sleep stage classification using single-channel EOG. *Comput. Biol. Med.*, **102**, 211–20.
28. Aboalayon KAI, Faezipour M, Almuhammadi WS and Moslehpour S. 2016. Sleep stage classification using EEG signal analysis: A comprehensive survey and new investigation. *Entropy*, **18**, 1–31.
29. Rostaghi M and Azami H. 2016. Dispersion entropy: A measure for time-series analysis. *IEEE Signal Process. Lett.*, **23**, 610–4.
30. Azami H and Escudero J. 2018. Amplitude- and fluctuation-based dispersion entropy. *Entropy*, **20**, 210.
31. Richman JS and Moorman JR. 2000. Physiological time-series analysis using approximate and sample entropy. *Am. J. Physiol.—Hear. Circ. Physiol.*, **278**, **H2039–H2049**.
32. Bandt C and Pompe B. 2002. Permutation entropy: A natural complexity measure for time series. *Phys. Rev. Lett.*, **88**(, 174102.
33. Kemp B, Zwinderman AH, Tuk B, Kamphuisen HAC and Oberyé JJL. 2000. Analysis of a sleep-dependent neuronal feedback loop: The slow-wave microcontinuity of the EEG. *IEEE Trans. Biomed. Eng.*, **47**, 1185–94.
34. Goldberger AL, Amaral LA, Glass L, Hausdorff JM, Ivanov PC, Mark RG, Mietus JE, Moody GB, Peng CK and Stanley HE. 2000. PhysioBank, PhysioToolkit, and PhysioNet: Components of a new research resource for complex physiologic signals. *Circulation*, **101**, e215–20.

35. Wu Z and Huang NE. 2009. Ensemble empirical mode decomposition: A noise-assisted data analysis method. *Adv. Adapt. Data Anal.*, **1**, 1–41.
36. Iranzo A, Santamaria J and Tolosa E. 2009. The clinical and pathophysiological relevance of REM sleep behavior disorder in neurodegenerative diseases. *Sleep Med. Rev.*, **13**, 385–401.
37. Iranzo A, Molinuevo JL, Santamaría J, Serradell M, Martí MJ, Valldeoriola F and Tolosa E. 2006. Rapid-eye-movement sleep behaviour disorder as an early marker for a neurodegenerative disorder: a descriptive study. *Lancet Neurol.*, **5**, 572–7.
38. Frei MG and Osorio I. 2007. Intrinsic time-scale decomposition: time–frequency–energy analysis and real-time filtering of non-stationary signals. *Proc. R. Soc. A Math. Phys. Eng. Sci.*, **463**, 321–42.
39. Kim H and Ling H. 1993. Wavelet analysis of radar echo from finite-size targets. *IEEE Trans. Antennas Propag.*, **41**, 200–7.
40. Rioul O and Duhamel P. 1992. Fast algorithms for discrete and continuous wavelet transforms. *IEEE Trans. Inf. Theory*, **38**, 569–86.
41. Crowley PM. 2007. A guide to wavelets for economists. *J. Econ. Surv.*, **21**, 207–67.
42. Griffel DH and Meyer Y. 1995. Wavelets and operators. *Math. Gaz.*, **79**, 227.
43. Breiman L. 2001. Random forests. *Mach. Learn.*, **45**, 5–32.
44. Hastie T, Tibshirani R, Friedman J. 2009. *The Elements of Statistical Learning The Elements of Statistical Learning Data Mining, Inference, and Prediction*, 2nd Springer-Verlag New York
45. Rodriguez-Galiano VF, Ghimire B, Rogan J, Chica-Olmo M and Rigol-Sanchez JP. 2012. An assessment of the effectiveness of a random forest classifier for land-cover classification. *ISPRS J. Photogramm. Remote Sens.*, **67**, 93–104.
46. Subasi A. 2007. EEG signal classification using wavelet feature extraction and a mixture of expert model. *Expert Syst. Appl.*, **32**, 1084–93.
47. Omerhodzic I, Avdakovic S, Nuhanovic A and Dizdarevic K. 2010. Energy distribution of EEG signals : EEG signal wavelet-neural network classifier World Academy of Science, Engineering and Technology, 61, 1190–1195.
48. Klimesch W. 1999. EEG alpha and theta oscillations reflect cognitive and memory performance: A review and analysis. *Brain Res. Rev.*, **29**, 169–95.
49. Sharma R, Pachori RB and Upadhyay A. 2017. Automatic sleep stages classification based on iterative filtering of electroencephalogram signals. *Neural Comput. Appl.*, **28** 2959–78.
50. Güneş S, Polat K and Yosunkaya Ş. 2010. Efficient sleep stage recognition system based on EEG signal using k-means clustering based feature weighting. *Expert Syst. Appl.*, **37**, 7922–8.
51. Miskovic V, MacDonald KJ, Rhodes LJ and Cote KA. 2019. Changes in EEG multiscale entropy and power-law frequency scaling during the human sleep cycle *Hum. Brain Mapp.*, **40**, 538–51.

3 Detection of Epileptic Electroencephalogram Signals Employing Visibility Graph Motifs

Sayanjit Singha Roy, Sudip Modak,
Kaniska Samanta, Soumya Chatterjee,
and Rohit Bose

CONTENTS

3.1 INTRODUCTION

Electroencephalography (EEG) signals are a type of bio-potential, which are non-stationary and aperiodic in nature, generated from neurons, i.e., nerve cells inside the human brain. The electrical discharges released from the neurons inside the brain become rapidly abrupt in the case of neurological disorders. Thus, analysis of EEG signals is considered to be the most common approach in the diagnosis of neurological diseases, since EEG signals inherently contain important character-istics regarding the different firing patterns of the neurons. Epilepsy is one of the most commonly occurring and severe kinds of neurodegenerative disorder, which affects roughly 1–2% of the worldwide human population, and people of all ages are prone to this disease [1, 2]. Epilepsy occurs because of the abnormal functioning of the neurons, which interrupt the regular transmission of information from one brain cell to another. Hence, transient disturbances occur in the discharge patterns of the neurons, which results in repeated epileptic seizures [3]. Epileptic seizures are generally characterized by recurrent high amplitude spikes observed in EEG recordings. In clinics, the analysis of EEG signals is considered the most viable method to locate and isolate the epileptogenic regions inside the human brain before removing them surgically [4]. During seizure activities, the neurons are observed to be discharging abnormally, and the flow of electrical signals within the human brain becomes highly anomalous, which subsequently develops into several symp-toms, such as the rapid declination of mental and physical performance, partial or complete loss of consciousness, and trembling movements in the arms and legs. [5]. It has also been reported in the literature that the onset of these seizure conditions are highly impulsive and can prove to be catastrophic if not appropriately diagnosed [6]. Thus, prompt and accurate diagnosis of epilepsy is necessary to preclude such detrimental physiological aberrations. Previously, in the pathology labs and clinics, the conventional methods of the EEG screening technique included minute visual inspection of the EEG recordings by experienced doctors. However, these methods are largely dependent on human interventions, which are obviously error-prone and time consuming. Apart from that, visual investigation of the EEG recordings, which are very long in length, is a tedious job [7]. Another method for the detection of epi-lepsy includes interpretation of functional magnetic resonance images (fMRI) and diffusion magnetic resonance images (dMRI) [8]. However, from a practical point of view, these imaging methods are not only expensive and have great complexity in their implementation, but they are also reported to produce poorer time resolu-tions compared to EEG-based analysis techniques [9]. Therefore, considering the above-mentioned facts, the aim to develop an automated and computerized disease diagnostic scheme for the accurate and prompt detection of epilepsy based on EEG signal analysis has been the foremost priority of researchers for the past few years all across the world.

In the literature, several studies have been reported, where various signal pro-cessing and machine learning algorithms were implemented to discriminate EEG signals into healthy, inter-ictal, and epileptic seizure categories [10–37]. In [10], a cross-correlation technique was implemented in the time domain to analyze and extract suitable features for classifying healthy and seizure EEG signals. In [11], frequency-domain-based spectral analysis of the EEG signals was carried out using the fast Fourier transform (FFT), and the identification of epileptic seizure EEG sig-nals was performed using a decision tree (DT) classifier. Both time and frequency-domain-based features were utilized in [12] to differentiate between the EEG signals for healthy and epileptic seizure categories using a special form of a recurrent neural network, i.e., known as the Elman network. Since, EEG signals typically exhibit non-stationary characteristics, applications of different joint time-frequency-domain-based techniques were reported in the literature for identifying the epileptic seizure category of EEG signals. In [13], the application of the wavelet transform was reported for separating healthy and seizure class EEG signals. Implementation of the discrete wavelet transform (DWT) method was reported in [14] to categorize the EEG signals into different frequency bands, and the corresponding sub-band frequencies were fed into a modular neural network architecture as inputs to classify the EEG signals into the healthy and epileptic seizure categories. The Stockwell transform (ST) was implemented in [15] to extract several statistical and energy domain features from the time-frequency spectrum of the EEG signals, and a detailed discriminative analysis was carried out based on the different lengths of the EEG signals using support vector machines (SVM) and a k-nearest neighbor (k-NN) classifier for distinguishing between the healthy, inter-ictal, and epileptic seizure class EEG signals. A modified implementation of the previously mentioned Stockwell transform, namely the cross-hyperbolic Stockwell transform (XHST) was applied in [16] to differentiate between healthy and seizure category EEG signals. Spectral features were extracted from the time-frequency bands of the EEG signals in [17], using the smoothed pseudo-Wigner-Ville distribution (SPWVD) technique, where the application of an artificial neural network (ANN) was reported to identify the EEG signals into normal, seizure-free, and epileptic seizure categories. Application of an ANN is also reported in [18] to identify the seizure class EEG signals based on the approximate entropy features extracted from the multi-wavelet transform of the respective EEG signals. The identification of the healthy and seizure class EEG signals was performed in [19], utilizing time-frequency localized three-band orthogonal wavelet filter banks. An improved representation of the decomposed eigenvalues, computed from the Hilbert transform and Henkel matrix, was reported in [20] for detecting the seizure category of EEG signals. The empirical mode decomposition (EMD) technique was implemented in [21] for discriminating between the healthy and epileptic seizure category of EEG signals. EMD iteratively decomposes the EEG signals into several band-limited analytic functions, namely intrinsic mode functions (IMFs). The application of the Fourier-Bessel (FB) expansion was reported in [22], for computing mean frequencies of the corresponding IMFs, which were subsequently used to classify between the pre-ictal (seizure-free) and seizure category intracranial EEG signals. In [23], rhythmic separation of the EEG signals was performed using the Fourier-Bessel series expansion (FBSE) technique, and the weighted multi-scale Renyi permutation entropy (WMRPE) based features extracted from the EEG spectrums were utilized to identify the epileptic seizure class of EEG signals. A similar experiment based on the extraction of the IMFs using the EMD technique was carried out in [24], where the immediate area of the IMFs was used as a feature to identify the seizure class of EEG signals. Implementation of the EMD technique was also reported in [25], where features based on the 2D and 3D phase-space representation (PSR) of the obtained IMFs were utilized for detection of seizure signals. The identification of healthy and seizure signals was performed in [26], utilizing features derived from the second-order differential plots (SODPs) of the IMFs, which were obtained using an EMD analysis of the corresponding EEG signals. Apart from that, implementation of the EMD technique was also reported in [27], where several features derived from the amplitude and frequency modulated bandwidths of the obtained IMFs were used for discrimination of the EEG signals. Statistical feature extraction from the amplitude-frequency modulated components of the EEG signals was reported in [28], using a hybrid implementation of optimum allocation sampling (OAS) and Teager energy operator (TEO) methods for detection of epileptic seizure EEG signals. In [29], the DWT technique was applied to decompose the EEG signals into several frequency subbands, and using Teager-Kaiser energy operator (TKEO) the instantaneous energy of the sub-frequency bands were measured. Following this, several statistical features were extracted from the TKEOs of different frequency bands, which were further utilized to discriminate between the healthy and epileptic seizure class of EEG signals. EEG signals manifest the anomalous firing patterns of the brain neurons during seizure activities, which in turn quantify the complex dynamics of the human brain; hence, apart from being highly non-stationary in nature, EEG signals also exhibit non-linear and chaotic characteristics [15]. Owing to this fact, several non-linear features, including fractal dimension (FD), approximate entropy, permutation entropy, largest Lyapunov exponent, etc., were used to classify the healthy and epileptic seizure class EEG signals [30–33]. The application of fractal mathematics implementing multi-fractal detrended fluctuation analysis (MFDFA), based on a non-linear feature extraction framework, was implemented in [5] to analyze the chaotic behavior of healthy and seizure class EEG signals in the fractal domain. A multiple hybrid fractal-dimension-based framework implementing the flexible analytic wavelet transform (FAWT) [34] and tunable-Q-factor wavelet transform (TQWT) [35] was also reported to identify seizure class EEG signals. Identification of epileptic seizure category EEG signals was performed in [36], utilizing histogram features computed from key point based 1D local binary patterns (LBP) of the respective EEG signals. In [37], a detailed discriminative analysis based on generalized and uniform LBP features was carried out for segregation of the healthy and epileptic seizure category EEG signals. Analysis of the epileptic EEG signals was done using the Hermite transform in [38], along with implementing a hybrid feature extraction framework, including several entropy, histogram, and statistical features, for detection of epileptic seizures. Recently, in [39], the application of a deep convolutional neural network (CNN) was reported for the programmed recognition of epileptic seizure class EEG signals. In [40], application of the SPWVD technique was reported for rhythmic separation of the EEG signals; using the obtained time-frequency images, identification of the epileptic EEG signal was then made using a least square-support vector machine (LS-SVM) classifier. Despite the fact that several methods were reported in the literature for detection and classification of epilepsy, with considerably good accuracies, the scope for improvement in the detection of epileptic EEG signals remains an important area of research.

The concept of transforming a time series into a graphical network and visualizing the underlying connectivity patterns between different nodes of the network was first introduced by Zhang et al. in [41]. In the recent literature, graph theory-based approaches to detect different brain-related disorders are gaining popularity, since human brain abnormalities can be thoroughly characterized by quantifying changes in the different network attributes [42]. In [43], a novel approach to mapping the time series data into complex networks was introduced using a visibility graph (VG) algorithm. Following this, several graph theory-based signal processing techniques using VG algorithms were reported for analyzing time-domain EEG signals for the detection of epilepsy [45–50]. A significant advantage of using VG aided graph-domain-based techniques to analyze the epileptic EEG signals is that the network topology can efficiently capture the random non-linear correlations present in the non-stationary EEG signals, which are caused due to the irregular firing patterns of the brain neurons during seizure activities [44]. Apart from that, the issue of signal adulteration, i.e., discriminating the EEG recordings from random and uncorrelated noise, is important for a better understanding of the dynamic discharge patterns of the brain neurons. In this context, graph-domain-based signal processing techniques based on visibility algorithms are particularly suitable to analyze epileptic EEG signals since they are reported to adeptly discriminate between chaotic signals and white noise. Moreover, VG based techniques are also immune to any vertical or horizontal transformation of the time series data, making it analytically more straightforward and flexible for implementation of real-time monitoring of the epileptic EEG signals. In [45], a conventional approach to extract complex network features employing the VG technique was reported for detection of epileptic seizure category EEG signals. An effective approach to construct VG networks from the higher frequency bands of EEG signals was reported in [46] for epileptic seizure identification. Application of a slightly modified and computationally simpler version of the conventional VG algorithm, namely horizontal visibility graph (HVG), was reported in [47] for detection of epilepsy, introducing a novel edge weight for constructing the corresponding HVG networks. A weighted VG (WVG) algorithm by introducing edge weights in terms of the gradient between two nodes was also reported in [48], for segregating the epileptic seizure-free EEG signals from the seizure class ones. Several other VG aided graph-domain-based feature extraction techniques were also reported in the recent literature on the detection of epilepsy [49, 50].

Taking into account the above-mentioned advantages and implementations of using VG based approaches to analyze epileptic EEG signals, yet one constraint of the VG based methods still remain, which is that they generally evaluate the EEG signals on a global scale without performing adequate tracing of the existing local fluctuations in the EEG signals. The epileptic EEG signals manifest the ephemeral discharge patterns of the brain neurons during seizure activities; therefore, analysis of the temporal characteristics quantifying local variations present in the EEG signals can provide considerably more significant information regarding various epileptic-seizure-related activities. Considering this fact, in this chapter, we are proposing a VG motif profile based feature extraction and classification framework for the detection of epileptic EEG signals. VG motifs are considered to be a subgraph of a larger graph. Besides, motifs can provide information regarding the underlying patterns present in a large complex network. It has also been reported in the existing literature that both sequential VG and HVG motifs can differentiate between the dynamics of different types of time series. However, the application of VG and HVG motifs to differentiate the underlying dynamics of the EEG time series has not yet been reported so far in the existing literature. In fact, to the best of the authors' knowledge, this the first study, where both VG and HVG sequential motifs have been used to quantify the differences between different categories of EEG signals.

To this end, we procured EEG signals corresponding to the healthy, inter-ictal, and seizure categories from an online benchmark data archive. After the acquisition of the EEG data, we computed the VG and HVG of the EEG time series to convert them into binary graphical networks. Following this, we computed sequential VG and HVG motifs considering a sliding window consisting of four nodes (data points) of the graph, for which six types of motif profiles could be obtained. The window is translated along the entire length of the signal, and occurrence of the six motifs for the entire signal length is computed. Finally, the frequency of the six VG and HVG sequential motifs were computed, and the process was repeated for all the categories of EEG signal. The frequency of occurrence of the conventional VG and HVG sequential motifs were utilized as features in this study for further discrimination of the EEG signals. Subsequently, these extracted features were made susceptible to the analysis of variance (ANOVA) test to measure their discriminative capabilities between different problem categories. In this study, we addressed eight binary and one multi-class classification problems by combining different sets of EEG signals for better identification of different categories of EEG signal. Finally, the frequency of occurrence of different motifs were served as inputs to the employed four highly efficient machine learning classifiers, namely a random forest (RF), support vector machine, k-nearest neighbor (k-NN), and naïve Bayes (NB). From the classification results, we observed a high degree of recognition accuracy in segregating different categories of EEG signals, utilizing both the conventional VG and the HVG sequential motif frequencies to distinguish between the features. The performance of the proposed methodology was also put against several models in the literature for comparison, from which comparable and even better results were achieved for some classification problems. A brief pictorial illustration of the proposed methodology is presented using a flowchart in Figure 3.1.

The rest of the chapter is arranged as follows. A brief description of the EEG data archive used in this study is provided in Section 3.2. Section 3.3, i.e., the methodology section, deals with the theoretical background of conventional VG, horizontal VG, and the concept of the VG motif profiles. The theoretical background of the machine learning classifiers is given in Section 3.4, and in Section 3.5 a brief description of the different classifiers is given. The results are discussed in Section 3.6, and lastly Section 3.7 concludes the chapter.

3.2   EEG DATASET

The EEG signals used in this study were provided by the University of Bonn, Germany [51]. The data archive contained five different sets of single-channel EEG signals, indicated as sets A, B, C, D and E, respectively. Monitored by expert neurologists, the recording of the EEG signals was done by implementing the standard 128-channel electrode system, while an average common reference was chosen for all the similar respective channels. After the initial recording process, the EEG recordings

```
┌─────────────────────────────────┐
│     EEG Signal acquisition      │
└─────────────────────────────────┘
                ↓
┌─────────────────────────────────┐
│  Convert EEG signals into Graph │
│        using VG and HVG         │
└─────────────────────────────────┘
                ↓
┌─────────────────────────────────┐
│ Select a sliding window consisting of │
│             4 nodes             │
└─────────────────────────────────┘
                ↓
┌─────────────────────────────────┐
│   Extract sequential VG and HVG │
│      Motifs within the window   │
└─────────────────────────────────┘
                ↓
┌─────────────────────────────────┐
│  Slide the window along the entire │
│        length of the signal     │
└─────────────────────────────────┘
                ↓
┌─────────────────────────────────┐
│ Compute the frequency of occurrence │
│ of each motifs for the entire signal │
└─────────────────────────────────┘
                ↓
┌─────────────────────────────────┐
│           ANOVA test            │
└─────────────────────────────────┘
                ↓
┌─────────────────────────────────┐
│         Classification          │
└─────────────────────────────────┘
```

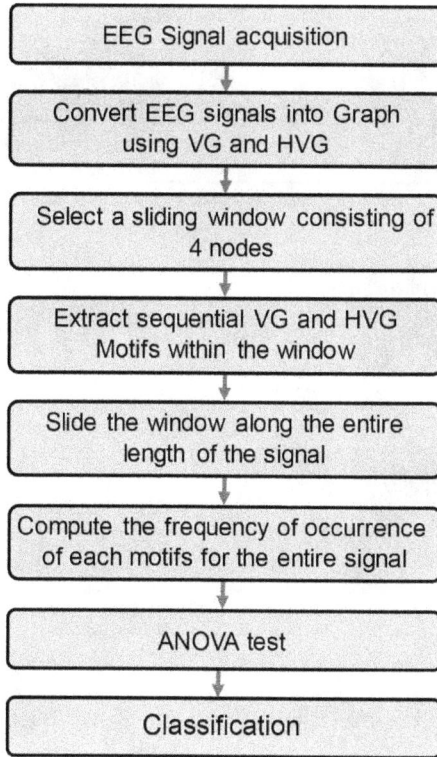

FIGURE 3.1   Flowchart showing the proposed seizure detection method.

were digitized at a sampling frequency of 173.6 Hz, using a 12-bit digitizer. Further, band-pass filtering was performed for noise and artifact removal purposes, with the filter setting kept at a range of between 0.53 and 40 Hz (12 dB/oct). Each set of the EEG recordings (from A–E) contained 100 single-channel EEG signals and each of those EEG recordings contained 4097 data points (out of which, only the first 4096 data points were considered for analysis in this study) and the duration of each recording process was 23.6 sec. It should be mentioned here that the recording of the first two sets of EEG signals, i.e., set A and set B, was done extra-cranially, while the rest, sets C–E, was done intra-cranially. Also, the EEG recording process for sets A and B was performed using surface electrodes, placed in accordance with the standard protocols of a 10–20 electrode system for five healthy participants, maintaining eyes open and eyes closed conditions, respectively. Similarly, EEG signals for sets C and D were recorded of five different volunteers who were previously diagnosed with epilepsy, while also maintaining eyes open and eyes closed conditions, respectively, and for this purpose recording of the EEG signals was performed during seizure-free interludes. EEG signals of set C were acquired from the hippocampal expansion of the opposite hemisphere, and those of set D were acquired from the epileptogenic region of the human brain, respectively. Set E consists of EEG signals corresponding

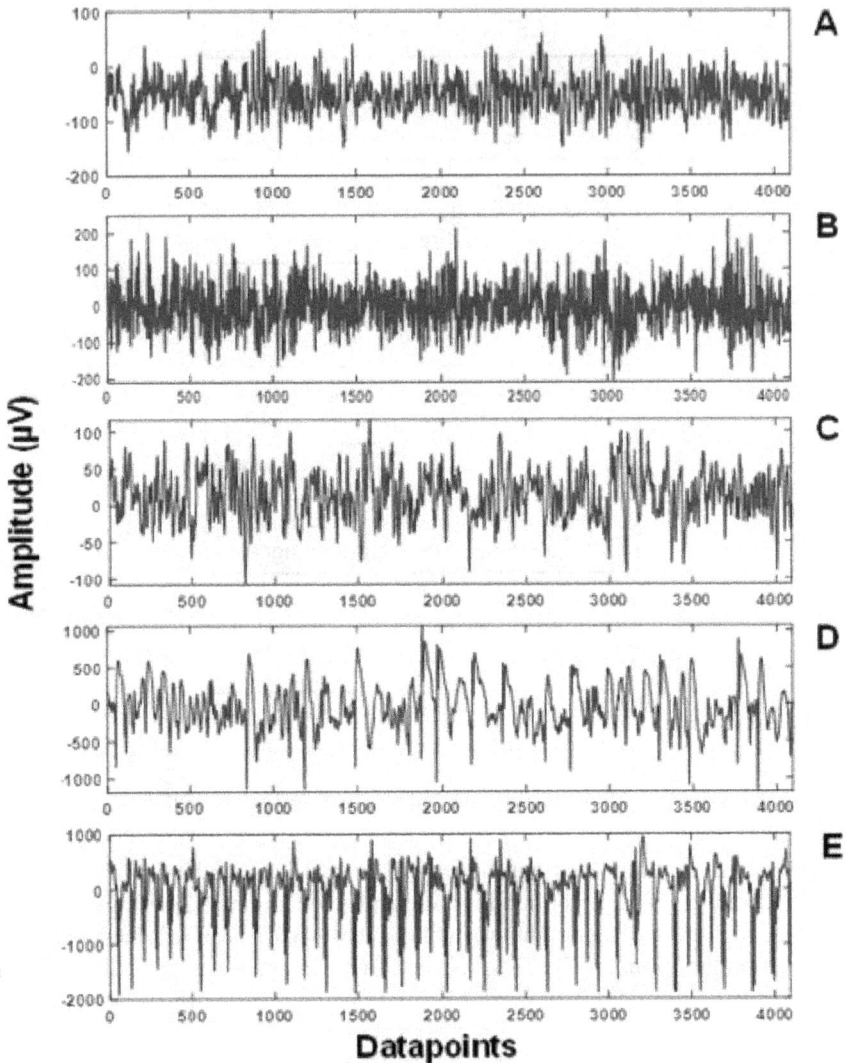

FIGURE 3.2 Time-domain EEG signals.

to epileptic seizure activities only. Sample time series representations of the different categories of EEG signals, i.e., from sets A–E, are shown below in Figure 3.2.

3.3 METHODOLOGY

3.3.1 CONCEPT OF A VISIBILITY GRAPH

The concept of a visibility graph algorithm was first proposed by Lacasa et al. in [43] to transform any time-domain signal into a graphical network, while retaining its temporal attributes. The VG algorithm enabled understanding of the structural connectivities between different nodes of the transformed time series network and to analyze them based on various topological properties of the graphical network. As proposed in [43], any time-domain signal can be converted into its equivalent topological network using a VG algorithm where a VG converted time series represents a binary and undirected graphical network without containing any self-loops, and the nodes of the graph denote the respective data points of the corresponding time series. According to a VG algorithm, the two data points "*a*" and "*b*" are considered to have connectivity or "visibility" between them if an intermediate datapoint "*c*" abides by the following mathematical criterion.

$$y_c < y_b + (y_a - y_b)\frac{(x_b - x_c)}{(x_b - x_a)} \qquad (3.1)$$

Where, the coordinates of the data points "*a*," "*b*," and "*c*" are denoted by (x_a, y_a), (x_b, y_b), and (x_c, y_c), respectively. Therefore, according to Equation (3.1), the mutual visibility among data points "*a*" and "*b*" depends on whether a straight line, i.e., otherwise called the visibility line, can be drawn between them without intersecting intermediate datapoint "*c*" in any manner. Considering a random time series, $x(t) = \{11,5,8,7,3,7,9\}$, the construction of the conventional VG network is shown in Figure 3.3, indicating the node connectivities with the help of a signal flow graph.

3.3.2 Concept of Horizontal Visibility Graph

Horizontal visibility graph theory was primarily developed as a modification of the conventional visibility algorithm. Proposed by Luque et al. in [44], HVG is considered a much simpler and analytically solvable subset of the original VG algorithm, and from the aspect of ease of implementation it is also computationally inexpensive.

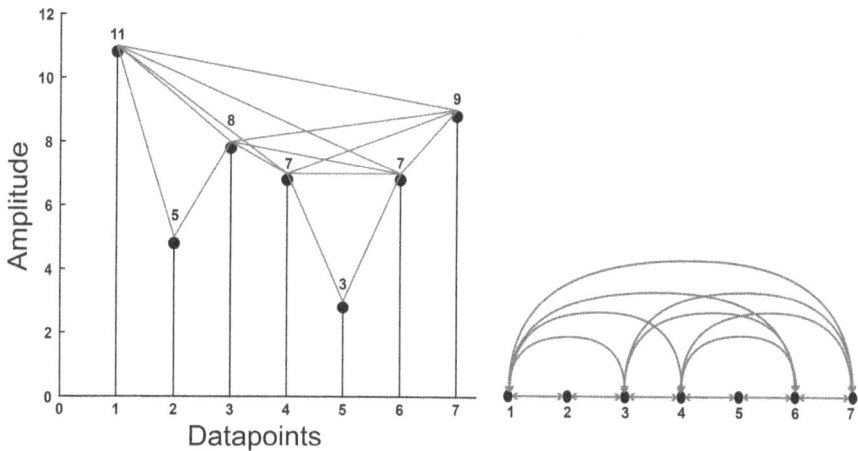

FIGURE 3.3 Conventional VG construction of a sample time series.

Similar to the conventional VG technique, HVG also maps a time-domain signal onto an indirect and binary graphical network, where the connectivity between the nodes of the network depends on the horizontal visibility condition. According to the proposed HVG algorithm in [44], presence of an intermediate node denoted by "c," between any two arbitrary nodes of the graphical network "a" and "b" will have mutual visibility between them if they satisfy the following mathematical criterion:

$$y_c < y_a, y_b \quad , \forall x_c \in (x_a, x_b) \tag{3.2}$$

Where, (x_a, y_a), (x_b, y_b), and (x_c, y_c) represent the coordinates of the nodes "a," "b," and "c," respectively. Abiding by the horizontal visibility criterion mentioned in Equation (3.2), two nodes of the graphical network are said to possess horizontal visibility among them. The obtained HVG adjacency matrix for a graph represented by $G(N, E)$ is also a "$N \times N$" binary symmetric matrix with an all-zero diagonal row-vector, indicating the corresponding graphical network does not contain any self-loops. Considering the previously mentioned example for computing a conventional VG network, HVG construction of the sample time series $x(t) = \{11,5,8,7,3,7,9\}$, is shown in Figure 3.4, also highlighting the nodal connectivities using a signal flow graph.

The output of both the conventional VG and HVG yields an adjacency matrix, also known as the graph shift matrix, which contains all the connectivity patterns and the edgewise information of the corresponding graphical network. As mentioned earlier, both the VG and HVG algorithm produces an undirected and connected binary network which in terms of graph theory, represent the different mutual connectivity patterns among the nodes of the network. For example, graph G (N, E) is considered where the nodes of the graph are denoted by N, and the number of edges between the corresponding nodes are denoted by E. The term "binary" network necessarily represents the fact that if any two arbitrary nodes N_i and N_j are visible to each other, or can be connected through a mutual visibility line, then the weight of the connecting

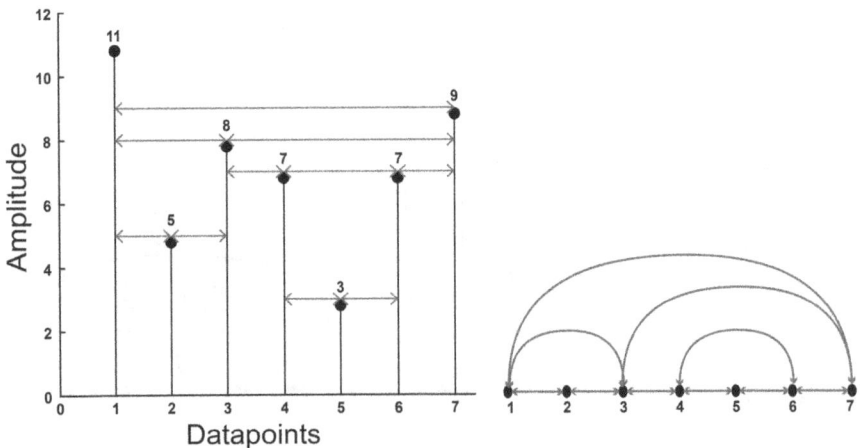

FIGURE 3.4   HVG construction of a sample time series.

edge will be reflected by "1" in the corresponding adjacency matrix; otherwise, the edge weight will be "0" if there exists no visibility between them. Thus, the obtained adjacency matrix is a binary one with the dimensions of $N \times N$ and is symmetric in nature as well since the graphical network is an undirected one. Again, the principal diagonal of the adjacency matrix is a $N \times 1$ dimensional zero row-vector, owing to the fact that the graphical network contains no self-loops. Therefore, it is imperative from the above discussion that as the number of nodes (data points of the time series) increases, the size of the adjacency matrix also increases, and computation of VG and HVG will become costly for a lengthy time series. Considering this fact, a new technique employing VG motifs was employed in this study to analyze an EEG time series that could reduce the complexity as well as capture the local variations present in the EEG time series effectively.

3.3.3 CONCEPT OF SEQUENTIAL VISIBILITY GRAPH MOTIFS

The concept of sequential VG motifs was first proposed by Iacovacci eta al. in [52] to distinguish between the dynamic patterns of different categories of time signals. Since then, different variants of VG motifs, such as limited penetrable VG motifs [53] and tetradic HVG motifs [54], have been proposed to analyze the dynamic fluctuations of a time series. Until now, the application of VG motifs to understand the non-linear and dynamic characteristics of the EEG time series during epileptic seizure onset was not well investigated. The concept of sequential VG motifs is briefly explained below.

The sequential motif of a VG (also true for HVG) can be defined as the set of all possible subgraphs consisting of M number of nodes in a consecutive manner along the Hamiltonian path of a VG, such that $M \in (2, 3, 4.....N)$, where N is the number of nodes of a graph. Thus, sequential motifs can be considered as a subset of the original VG. Interestingly, each sequential motif is also a VG [49]. In order to extract the sequential VG motifs, consider a window of size M that can slide along the along the Hamiltonian path of the VG such that it has $N - M$ number of consecutive overlapping steps. For each overlapping step, a particular motif is detected within the window. It was shown in [49] that for $M = 4$, a total 6 VG motifs (also HVG motifs) could be obtained. The variation of six motif profiles for window size $M = 4$ is shown in Figure 3.5. As the window slides across all the nodes of the Hamiltonian path of the graph, the total number of six VG (HVG) motifs, as indicated in Figure 3.5, present in the graph, can be obtained. In Figure 3.6, the extraction of a VG motif from a time series is shown. It is important to mention here that since the signal is analyzed within a window, the local fluctuations present in the time series is better reflected in the VG (HVG) motif profile. After the extraction of all the motifs, the next step is to compute the frequency of a particular VG (HVG) motif. Let the occurrence of a particular motif M_i be N_i, then the frequency of occurrence of that particular motif can be expressed by the following equation as:

$$f_i = \frac{N_i}{\sum_{i=1,2...6} N_i} \tag{3.3}$$

Motif number	Motif profile
1	
2	
3	
4	
5	
6	

FIGURE 3.5 Description of six sequential VG motif profiles for $M = 4$.

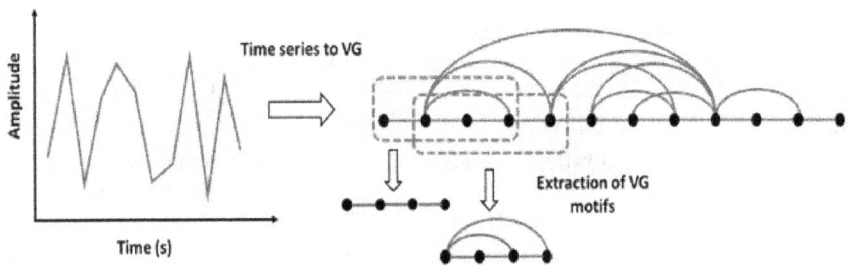

FIGURE 3.6 Extraction of sequential motifs from a time series converted to VG.

Using Equation (3.3), the frequency of six sequential motifs can be obtained. In the present work, the frequencies of different motifs were selected as features to quantify the differences between different categories of EEG signal.

3.4   MACHINE LEARNING CLASSIFIERS

3.4.1   RANDOM FOREST CLASSIFIER

Proposed by Brieman et al. in [55], a random forest algorithm is a tree-based ensemble classification algorithm where every tree is assigned a particular weight to produce an individual output. The final classifier output is based on the decision gained from the majority of votes of the ensemble tree classifiers. The sequence of steps for building the RF classifier involves a training input, namely the in-bag data-set, formed by random selection with the substitute from the entire input data and this process is known as the bootstrap bagging technique. In this way, approximately one-third of the in-bag dataset is left out, which is termed the testing dataset, also often referred to as the out-bag dataset. Input into each tree in the forest is formed by assigning random subsets of the previously mentioned in-bag dataset, and this process continues until feature input at every split node has grown without pruning [56]. The training dataset I_n at node n is randomly split into a left and right subset of attributes I_l and I_r to form split nodes by using the split function $f(v_i)$ for feature vector v, depending upon the maximum information gain selection at the root node. The information gain is estimated by the entropy measures shown in Equation (3.4)

$$\Delta E = -\frac{|I_l|}{|I_n|}E(I_l) - \frac{|I_r|}{|I_n|}E(I_r)$$

(3.4)

The splitting of the nodes continues as long as the information gain is positive; oth-erwise, it terminates into a leaf node. The overall classification model of the RF clas-sifier depends upon the votes obtained from the output of each classifier, weighted according to their performance to make them combine linearly [57]. Thereafter, a random selection of the training samples is made from the entire training dataset with the help of an adaboost classifier to construct inputs for each individual tree in the forest. After completing the training of all the adaboost classifiers, majority voting is done to draw the inference. For testing purposes, sub-sampled features are used in each boosting tree. Each adaboost classifier produces individual probability values, and probabilistic mapping is performed based on the average output from all the classifiers to get the final classifier result [58], as defined in Equation (3.5):

$$P_{final}(c_i / x) = \frac{1}{T}\sum_{i=1}^{T} P(c_i / x)$$

(3.5)

Where, the total number of trees in the random forest is indicated by T and c_i denotes the respective classes of the validation data x. As mentioned earlier, the process of training the tree classifiers is cumulative in nature, and the final output is delivered using the majority voting technique. In this study, T was set to 50 as it was found to deliver better performance.

3.4.2 Support Vector Machine Classifier

A support vector machine classifier is a well-known machine learning algorithm where labeled class inputs are learned in the training phase of the classifier, which helps in classification of the validation dataset between the different respective cat-egories of the input data. A detailed description of the formulation of a SVM algo-rithm is reported in [59]. A SVM model maps the input feature data into a higher dimensional feature space using several non-linear mapping functions. The aim is to find the optimal separating hyperplane (OSH) where data vectors close to the hyperplane are called support vectors, which are solely responsible for classification. However, in practical situations, perfect linear mapping of the input data is not possible, and it leads to misclassification of data points around the margin. Therefore, a cost function is evaluated to maximize the marginal boundary and to minimize the risk of error as well, which is expressed as the following:

$$J(w,b,\xi) = \frac{1}{2}\|w\|^2 + C\sum_{i=1}^{N}\xi_i \qquad (3.6)$$

where, C represents the positive regularization constant and ξ_i represents the positive set of variables, also known as the slack variables. The first term in Equation (3.6) is used to balance the model complexities, and the second term is used to reduce misclassifications with the help of the regularization technique [60]. The magnitude of ξ_i indicates the distance of x_i from the decision boundary. The OSH is found by converting linearly non-separable data into linearly separable data with the help of several kernel functions denoted by $K(x_i, x_j)$, which shows the product of the input vectors according to Mercer's condition [61]. This technique helps the SVM model to perform in a higher dimensional feature space without investing in much complexity. The linear, polynomial, and radial basis function (RBF) are some of the popularly utilized kernel functions for enhancing the classification of input feature data in a high dimensional feature space, which is represented by the following mathematical equations, respectively.

$$f(x_i, x_j) = x_i^T x_j \qquad (3.7)$$

$$f(x_i, x_j) = (1 + x_i^T x_j)^m \qquad (3.8)$$

$$f(x_i, x_j) = e^{\frac{-(\|x_i - x_j\|^2)}{2\sigma^2}} \qquad (3.9)$$

Where, "a" and "b" signify the random input training samples, respectively. Also, the index of the polynomial and width of the RBF kernel are denoted by m and σ, respectively. It is worth mentioning here that the choice of kernel functions, depending on the classification problem at hand, significantly affect the classifier's performance. Keeping this in mind, in this study, we have selected polynomial kernel functions using polynomial index index "$m = 3$" and varied the RBF kernel width within a range of 0.5 to 100 with a step size of 0.1 for obtaining the optimum performance of the SVM classifier.

3.4.3   K-NEAREST NEIGHBOR CLASSIFIER

A k-nearest neighbor classifier, popularly known as the k-NN, is a non-parametric, supervised machine learning classifier. A detailed formulation of a k-NN algorithm was reported in [59]. In the training phase of a k-NN classifier, feature vectors are sorted based on the most frequently occurring class of the input feature data. After that in the testing phase, feature vectors with unknown class labels are categorized based on the majority voting technique contingent to the k number of nearest neighbors among the previously mentioned training dataset [62]. The k value depends on the user-defined number of the nearest neighbors in the training dataset. Selection of the k value has a direct impact on classification performance as small values of k might result in overfitting, whereas for larger values of k, the decision-making process gets adultered with the influence of noise. Thus, proper selection of the k value is necessary for ensuring optimum performance of a k-NN classifier, and it is generally an odd number to avoid tied votes [63]. Apart from that, an appropriate distance metric is also chosen to measure the distance between training feature samples for finding the similarity between class labels. Different distance metrics are used in a k-NN algorithm, such as city block, Euclidean, etc. Let us consider two points in a k dimensional space such that $x(i) = x_1, x_2,..., x_k$ and $y(i) = y_1, y_2,..., y_k$. The Euclidean distance measures the shortest path between two given points. The Euclidean distance between two points is given by Equation (3.10):

$$d(x,y) = \sqrt{\sum_{i=1}^{k} (y_i - x_i)^2} \qquad (3.10)$$

Using Equation (3.10), the Euclidean distance parameter was calculated by computing the square root of the linear distance between the two points. In this work, the Euclidean distance parameter was chosen as the distance parameter, and the value of k was varied within a range of 1–15 with a step size of two to obtain the optimum performance of the k-NN classifier.

3.4.4 NAÏVE-BAYES CLASSIFIER

A naïve Bayes is another supervised machine learning algorithm, the concept of which originated based on the Bayes theorem of probability [64]. The significance of the term "naïve" in the NB algorithm is that it considers the independency of the features. The working principle of a NB classifier can be expressed using the following mathematical expression:

$$P(H/E) = \frac{P(E/H) \times P(H)}{P(E)} \qquad (3.11)$$

Where, E is the data sample with an unknown class label, $P(E)$ is the probability of data sample E, and H is the hypothesis that E belongs to a particular class C_i where, $P(H)$ is the prior hypothesis that E fits into the previously mentioned class C_i. $P(H/E)$ and $P(E/H)$ are the probabilities given that there is evidence that the corresponding hypothesizes are true, respectively. Multiple input features are used for classification using the NB classifier; therefore a probability chain is made to find out the resulting probability, based on a given set of provided features that are utilized as an input

[65]. The formation of the probability chain can be understood using the following equation:

$$P(H/E) = \prod_{i=1}^{n} P(H_i/E) \qquad (3.12)$$

Here, $H(i) = H_1, H_2, ..., H_n$ is the corresponding feature input to the NB classifier, based on which the membership probabilities are predicted. Since, the feature inputs are considered unrelated to each other, the outputs therefore become significant while finding the probability.

3.5 DESCRIPTION OF CLASSIFICATION PROBLEMS

In this study, several classification problems were addressed to differentiate between healthy and epileptic seizure category EEG signals. The data archive used in this study consisted of five different sets of EEG signals, namely sets A, B, C, D and E, with 100 signals belonging to each set. As mentioned earlier in Section 3.2, the EEG signals in the first two sets, i.e., sets A and B, corresponded to the healthy categories, which were also termed the pre-ictal categories of EEG signals. The latter two sets of EEG signals, i.e., sets C and D were called the inter-ictal categories, which also corresponded to the seizure-free categories. However, EEG signals belonging to the same category were extracted from epilepsy patients but within the interludes of seizure onsets. EEG signals of the last remaining set E belonged to the seizure category. Hence, based on the occurrence of the seizure onsets and depending on the method of recording the EEG signals, eight binary and one multi-class classification tasks were carried out in this work, by accumulating the EEG signals of different sets together. Descriptions of the classification problems are presented in Table. 3.1, along with their significances.

From Table 3.1, we can see that C-I through C-VIII describes the binary classification problems, and the last one, i.e., C-IX, represents the multi-class classification

TABLE 3.1
Description of Classification Problem

Serial no	Combination of EEG signals	Classification type
C-I	A vs E	Binary
C-II	B vs E	Binary
C-III	C vs E	Binary
C-IV	D vs E	Binary
C-V	AB vs E	Binary
C-VI	CD vs E	Binary
C-VII	ABCD vs E	Binary
C-VIII	AB vs CD	Binary
C-IX	AB vs CD vs E	Multi-class

problem. The first four classification tasks, i.e., from C-I to C-IV, were performed with the intention to segregate the seizure-free EEG signals from the seizure class ones, using only single sets of EEG signals. Apart from that, EEG signals of the pre-ictal (AB) and the inter-ictal (CD) categories were also accumulated separately by combining the EEG signals recorded for eyes open and eyes closed conditions together. Using combined sets of the pre-ictal (set AB) and inter-ictal (set CD) categories, identification of the epileptic seizure EEG signals was performed with respect to set E, which are depicted by classification problems C-V and C-VI, respectively. Classification of the seizure-free and seizure class EEG signals was also carried out by combining the pre-ictal and inter-ictal categories together (set ABCD), and CP-VII denotes the corresponding classification task performed between set ABCD and set E. The following penultimate classification problem, i.e., C-VIII, represents the discrimination of EEG signals between the pre-ictal and inter-ictal categories, which was performed using EEG signals from the seizure-free category only but with the purpose to identify the healthy EEG signals from the inter-ictal ones. The final multi-class classification task, denoted by C-IX, was performed to classify any unknown EEG signals into either the pre-ictal, inter-ictal, or the seizure category. Using the conventional VG and HVG motif profiles, the above-mentioned classification tasks were performed in this study and the results are discussed in detail in the following section.

3.6 RESULTS AND DISCUSSIONS

3.6.1 EEG SIGNAL ANALYSIS USING VG AND HVG MOTIFS

For this work, we procured five differents categories of EEG signals with 100 signals belonging to each category, resulting in a total of 500 EEG signals. Initially, the EEG signals were converted into undirected and binary graphs using VG and HVG, as described by Equations (3.1) and (3.2), respectively. Then, we considered a sliding window of size M that slides along the graph sequentially; each time the window slides along the Hamiltonian path of the graph, it extracts a sequential motif profile, as described in Section 3.3.3. It should be mentioned here that this window size was selected empirically, and the choice of window length depends on the type of the signal to be analyzed. In the present problem, which deals with epilepsy detection, EEG recordings reveal periodic spike patterns during seizure inception. Therefore, instead of considering the total graph obtained for the entire length of the signal, analysis of the EEG signals on the local scale may provide better information regarding the instantaneous change in dynamic patterns. Also, as mentioned earlier, each motif is a subgraph (VG or HVG) of the original graph. Therefore, in this chapter, we considered a sliding window consisting of $M = 4$ nodes (EEG sample points) and traversed the sliding window along the entire length of the EEG signal to identify the motifs lying within the specified window. Now, as explained in Section 3.3.3, computation of VG and HVG motif profiles using $M = 4$ can result in six types of graph motifs. In this way, we computed the total number of occurrences of each type of motif profile for the entire signal length. Finally, the frequency of occurrence of different types of motifs for the particular EEG signal was computed using Equation (3.3). This procedure was repeated for all five sets of EEG signals. The frequencies of the different types of motifs were used as features in this study to categorize different types of EEG signals.

Before feeding the VG and HVG motif features into the employed machine learning classifiers, we evaluated the discriminative capabilities of the same using the one-way ANOVA test. The ANOVA test is performed to determine the statistical distinguishability of the feature inputs between different problem classes. ANOVA analysis essentially yields a probabilistic value, i.e., commonly termed as the "p-value," which is used to measure the distinguishability of any feature parameter between the corresponding problem classes. A lower "p-value" ("p" $< 10^{-5}$ for this study) signifies a lower probable rejection of a null hypothesis being false in the ANOVA analysis of the respective feature parameter. Apart from that, a lower "p-value" also indicates a lesser probability of occurrence of a particular feature value between the problem classes [66]. Hence, the obtained "p-values" from the ANOVA test for the VG and HVG motif features was used as a metric of distinguishability in this study. The variation of the frequency of occurrences of sequential motif features computed for conventional VG and HVG along with their "p-values" for different sets are shown in Table 3.2–3.3, respectively.

The box-whisker plot was also analyzed in this chapter to further visualize the class separation between the extracted VG and HVG motif features. The obtained boxplots corresponding to the conventional VG and HVG motif features, respectively, are presented in Figure 3.7 and Figure 3.8, respectively. From Figures 3.7–3.8, it can be seen that extracted motif features corresponding to different problem categories possess sufficient class separation between them in terms of their median and quantile values. Finally, the extracted conventional VG and HVG motif features were fed into four efficient machine learning classifiers to perform the classification problems, as mentioned in Section 3.5. We calculated three performance metrics, i.e., accuracy, specificity, and sensitivity, from the confusion matrices obtained for the respective classification problems. In addition, to avoid the possible issues of overfitting and also for ensuring the reliable classifiers' performance, we employed

TABLE 3.2

Variation (Mean ± Deviation) of Frequency of Six VG Motifs along with "p-Values"

Motif number	A	B	C	D	E	"p-value"
1	0.382 ± 0.102	0.493 ± 0.121	0.262 ± 0.093	0.308 ± 0.087	0.64 ± 0.172	7.454×10^{-89}
2	0.497 ± 0.135	0.326 ± 0.154	0.705 ± 0.125	0.659 ± 0.110	0.201 ± 0.146	3.616×10^{-118}
3	0.495 ± 0.133	0.320 ± 0.150	0.704 ± 0.125	0.654 ± 0.11	0.195 ± 0.148	3.136×10^{-120}
4	0.631 ± 0.122	0.459 ± 0.157	0.784 ± 0.079	0.765 ± 0.098	0.261 ± 0.183	1.14×10^{-123}
5	0.663 ± 0.111	0.502 ± 0.167	0.808 ± 0.078	0.797 ± 0.098	0.312 ± 0.187	1.589×10^{-115}
6	0.403 ± 0.094	0.561 ± 0.122	0.220 ± 0.087	0.236 ± 0.084	0.637 ± 0.152	3.083×10^{-127}

TABLE 3.3

Variation (Mean ± Deviation) of Frequency of Six HVG Motifs along with "p-Values"

Motif number	A	B	C	D	E	"p-value"
1	0.394 ± 0.107	0.548 ± 0.152	0.385 ± 0.150	0.472 ± 0.137	0.738 ± 0.095	7.186×10^{-73}
2	0.389 ± 0.109	0.273 ± 0.097	0.520 ± 0.205	0.510 ± 0.184	0.143 ± 0.102	3.299×10^{-71}
3	0.530 ± 0.115	0.369 ± 0.142	0.620 ± 0.157	0.535 ± 0.140	0.215 ± 0.094	8.426×10^{-84}
4	0.514 ± 0.104	0.358 ± 0.128	0.629 ± 0.155	0554 ± 0.136	0.202 ± 0.097	1.747×10^{-96}
5	0.766 ± 0.106	0.638 ± 0.155	0.587 ± 0.117	0.524 ± 0.127	0.359 ± 0.128	1.401×10^{-78}
6	0.698 ± 0.102	0.578 ± 0.147	0.517 ± 0.102	0.451 ± 0.109	0.352 ± 0.108	3.991×10^{-75}

a ten-fold cross-validation technique in the training and validation process. In this way, we split the input feature data into the ratio of 9:1, out of which nine parts were used for training the classifiers for each fold, and the single remaining parts were used to validate the outputs of the classifiers. The whole classification process was carried out for ten consecutive iterations, and the performance parameters were reported in this study in terms of their mean and standard deviation values.

3.6.2 CLASSIFICATION PERFORMANCE

The performance of the four machine learning classifiers, namely RF, SVM, k-NN, and NB classifiers, to discriminate between the different categories of EEG signals is reported in brief in this section. Table 3.4 through Table 3.7 reports the classification results obtained for RF, SVM, k-NN, and NB classifiers, respectively. In the case of the SVM and k-NN classifiers, the details of the kernel function and the k values that yielded the best classification performances for both VG and HVG motifs are given in Table 3.5 and 3.6, respectively. From Tables 3.4–3.7, it can be observed that very high recognition was obtained for all classification

problems using the four machine learning classifiers. Considering the performance of different classifiers, it was observed that among the four classifiers RF and SVM delivered better performances compared to the k-NN and NB classifiers. Between NB and k-NN, the performance of the NB classifier was found to be inferior for all nine classification problems. Between conventional VG and HVG motifs, the performance of the HVG motifs was found to be superior for all the cases. For C-I to C-IV, C-VI, and C-IX, the highest classification accuracy was obtained for the RF classifier, whereas for C-VII, C-VII, and

C-VIII, the SVM classifier delivered the best classification performance, yielding very high classification accuracies. Also, we can observe from the results presented in Table 3.4–3.7, that the specificity and the sensitivity values obtained for all classification problems were reasonably satisfactory. The most important observation was that for all classification problems the standard deviation value obtained for each problem was minimal, which signified the robustness of the proposed model.

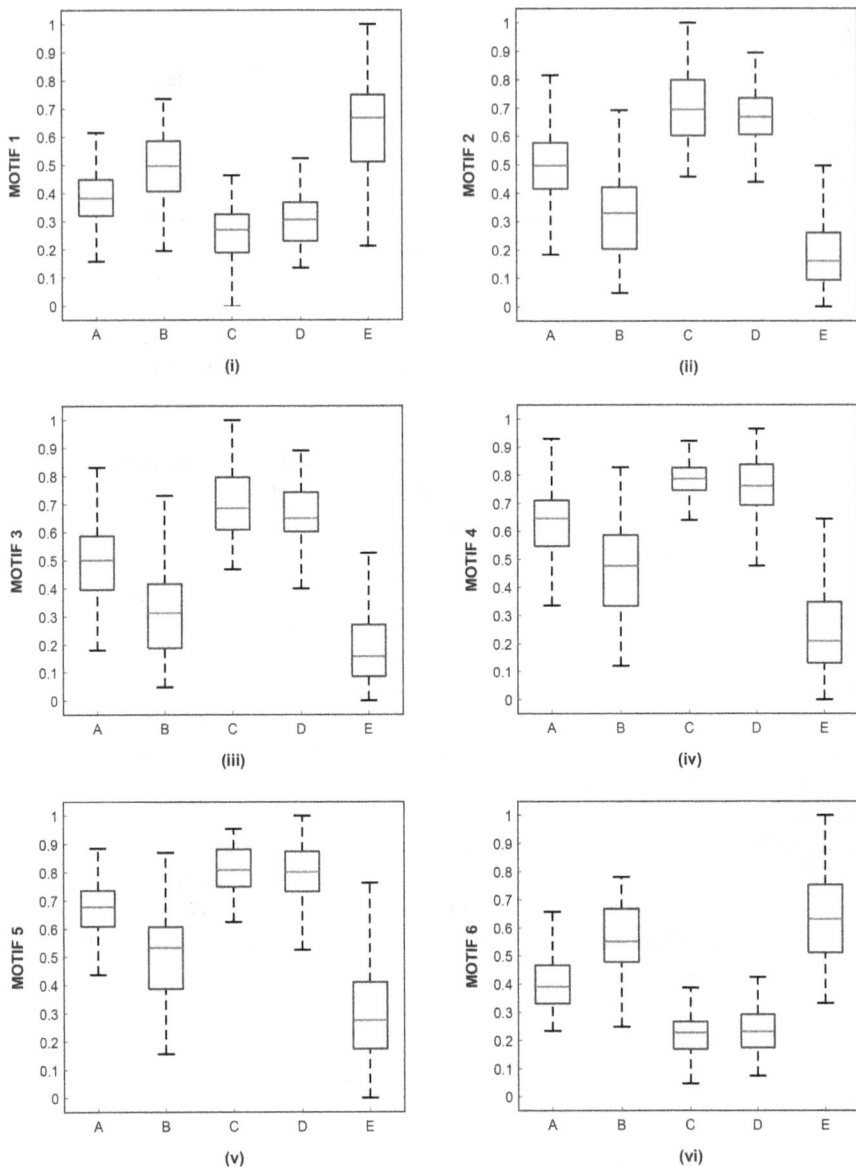

FIGURE 3.7   Boxplot analysis of the frequency of six VG motifs for five sets.

3.6.3 COMPARISON WITH EXISTING LITERATURES

We also compared the performance of our proposed technique with some existing methods, which were studied on the same dataset. It should be mentioned here that the comparative study was done for all nine classification problems, which are reported in Table 3.8. Along with the classification accuracy, sensitivity and specificity are

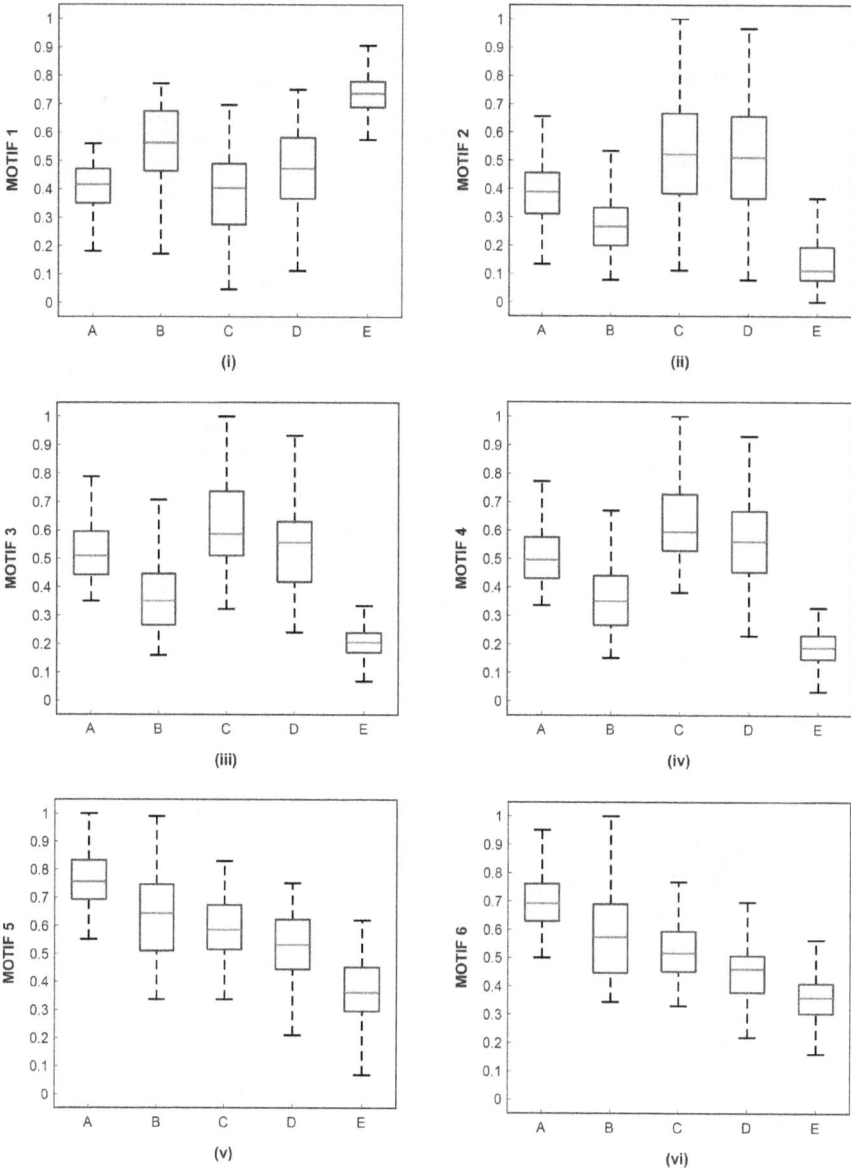

FIGURE 3.8 Box-polt analysis of the frequency of six HVG motifs for five sets.

also reported in Table 3.8, in order to have a comprehensive overview of the performance of the proposed method in comparison with the existing state of the art techniques. From Table 3.8, it can be observed that for the majority of the classification problems, the proposed method delivered a comparable performance while for some problems the performance was found to be even better than the existing methods.

TABLE 3.4
Classification Report of a RF Classifier

Classification	VG motifs			HVG motifs		
	Accuracy (%)	Specificity (%)	Sensitivity (%)	Accuracy (%)	Specificity (%)	Sensitivity (%)
C-I	99.25 ± 0.32	98.50 ± 0.35	99.08 ± 0.28	**100 ± 0**	100 ± 0	100 ± 0
C-II	93.85 ± 0.67	91.83 ± 0.53	90.62 ± 0.27	**95.93 ± 0.41**	95.25 ± 0.33	93.50 ± 0.37
C-III	98.16 ± 0.25	97.78 ± 0.19	95.93 ± 0.31	**100 ± 0**	97.32 ± 0.25	98.27 ± 0.22
C-IV	97.32 ± 0.46	96.90 ± 0.28	96.29 ± 0.31	**98.38 ± 0.27**	98.69 ± 0.16	97.65 ± 0.21
C-V	94.92 ± 0.62	95.31 ± 0.25	94.67 ± 0.34	95.62 ± 0.37	95.91 ± 0.19	96.14 ± 0.23
C-VI	97.56 ± 0.34	96.68 ± 0.21	97.16 ± 0.29	**99.25 ± 0.43**	97.25 ± 0.72	98.23 ± 0.53
C-VII	95.44 ± 0.24	94.81 ± 0.26	93.95 ± 0.59	96.62 ± 0.35	97.06 ± 0.14	96.84 ± 0.27
C-VIII	94.96 ± 0.41	95.32 ± 0.34	94.29 ± 0.33	96.44 ± 0.16	96.94 ± 0.25	97.17 ± 0.31
C-IX	97.69 ± 0.32	97.43 ± 0.17	98.23 ± 0.25	**98.90 ± 0.34**	98.94 ± 0.29	99.52 ± 0.37

This is because most of the existing studies have done an analysis of EEG signals on a global scale, while the present study focused on capturing the instantaneous changes in the dynamics of the EEG patterns in terms of the graph motifs. Thus, the local fluctuations that occur during seizure onsets were better characterized in this chapter using the VG and HVG motifs than the majority of the existing models.

3.7 CONCLUSIONS

In this chapter, a novel framework employing a graph theory aided conventional VG and HVG sequential motifs based on the analysis of EEG signals was proposed for the accurate detection of epilepsy. For this purpose, we procured EEG signals corresponding to the healthy, inter-ictal, and epileptic seizure categories from an available online benchmark dataset. Initially, the EEG signals were transformed from the time domain into the graph domain using VG and HVG. Now, instead of analyzing the EEG signals on a global scale, we selected a sliding window of a fixed length that translated along the Hamiltonian path of the graph (i.e., the entire length of the signal). In this study, the size of the sliding window was kept at four and we observed that for four nodes (data points), six different motif profiles were obtained for both VG and HVG. The window was translated along the entire length of the EEG signals to extract the different sequential motif profiles within the window and the number of occurrences of six different motif profiles for the entire signal length was obtained. Finally, we computed the frequency of occurrence of six different motifs for a particular EEG signal, and the process was repeated for the different classes of EEG signals. In this chapter, the extracted frequencies of occurrences of conventional VG and HVG motif profiles were utilized as feature attributes to differentiate between the various categories of healthy and epileptic signals. Further, an ANOVA analysis was performed to measure the statistical distinguishability of the extracted features. Finally, extensive discrimination between pre-ictal, inter-ictal, and seizure

TABLE 3.5

Classification Report of SVM Classifier

Classification	VG motifs				HVG motifs			
	Accuracy (%)	Specificity (%)	Sensitivity (%)	Kernel type	Accuracy (%)	Specificity (%)	Sensitivity (%)	Kernel type
C-I	98.99 ± 0.25	99.07 ± 0.55	98.25 ± 0.48	Linear	99.46 ± 0.26	98.87 ± 0.34	99.16 ± 0.30	Linear
C-II	92.17 ± 0.37	93.19 ± 0.43	92.59 ± 0.40	Linear	94.02 ± 0.43	95.53 ± 0.33	95.27 ± 0.32	Linear
C-III	97.06 ± 0.43	98.16 ± 0.24	97.61 ± 0.21	Linear	98.82 ± 0.39	99.25 ± 0.27	98.92 ± 0.29	Linear
C-IV	96.24 ± 0.41	95.67 ± 0.49	96.49 ± 0.36	Linear	97.96 ± 0.32	98.45 ± 0.36	97.13 ± 0.33	Polynomial, "$m = 3$"
C-V	94.77 ± 0.65	95.02 ± 0.27	94.56 ± 0.31	Polynomial "$m = 3$"	97.42 ± **0.45**	97.15 ± 0.42	96.34 ± 0.40	RBF $\sigma = 1.8$
C-VI	97.14 ± 0.32	98.43 ± 0.44	98.11 ± 0.26	RBF, $\sigma = 1.5$	98.49 ± 0.50	98.96 ± 0.52	99.14 ± 0.49	Polynomial "$m = 3$"
C-VII	95.73 ± 0.53	96.47 ± 0.22	95.46 ± 0.39	Linear	**97.35 ± 0.33**	97.26 ± 0.30	97.02 ± 0.36	RBF, $\sigma = 2.2$
C-VIII	96.78 ± 0.49	95.44 ± 0.37	96.28 ± 0.32	Polynomial "$m = 3$"	**97.82 ± 0.42**	97.50 ± 0.45	98.41 ± 0.47	RBF $\sigma = 1.7$
C-IX	97.49 ± 0.36	98.25 ± 0.33	97.68 ± 0.35	RBF $\Sigma = 2.1$	98.45 ± 0.28	99.05 ± 0.25	98.82 ± 0.23	RBF $\Sigma = 2.4$

TABLE 3.6
Classification Report of *k*-NN Classifier

Classification	VG motifs				HVG motifs			
	Accuracy (%)	Specificity (%)	Sensitivity (%)	k value	Accuracy (%)	Specificity (%)	Sensitivity (%)	k- value
C-I	98.47 ± 0.45	97.86 ± 0.40	98.11 ± 0.43	k = 3	99.65 ± 0.25	99.20 ± 0.27	98.87 ± 0.22	k = 5
C-II	93.15 ± 0.58	94.42 ± 0.49	94.05 ± 0.53	k = 5	94.83 ± 0.45	95.27 ± 0.43	95.48 ± 0.47	k = 7
C-III	97.90 ± 0.39	98.25 ± 0.32	98.67 ± 0.35	k = 5	99.36 ± 0.37	98.79 ± 0.32	99.25 ± 0.33	k = 3
C-IV	96.38 ± 0.34	97.31 ± 0.36	96.95 ± 0.39	k = 3	97.17 ± 0.35	98.24 ± 0.44	97.89 ± 0.30	k = 5
C-V	94.50 ± 0.41	94.67 ± 0.46	95.50 ± 0.42	k = 5	96.04 ± 0.49	95.67 ± 0.51	96.47 ± 0.55	k = 5
C-VI	97.11 ± 0.35	98.40 ± 0.30	97.76 ± 0.32	k = 7	98.97 ± 0.39	99.23 ± 0.30	98.46 ± 0.41	k = 7
C-VII	94.73 ± 0.55	95.11 ± 0.52	95.53 ± 0.56	k = 9	96.17 ± 0.45	97.20 ± 0.52	96.79 ± 0.55	k = 9
C-VIII	95.87 ± 0.29	96.68 ± 0.26	96.42 ± 0.24	k = 7	97.19 ± 0.24	98.15 ± 0.29	98.05 ± 0.22	k = 7
C-IX	96.22 ± 0.31	97.29 ± 0.34	96.94 ± 0.35	k = 5	98 ± 0.42	98.47 ± 0.38	98.25 ± 0.45	k = 5

TABLE 3.7
Classification Report of NB Classifier

	VG motifs			HVG motifs		
Classification	Accuracy (%)	Specificity (%)	Sensitivity (%)	Accuracy (%)	Specificity (%)	Sensitivity (%)
C-I	97.24 ± 0.37	97.69 ± 0.43	98.13 ± 0.40	98.92 ± 0.34	99.25 ± 0.26	98.57 ± 0.38
C-II	91.16 ± 0.65	92.64 ± 0.69	92.19 ± 0.72	92.73 ± 0.79	93.69 ± 0.75	93.44 ± 0.69
C-III	94.29 ± 0.43	95.27 ± 0.46	94.89 ± 0.51	96.54 ± 0.59	97.79 ± 0.63	97.15 ± 0.45
C-IV	94.63 ± 0.38	94.87 ± 0.35	95.22 ± 0.31	96.21 ± 0.45	96.89 ± 0.38	97.25 ± 0.34
C-V	93.05 ± 0.49	94.08 ± 0.39	93.79 ± 0.45	94.23 ± 0.47	95.27 ± 0.41	94.69 ± 0.46
C-VI	95.60 ± 0.27	96.22 ± 0.26	96.45 ± 0.25	97.85 ± 0.32	97.29 ± 0.38	98.15 ± 0.30
C-VII	93.69 ± 0.50	93.73 ± 0.53	94.07 ± 0.44	95.22 ± 0.36	94.69 ± 0.43	95.49 ± 0.39
C-VIII	93.66 ± 0.59	93.29 ± 0.62	94.15 ± 0.55	95.28 ± 0.49	95.76 ± 0.44	96.26 ± 0.37
C-IX	95.47 ± 0.53	96.29 ± 0.45	96.44 ± 0.48	97.32 ± 0.30	97.19 ± 0.35	98.50 ± 0.28

category EEG signals was performed using four well-known machine learning classifiers, namely, RF, SVM, k-NN, and NB classifiers. A total of nine classification problems were addressed in this study based on the onset of seizure activities as well as the source of the EEG signals. The most important outcomes of the investigations are summarized below.

We observed that the proposed VG and HVG motif profile based epilepsy recognition framework performed significantly well in distinguishing between healthy, inter-ictal, and epileptic EEG signals. From the classification performance, we saw that the binary classification problems for discerning healthy, inter-ictal, and seizure-free (by combining healthy and inter-ictal) EEG signals from seizure signals delivered better recognition accuracies compared to the other classification problems. Also, the performance of different classifiers for the multi-class classification task for identifying any unknown EEG signal into the mentioned pre-ictal, inter-ictal, and seizure categories were also observed to be reasonably satisfactory, which indicates the superiority of the classification framework. It was also observed that the overall better classification performance was obtained utilizing the HVG motif features compared to utilizing the conventional VG motif features. Further, comparative analysis of the literature revealed that the proposed disease detection framework delivered similar classification performance with respect to the existing ones and was also found to perform exceptionally better for some of the classification problems. Thus, the proposed framework can be potentially implemented for the clinical diagnosis of seizures.

However, the present study has certain limitations. We validated our proposed disease diagnostic scheme on a relatively small number of EEG signals (100 per class). Prior to real-time applications and clinical trials, further experiments are needed on a more extensive database consisting of EEG signals from a larger number of patients, recorded using multiple electrode channels. This could be considered as a future extension of the present work of this study. In addition to that, the proposed

TABLE 3.8

Comparison with Existing Methods

Problem	Literature	Methods	Accuracy (%)	Sensitivity (%)	Specificity (%)
C-I	[23]	FBSE-WMRPE + LS-SVM	99.5	—	—
	[67]	Dual tree complex wavelet transform (DTCWT) + complex-valued neural networks (CVANN)	99.50	99	100
	[29]	DWT-TKEO + SVM	99.56	99.12	100
	[16]	XHST + k-NN	100	100	100
	[48]	Weighted complex network features + (SVM + k-NN)	100	100	100
	[5]	MFDFA + SVM	100	100	100
	This work	**Sequential HVG motif profiles + RF**	**100**	**100**	**100**
C-II	[68]	EMD + power spectral density (PSD)	83.68	78.38	76.66
	[32]	Permutation entropy + SVM	82.88	—	—
	This work	**Sequential HVG motif profiles + RF**	**95.93**	**95.25**	**93.5**
C-III	[69]	Discrete cosine transform (DCT)-based EEG rhythms-Hurst exponent + SVM	97.50	98	97
	[68]	EMD + power spectral density (PSD)	96.39	97.88	94.11
	[48]	Weighted complex network features + SVM	98.25	98	98.49
	[70]	Clustering technique (CT) + LS-SVM	96.20	96.20	96.20
	[34]	FAWT - FD + LS-SVM	99	100	98
	[71]	Multi-scale radial basis functions (MRBF) - modified particle swarm optimization (MPSO) + SVM	99.80	99.60	100
	[5]	MFDFA + SVM	100	100	100
	[35]	TQWT - FD + LS-SVM	100	100	100
	This work	**Sequential HVG motif profiles + RF**	**100**	**100**	**100**
C-IV	[23]	FBSE - WMRPE + LS-SVM	97.5	—	—
	[69]	DCT-based EEG rhythms-Hurst exponent + SVM	96.35	96.50	96.20
	[68]	EMD + power spectral density (PSD)	93	97.88	88.55
	[71]	MRBF - MPSO + SVM	97.60	96.80	97.20
	[48]	Weighted complex network features + SVM	93.25	90.6	96.25
	[72]	Local mean decomposition (LMD) based hybrid features + SVM optimized by genetic algorithm (GA-SVM)	98.10	98.80	97.40
	[73]	Complete ensemble empirical mode decomposition with adaptive noise (CEEMDAN) based spectral features + linear programming boosting (LPBoost)	97	97.4	98.25

(Continued)

TABLE 3.8 (CONTINUED)
Comparison with Existing Methods

Problem	Literature	Methods	Accuracy (%)	Sensitivity (%)	Specificity (%)
	This work	**Sequential HVG motif profiles + RF**	**98.38**	**98.69**	**97.65**
C-V	[69]	DCT-based EEG rhythms-Hurst exponent + SVM	97.27	97.40	97.15
	This work	**Sequential HVG motif profiles + SVM**	**97.42**	**97.15**	**96.34**
C-VI	[23]	FBSE - WMRPE + LS-SVM	99	—	—
	[69]	DCT based EEG rhythms - Hurst exponent + SVM	96.92	96.85	97
	[71]	MRBF-MPSO + SVM	98.73	98	99.10
	[34]	FAWT-FD + LS-SVM	98.67	100	96
	[74]	DTCWT + general regression neural network (GRNN)	98.67	97.92	98.92
	[75]	Statistical features + LS-SVM	97.19	96.96	99.66
	[76]	TQWT + Kraskov entropy based features + LS-SVM	97.75	97	99
	This work	**Sequential HVG motif profiles + RF**	**99.25**	**97.25**	**98.23**
C-VII	[74]	DTCWT + GRNN	95.24	—	—
	[47]	Fast weighted horizontal visibility graph constructing algorithm (FWHVA) based complex network features + k-NN	95.4	—	—
	[77]	TQWT quality factor (Q) based multi-scale entropy features + k-NN	96.4	91	97.8
	This work	**Frequency of ocuurence of Sequential HVG motif +SVM**	**97.35**	**97.26**	**97.02**
C-VIII	[34]	FAWT-FD + LS-SVM	92.50	90.5	94.5
	[5]	MFDFA + SVM	95.50	94.75	95.20
	[69]	DCT based EEG rhythms - Hurst exponent + SVM	97.70	97.45	97.95
	This work	**Sequential HVG motif profiles + SVM**	**97.82**	**99.30**	**98.91**
C-IX	[77]	TQWT - Q-based entropy features + k-NN	98.6	96	99.75
	[78]	Variational mode decomposition + autoregression features + RF	97.352	—	—
	[36]	Key point-1D-LBP based histogram features + SVM	98.80	—	—
	[72]	LMD + GA-SVM	98.40	98.33	—
	[67]	DTCWT + CVANN	97.79	97.04	98.01
	[73]	CEEMDAN + LPBoost	97.6	98.11	96.93
	This work	**Sequential HVG motif profiles +RF**	**98.90**	**98.94**	**99.52**

disease diagnostic scheme employing the frequency of occurrence of conventional VG and HVG motif profile based features will be implemented for detection of other neurological disorders, such as Alzheimer's disease, schizophrenia, attention deficit hyperactivity disorder (ADHD), etc.

REFERENCES

1. Acharya U R, Sree S V, Chattopadhyay S, Yu W and Ang P C 2011 Application of recurrence quantification analysis for the automated identification of epileptic EEG signals, *Int. J. Neural Syst.* **21**(03), 199–211.
2. Guo L, Rivero D, Dorado J, Munteanu C R and Pazos A 2011 Automatic feature extraction using genetic programming: An application to epileptic EEG classification,*Expert. Syst. Appl.* **38**(38), 10425–10436.
3. Pachori R B and Patidar S 2014 Epileptic seizure classification in EEG signals using second-order difference plot of intrinsic mode functions, *Comput. Methods Prog. Biomed.* **113**(2), 494–502.
4. Sharma R, Pachori R B and Acharya U R 2015 An integrated index for the identification of focal electroencephalogram signals using discrete wavelet transform and entropy measures, *Entropy* **17**(8), 5218–5240.
5. Bose R, Pratiher S and Chatterjee S 2018 Detection of epileptic seizure employing a novel set of features extracted from multi-fractal spectrum of electroencephalogram signals, *IET Signal Process.* **13**(2), 157–164.
6. Pati S and Alexopoulos A V 2010 Pharmacoresistant epilepsy: From pathogenesis to current and emerging therapies, *Cleve Clin J Med* **7**(7), 457–567.
7. Li P, Karmakar C and Yan C 2016 Classification of 5-S epileptic EEG recordings using distribution entropy and sample entropy, *Front. Physiol.* **7**(66), 136.
8. Seeck M, Lazeyras F, Michel C M, Blanke O, Gericke C A, Ives J, Delavelle J, Golay X, Haenggeli C A, De Tribolet N and Landis T 1998 Non-invasive epileptic focus localization using EEG-triggered functional MRI and electromagnetic tomography, *Electroencephalogr. Clin. Neurophysiol.* **106**(6), 508–512.
9. Sharma M, Dhere A and Pachori R B 2017 An automatic detection of focal EEG signals using new class of time–frequency localized orthogonal wavelet filter banks, *Knowl.-Based Syst.* **118**, 217–227.
10. Chandaka S, Chatterjee A and Munshi S 2009 Cross-correlation aided support vector machine classifier for classification of EEG signals, *Expert Syst. Appl.* **36**(2), 1329–1336.
11. Polat K and Güneş S 2007 Classification of epileptiform EEG using a hybrid system based on decision tree classifier and fast Fourier transform, *Appl. Math. Comput.* **187**(2), 1017–1026.
12. Srinivasan V, Eswaran C and Sriraam N 2005 Artificial neural network based epileptic detection using time-domain and frequency-domain features, *J. Med. Syst.* **29**(6), 647–660.
13. Gandhi T, Panigrahi B K and Anand S 2011 A comparative study of wavelet families for EEG signal classification, *Neurocomputing* **74**(17), 3051–3057.
14. Subasi A 2007 EEG signal classification using wavelet feature extraction and a mixture of expert model, *Expert Syst. Appl.* **32**(4), 1084–1093.
15. Chatterjee S, Choudhury N R and Bose R 2017 Detection of epileptic seizure and seizure-free EEG signals employing generalized S-transform, *IET Sci. Meas. Technol.* **11**(7), 847–855.

16. Choudhury N R, Roy S S, Pal A, Chatterjee S and Bose R 2019 Epileptic seizure detection employing cross-hyperbolic Stockwell transform. *Proc. Fourth International Conference on Research in Computational Intelligence and Communication Networks (ICRCICN), (Kolkata, West Bengal, India, Nov. 2018),* IEEE, pp. 70–74.
17. Tzallas A, Tsipouras M and Fotiadis D A 2007 A time-frequency based method for the detection of epileptic seizures in EEG recordings. *Proc. 20th IEEE International Symposium on Computer-Based Medical Systems (Maribor, Slovenia, Jun. 2007),* IEEE, pp. 135–140.
18. Guo L, Rivero D and Pazos A 2010 Epileptic seizure detection using multi-wavelet transform based approximate entropy and artificial neural networks, *J. Neurosci. Methods.* **193**(1), 156–163.
19. Bhati D, Pachori R B and Gadre V M 2017 A novel approach for time-frequency localization of scaling functions and design of three-band biorthogonal linear phase wavelet filter banks, *Digit. Signal Process.* **69**, 309–322.
20. Sharma R R and Pachori R B 2018 Time-frequency representation using IEVDHM-HT with application to classification of epileptic EEG signals, *IET Sci. Meas. Technol.* **12**(1), 72–82.
21. Pachori R B 2008 Discrimination between ictal and seizure-free EEG signals using empirical mode decomposition, *Res. Lett. Signal Process.* **2008**, 293056.
22. Bajaj V and Pachori R B 2011 EEG signal classification using empirical mode decomposition and support vector machine. *Proc. International Conference on Soft Computing for Problem Solving (SocProS)(New Delhi, India. Dec. 2011),* Springer, pp. 623–635.
23. Gupta V and Pachori R B 2019 Epileptic seizure identification using entropy of FBSE based EEG rhythms, *Biomed. Signal Process. Control.* **53**, 101569.
24. Bajaj V and Pachori R B 2013 Epileptic seizure detection based on the instantaneous area of analytic intrinsic mode functions of EEG signals, *Biomed. Eng. Lett.* **3**(1), 17–21.
25. Sharma R and Pachori R B 2015 Classification of epileptic seizures in EEG signals based on phase space representation of intrinsic mode functions, *Expert Syst. Appl.* **42**(3), 1106–1117.
26. Pachori R B and Patidar S 2014 Epileptic seizure classification in EEG signals using second-order difference plot of intrinsic mode functions, *Comput. Meth. Prog. Bio.* **113**(2), 494–502.
27. Bajaj V and Pachori R B 2012 Classification of seizure and nonseizure EEG signals using empirical mode decomposition, *Health Inf. Sci. Syst.* **5**(1), 7.
28. Taran S, Bajaj V and Siuly S 2017 An optimum allocation sampling based feature extraction scheme for distinguishing seizure and seizure-free EEG signals, *IEEE Trans. Inf. Technol. Biomed.* **16**(6), 1135–1142.
29. Badani S, Saha S, Kumar A, Chatterjee S and Bose R 2017 Detection of epilepsy based on discrete wavelet transform and Teager-Kaiser energy operator. *Proc. IEEE Calcutta Conference (CALCON)(Kolkata, West Bengal, India, Dec. 2017),* IEEE, pp. 164–167.
30. Accardo A, Affinito M, Carrozzi M and Bouquet F 1997 Use of the fractal dimension for the analysis of electroencephalographic time series, *Biol. Cybern.* **77**(5), 339–350.
31. Srinivasan V, Eswaran C and Sriraam N N 2007 Approximate entropy-based epileptic EEG detection using artificial neural networks, *IEEE Trans. Inf. Technol. Biomed.* **11**(3), 288–295.
32. Nicolaou N and Georgiou J 2012 Detection of epileptic electroencephalogram based on permutation entropy and support vector machines, *Expert Syst. Appl.* **39**(1), 202–209.
33. Güler N F, Übeyli E D and Güleri 2005 Recurrent neural networks employing Lyapunov exponents for EEG signals classification, *Expert Syst. Appl.* **29**(3), 506–514.

34. Sharma M, Pachori R B and Acharya U R 2017 A new approach to characterize epileptic seizures using analytic time-frequency flexible wavelet transform and fractal dimension, *Pattern Recognit. Lett.* **94**, 172–179.
35. Sharma M and Pachori R B 2017 A novel approach to detect epileptic seizures using a combination of tunable-Q wavelet transform and fractal dimension, *J. Mech. Med. Biol.* **17**(4), 1740003.
36. Tiwari A K, Pachori R B, Kanhangad V and Panigrahi B K 2016 Automated diagnosis of epilepsy using key-point based local binary pattern of EEG signals, *IEEE J. Biomed. Health Inform.* **21**(4), 888–896.
37. Kaya Y, Uyar M, Tekin R and Yıldırım S 2014 1D-local binary pattern based feature extraction for classification of epileptic EEG signals, *Appl. Math. Comput.* **243**, 209–219.
38. Siuly S, Alcin O F, Bajaj V, Sengur A and Zhang Y 2018 Exploring hermite transformation in brain signal analysis for the detection of epileptic seizure, *IET Sci. Meas. Technol.* **13**(1), 35–41.
39. Acharya U R, Oh S L, Hagiwara Y, Tan J H and Adeli H 2017 Deep convolutional neural network for the automated detection and diagnosis of seizure using EEG signals, *Comput. Biol. Med.* **100**, 270–278.
40. Bajaj V, Rai K, Kumar A and Sharma D 2017 Time-frequency image based features for classification of epileptic seizures from EEG signals, *Biomed. Phys. Eng. Express* **3**(1), 015012.
41. Zhang J and Michael S 2006 Complex network from pseudoperiodic time series: Topology versus dynamics, *Phys. Rev. Lett.* **96**(23), 238701.
42. Stam C J and Van Straaten E C 2012 The organization of physiological brain networks, *Clin. Neurophysiol.* **123**(6), 1067–1087.
43. Lacasa L, Luque B, Ballesteros F, Luque J and Nuno J C 2008 From time series to complex networks: The visibility graph, *Proc. Natl. Acad. Sci. U. S. A.* **105**(13), 4972–4975.
44. Luque B, Lacasa L, Ballesteros F and Luque J 2009 Horizontal visibility graphs: Exact results for random time series, *Phys. Rev. E.* **80**(4), 046103-1–046103-11.
45. Ni Y, Wang Y, Yu T and Li X 2014 Analysis of epileptic seizures with complex network, *Comput. Math. Methods Med.* **2014**, 1–6.
46. Tang X, Xia L, Liao Y, Liu W, Peng Y, Gao T and Zeng Y 2013 New approach to epileptic diagnosis using visibility graph of high-frequency signal, *Clin. EEG Neurosci.* **44**(2), 150–156.
47. Zhu G, Li Y and Wen P 2014 Epileptic seizure detection in EEGs signals using a fast weighted horizontal visibility algorithm, *Comput. Meth. Program Biol.* **115**(2), 64–75.
48. Supriya S, Siuly S, Wang H, Cao J and Zhang Y 2016 Weighted visibility graph with complex network features in the detection of epilepsy, *IEEE Access* **4**, 6554–6566.
49. Mohammadpoory Z, Nasrolahzadeh M and Haddadnia J 2017 Epileptic seizure detection in EEGs signals based on the weighted visibility graph entropy, *Seizure***50**, 202–208.
50. Zeng M, Zhao C Y and Meng Q H 2019 Detecting seizures from EEG signals using the entropy of visibility heights of hierarchical neighbors, *IEEE Access* **7**, 7889–7896.
51. Andrzejak R G, Lehnertz K, Mormann F, Rieke C, David P and Elger C E 2001 Indications of non-linear deterministic and finite dimensional structures in time series of brain electrical activity: Dependence on recording region and brain state, *Phys. Rev. E.* **64**(6), 061907.
52. Iacovacci J and Lacasa L 2016 Sequential visibility-graph motifs, *Phys. Rev. E.* **93**(4), 042309.
53. Ren W and Jin N 2020 Sequential limited penetrable visibility-graph motifs. *Nonlinear Dynamics*, **155**, 1–10.

54. Xie W J, Han R Q and Zhou W X 2019 Tetradic motif profiles of horizontal visibility graphs, *Commun. Nonlinear Sci. Numer. Simul.* **72**, 544–551.

55. Breiman L 2001 Random forests, *Machine Learning* **45**(1), 5–32.

56. Roy S S, Dey S and Chatterjee S 2020 Autocorrelation aided random forest classifier based bearing fault detection framework, *IEEE Sensors J.* **20**(18), 10792 –10800.

57. Oshiro T M, Perez P S and Baranauskas J A 2012 How many trees in a random forest?. *Proc. International Workshop on Machine Learning and Data Mining in Pattern Recognition (Heidelberg, Berlin, July 2012)*, Springer, pp. 154–168.

58. Cheng E, McLaughlin S, Megalooikonomou V, Bakic P R, Maidment A D and Ling H 2011 Learning-based vessel segmentation in mammographic images. *Proc. IEEE First International Conference on Healthcare Informatics, Imaging and Systems Biology (San Jose, CA, USA, July 2011)*, IEEE, pp. 315–322.

59. Chatterjee S, Roy S S, Bose R and Pratiher S 2020 Feature extraction from multifractal spectrum of electromyograms for diagnosis of neuromuscular disorders, *IET Sci. Meas. Technol.* **14**(7),817–824.

60. Luts J, Ojeda F, Van de Plas R, De Moor B, Van Huffel S and Suykens J A 2010 A tutorial on support vector machine-based methods for classification problems in chemometrics, *Analytica Chimica Acta.* **665**(2), 2010.

61. Battineni G, Chintalapudi N and Amenta F 2019 Machine learning in medicine: Performance calculation of dementia prediction by support vector machines (SVM), *Informatics in Medicine Unlocked* **16**, 100200.

62. Ganesan K and Rajaguru H 2019 Performance analysis of KNN classifier with various distance metrics method for MRI images, *Soft Computing and Signal Processing. Advances in Intelligent Systems and Computing*, vol 900, eds. Wang J, Reddy G and Prasad V (Springer: Singapore) pp. 673–682.

63. Supardi N Z, Mashor M Y, Harun N H, Bakri F A and Hassan R 2012 Classification of blasts in acute leukemia blood samples using k-nearest neighbour. *Proc. IEEE 8th International Colloquium on Signal Processing and its Applications (Melaka, Malaysia, March 2012)*, IEEE, pp. 461–465.

64. Zaw H T, Maneerat N and Win K Y 2019 Brain tumor detection based on naïve Bayes classification. *Proc. 5th International Conference on Engineering, Applied Sciences and Technology (ICEAST) (Luang Prabang, Laos, July 2019)*, IEEE, pp. 1–4.

65. Erickson B J, Korfiatis P, Akkus Z and Kline T L 2017 Machine learning for medical imaging, *Radiographics* **37**(2), 505–515.

66. Dey S, Roy S S, Samanta K, Modak S and Chatterjee S 2019 Autocorrelation based feature extraction for bearing fault detection in induction motors. *Proc. International Conference on Electrical, Electronics and Computer Engineering (UPCON)(Uttar Pradesh, India, Nov. 2019)*, IEEE, pp. 1–5.

67. Peker M, Baha S and Dursun D 2015 A novel method for automated diagnosis of epilepsy using complex-valued classifiers, *IEEE J. Biomed. Health Inform.* **20**(1), 108–118.

68. Mert A and Akan A 2018 Seizure onset detection based on frequency domain metric of empirical mode decomposition, *Signal Image Video Process.* **12**(8), 1489–1496.

69. Gupta, A, Singh P and Karlekar M 2018 A novel signal modeling approach for classification of seizure and seizure-free EEG signals, *IEEE Trans. Neural Syst. Rehabil. Eng.* **26**(5), 925–935.

70. Siuly S, Li Y and Zhang Y 2016 A novel clustering technique for the detection of epileptic seizures, *EEG Signal Analysis and Classification. Health Information Science* (Springer: Cham) pp. 83–97.

71. Li Y, Wang X D, Luo M L, Li K, Yang X F and Guo Q 2017 Epileptic seizure classification of EEGs using time–frequency analysis based multi-scale radial basis functions, *IEEE J. Biomed. Health Inform.* **22**(2), 386–397.

72. Zhang T and Chen W 2016 LMD based features for the automatic seizure detection of EEG signals using SVM, *IEEE Trans. Neural Syst. Rehabil. Eng.* **25**(8), 1100–1108.

73. Hassan A R and Subasi A 2016 Automatic identification of epileptic seizures from EEG signals using linear programming boosting, *Comput. Methods Programs Biomed.* **136**, 65–77.

74. Swami P, Gandhi T K, Panigrahi B K, Tripathi M and Anand S A 2016 A novel robust diagnostic model to detect seizures in electroencephalography, *Expert Syst. Appl.* **56**, 116–130.

75. Behara D S, Kumar A, Swami P, Panigrahi B K and Gandhi T K 2016 Detection of epileptic seizure patterns in EEG through fragmented feature extraction. *Proc. 3rd International Conference on Computing for Sustainable Global Development (INDIACom)(New Delhi, India, March 2016)*, IEEE, pp. 2539–2542.

76. Patidar S and Panigrahi T 2017 Detection of epileptic seizure using Kraskov entropy applied on tunable-Q wavelet transform of EEG signals, *Biomed. Signal Process. Control.* **34**, 74–80.

77. Bhattacharyya A, Pachori R B, Upadhyay A and Acharya U R 2017 Tunable-Q wavelet transform based multi-scale entropy measure for automated classification of epileptic EEG signals, *Appl. Sci.* **7**(4), 385.

78. Zhang T, Chen W and Li M 2017 AR based quadratic feature extraction in the VMD domain for the automated seizure detection of EEG using random forest classifier, *Biomed. Signal Process. Control.* **31**, 550–559.

4 Effect of Various Standing Poses of Yoga on the Musculoskeletal System Using EMG

*Padmini Sahu, Bikesh Kumar Singh,
and Neelamshobha Nirala*

CONTENTS

4.1 INTRODUCTION

Yoga is an ancient Indian science that is useful for maintaining the physical fitness of the human body. The origin of yoga comes from the pre-Vedic period, i.e., it is more than 5000 years old. Yoga includes various postures, breathing exercises, cleansing techniques, meditations, pranayamas, and healthcare modifications [1]. Yoga is very effective for staying physically and mentally healthy. It helps the muscles to gain strength, tone the body, and increase fitness levels. Modern yoga studies include the prevention and cure of diseases, recovery from injuries, and lifestyle science [2]. Yoga is also important for controlling and calming the human body; it also helps people to stay healthy, to keep the musculoskeletal system in shape, and to heal cardiopulmonary diseases [3]. By practicing yoga regularly, it helps to develop healthy hormone levels in the body, which supports and maintains physical fitness [4]. Yoga poses (asanas) are performed using the movement of various parts of the human body. The yoga asanas consist of physical postures that strengthen and stretch muscles and joints to balance the musculoskeletal system of the human body. The various benefits of yoga are stress reduction, mind relaxation, body relief, maintaining health and fitness, and healthcare management [5]. Yoga has been proven to improve the muscular activity of orphaned children [6] and Parkinson's patients [7]; it is also used in physiotherapy to relieve joints, muscles, and the entire skeletal system [8]. In [9], Dev *et al.* demonstrated that there was an increased oxyhemoglobin (HbO) concentration in the prefrontal cortex (PFC) of a yoga group compared to a non-yoga group during an attention-based task. In [10], it was shown that yoga had a positive effect on the treatment of high blood pressure. Sharma *et al.* [11] showed that yoga therapy could improve chronic pain and psychological impacts significantly. Kaminoff *et al.* [12] beautifully explained the science of the human anatomy and yoga, and also defined the relationship between the two.

The various yoga poses are categorized as standing poses, sitting poses, arm support poses, kneeling poses, supine poses, and prone poses. Each pose has various sub-poses, such as standing poses containing a tree pose, triangle pose, and chair pose, with specific joint and muscle actions. Ni *et al.* [13] investigated specific muscle activation patterns for training and rehabilitation during different standing poses. The investigation was performed on muscles of the trunk and hip (to address lower back pain), which in the study declined in performance during the various standing poses, such as chair pose, high plank, and mountain pose. The EMG signal of the muscles was analyzed using Labview software in which the muscle activity was measured using the root mean square of the EMG signal acquired during the yoga poses. The standing poses help to improve and strengthen the movement of the knee muscles [14]. Standing poses are very important for children, students, teachers, coaches, practitioners, and older adults. as they can easily learn various postures like standing upright, forward bending, back bending, balancing, and twisting of the body [15]. Standing poses are widely used in the biomedical field, and they are highly beneficial for older adults as they can help improve joint pain, muscle activity, balance, and the control of body posture [16]. Ekstrom *et al.* [17] analyzed the

EMG signal of the activation of various muscles, such as the spine, hip, and knee so that clinicians could be aware of the impact of a particular exercise on strength, endurance, or stabilization. They used a maximum voluntary isometric contraction (MVIC) for normalization of the EMG data recorded for each muscle during nine exercises.

Electromyography (EMG) is a technique for recording and evaluating electrical signals generated during muscle relaxation and contraction in the skeletal system. These generated electrical signals are called EMG signals [18]. An EMG signal is a biomedical signal that can be recorded by placing an electrode onto a muscle group that provides information about muscle force or movement of the body [19]. In the field of biomechanics, De Luca [20] explored the uses of sEMG in three groups of applications: the relationship of force/EMG signal, activation timing of muscle contractions, and EMG signals of muscle fatigue. He explained how to detect and process an EMG signal, the factors that can affect an EMG signal, the force produced by a muscle and its relationship with an EMG signal, the activation timing of the muscles, an EMG signal as a fatigue index, and issues related with the EMG signal for international agreement. Ekstrom *et al.* [21] described and identified various exercises for trapezius and serratus anterior muscles using EMG signals that could help practitioners to develop exercise programs for these muscles. Sadikoglu *et al.* [22] presented detailed knowledge of the EMG signal and its analysis. Bose *et al.* [23] presented a feature extraction scheme for the classification of EMG signals into three classes: autophagy, myopathy, and neuropathy. The deep feature extraction and transfer learning methods were explored for the classification of physical actions based on surface EMG signals [24]. Analytical features with least square support vector machine (LS-SVM) classifiers were explored for the classification of amyotrophic lateral sclerosis (ALS) and standard EMG signals [25]. Chadha *et al.* [26] analyzed an EMG signal using the tunable-Q factor wavelet transform (TQWT) in order to classify physical actions through a multi-class least square support vector machine (MC-LS-SVM). Wu *et al.* [27] proposed a real-time multimodal American Sign Language recognition system from a feature-level fusion scheme using surface electromyography (sEMG). Atzori *et al.* [28] worked to control a robotic prosthetic hand with the help of a non-invasive method using sEMG, hand kinematics, and hand forces by applying machine learning algorithms. Xi *et al.* [29] proposed a feature-level fusion method for processing and analyzing the EMG signals of eight activities in daily life (ADLs). Salem *et al.* [30] used 21 Hatha yoga postures using sEMG for biomechanical analysis and evaluated muscle movements, joint angles, and forces on senior citizens, and found that it was very effective for balancing their musculoskeletal system.

The rest of the chapter is organized as follows: Section 4.2 describes the different standing poses of yoga. Section 4.3 briefly explains the effect of standing poses on muscles. Section 4.4 describes the benefits of yoga standing poses according to the literature. Section 4.5 gives a detailed description of EMG signals. Section 4.6 reviews the methodology adopted in existing studies for analyzing the EMG signals. Conclusions and future scope are summarized in Section 4.7.

4.2   STANDING POSES OF YOGA

Tirumalai Krishnamacharya (aged 100 between November 18, 1888, to February 28, 1989) was "The father of modern yoga." He was a great ayurvedic restorer, yoga teacher, and scholar [31]. A standing pose is a pose where you are standing on the soles of the feet. Standing poses are done by standing on one or both legs, i.e., at least one leg should be in contact with the ground. Yoga starts with standing poses that help to do the essential process in the human body like inhaling and exhaling, and support and relaxing [12]. Standing poses have the ability to maintain and control the posture of the human body as they have the highest center of gravity among all the yoga poses [32]. Various standing poses are there in the literature that is based on a combination of standing postures and balance postures. Examples of standing postures include the mountain, triangle, side angle, and king of the dancer postures. Examples of balance postures include the chair, eagle, tree, and arm balance postures. The following sub-sections briefly describe the various standing poses.

4.2.1   TADASANA (MOUNTAIN POSE)

Tadasana is an important pose that requires concentration to balance the human body (also known as a balancing pose). Tadasana is a combination of two words, "Tada" and "sana," in which "tada" means "mountain" and "sana" means "straight"; so tadasana means standing straight like a mountain. Various muscles like the erector spine or back muscles, abdominal muscles, and trapezius muscles, are affected during the tadasana [13]. The benefits of tadasana are to increase height; to strengthen arms, thighs, knees, ankles, and legs; and to improve muscle toning, the digestive system, and body postures [12, 33]. The mountain pose is shown in Figure 4.1(a).

4.2.2   UTKATASANA (CHAIR POSE)

Utkatasana is a basic standing pose. It is a combination of "utkata" and "sana." The word "utkata" is a Sanskrit word which means superior, immense, and large. It is also known as the "chair pose" because when doing this pose it looks like you are sitting on an invisible chair [32]. Various muscles, such as rectus femoris muscles (thigh muscle), and iliopsoas muscles (a deep hip flexor muscle), are affected during utkatasana [13]. The benefits of utkatasana are to strengthen the thighs, ankles, and legs; to improve the hips, stomach, and shoulders; and to tone the knees and ankles [12]. The chair pose is shown in Figure 4.1(b).

4.2.3   UTTANASANA (STANDING FORWARD BEND)

Uttanasana is an easy standing posture that includes bending forward, which is why it is also known as the standing forward bend. It is a combination of three words "ut," "tan," and "asana" in which "ut" means "intense" "tan" means "stretch," and "asana" means "posture." Various muscles, such as the hamstrings, and calf muscles are affected with this asana. The benefits of Uttanasana are to stimulate the brain,

(a) Mountain Pose

(b) Chair Pose

(c) Standing Forward Bend Pose

(d) Tree Pose

(e) Extended Hand–Toe Pose

(f) Eagle Pose

(g) Warrior Pose I

(h) Warrior Pose II

(i) Warrior Pose III

(j) Triangle Pose

(k) King of dancers pose

(l) Wide Legged Forward Bend

(m) Extended Side Angle Pose

(n) Revolved Side Angle Pose

(o) Half moon pose

(p) Plank Pose

FIGURE 4.1   Various yoga standing poses [52].

strengthen the knee, thigh, hamstrings, hips, and calves; and to relax the nerves [12, 34]. The standing forward bend pose is shown in Figure 4.1(c).

4.2.4 VRIKSHASANA (TREE POSE)

Vrikshasana is an asymmetrical standing balancing pose. It is a combination of two words "Vriksh" and "asana" in which "Vriksh" means "tree" and "asana" means "pose." so Vrikshasana means "tree pose." It is also known as Vrksasana. It is a part of hatha yoga, and it is a very popular exercise in modern yoga. The various benefits of Vrikshasana are to stretch the thigh and shoulders; strengthen the torso, thigh, and ankles; thin the legs, and opens the hips. It also helps to improve flat feet; increase self-confidence; concentrate the mind; control balance [1, 12, 30, 32, 35]. The tree pose is shown in Figure 4.1(d).

4.2.5   Utthita Hasta Padangusthasana (Extended Hand–Toe Pose or Standing Big Toe Hold)

Utthita Hasta Padangusthasana is a combination of five words "utthita," "hasta," "pada," "augusta" and "asana" in which "utthita" means "extended," "hasta" means "hand," "pada" means "foot," "augusta" means "big toe," and "asana" means "pose," so uttita hasta padangusthasana means the "extended hand–toe pose" or "standing big toe hold." The benefits of this pose are to improve physical and mental balance; strengthen core muscles and legs; produce a stretch in the hip and knee muscles, and to maintain stability during exercises. It also helps to relieve back pain and depression, especially for the younger generation [12, 36, 37]. The extended hand–toe pose is shown in Figure 4.1(e).

4.2.6   Garudasana (Eagle Pose)

Garudasana is a type of balancing pose and a part of hatha yoga. It is a combination of two words "garud" and "asana" in which "garud" means "eagle" and "asana" means "pose," so Garudasana is known as the "eagle pose." The word "garud" refers to a god in Indian tradition, i.e., Hindu god "Vishnu." The benefits of garudasana are to improve concentration and balance; make the shoulders flexible; strengthen ankles; stretch the shoulders, upper back, hip, thigh, and ankles [12, 32, 34]. It is beneficial for musicians to reduce music performance anxiety (MPA), which is possible only when the musicians do regular and strict yoga [38]. It is also beneficial for older adults to maintain and balance the body [39]. The eagle pose is shown in Figure 4.1(f).

4.2.7   Virabhadrasana (Warrior)

The word "virabhadrasana" is a Sanskrit name which is related to Indian mythology and is also known as the warrior pose. Virabhadrasana is a combination of three words "vira," "bhadra," and "asana" in which "vira" means "hero," "bhadra" means "auspicious," and "asana" means "pose," so "virabhadrasana" means "heroic auspicious pose." There are three variations of virabhadrasana: (i) virabhadrasana I (warrior I), (ii) virabhadrasana II (warrior II), and (iii) virabhadrasana III (warrior III). The various benefits of virabhadrasana are to strengthen the shoulders, back muscles, and arms; stretch the neck, chest, lungs, shoulders, and abdomen; and improve physical fitness. These stretching andstrengthening exercises are to maintain the back bends [39]. They also help to improve body balance and mobility-related issues in older adults [40].

4.2.7.1   Virabhadrasana I (Warrior I)

Virabhadrasana I helps to relax a stiff neck and shoulders; strengthen the shoulders, arms, thighs, and ankles; helps with deep breathing; helps stretch the neck, shoulders, chest, lungs, thighs, inguinal region, and ankles; improves concentration; develops more power; and most importantly brings stability [5, 12, 32, 33]. It also helps to balance metabolic risk and healthy life in adults [41]. The warrior pose I is shown in Figure 4.1(g).

4.2.7.2   Virabhadrasana II (Warrior II)

Virabhadrasana II has almost the same features and benefits as virabhadrasana I, as the first few steps of virabhadrasana II are the same as virabhadrasana I. However, there is a difference in the later steps where virabhadrasana II tightens the arms and ankles; strengthens and shapes the legs; tones the stomach; and relieves back and legs cramps. Various muscles like the gluteus medius, lateral rotators of the hip, anterior thighs, deltoids, and adductors are affected during this pose [5, 12, 32, 41]. Warrior pose II is shown in Figure 4.1(h).

4.2.7.3   Virabhadrasana III (Warrior III)

The virabhadrasana III pose is more difficult than the previous two virabhadrasanas. The benefits of virabhadrasana III are to strengthen the thigh; develop concentration; increase strength and size of the legs and abdominal muscles; reduce fat in the hip; and improve physical postures, especially in adults [5, 12, 32, 35, 41]. warrior pose III is shown in Figure 4.1(i).

4.2.8   TRIKONASANA (TRIANGLE POSE)

Trikonasana is a hip opening standing pose. It is a combination of three words "tri," "kona" and "asana" in which "tri" means "three," "kona" means "angle," and "asana" means "pose," so trikonasana is called the "triangle pose." It is also known as the "happy pose" because when doing this pose, the Venus chakra opens, and the feeling is very joyful [1, 12, 32, 34]. The benefits of trikonasana are to shape the trunk muscles, waist, and legs; improve blood circulation and the digestion system; reduce fat; strengthen ankles [5]. Various muscles like the right oblique muscles, rectus abdominis, gluteus maximus, and gluteus medius muscles are affected during trikonasana [42]. Trikonasana has been used by many authors in the field of disease, medicine, and sports. It helps to improve rheumatoid arthritis (chronic inflammatory disorder) [43] and diabetes [44]. It also helps pregnant women during the prenatal and postnatal periods [45]. The triangle pose is shown in Figure 4.1(j).

4.2.9   NATARAJASANA (KING OF THE DANCE POSE)

Natarajasana is a very old and back bend standing pose from Indian civilization. Some features of the pose have been taken from the classical Indian dance "bharatnatyam." It is a combination of three words "nata," "raja" and "asana" in which "Nata" means "dance," "raja" means "king," and "asana" means "pose," so natarajasana means "king of the dance pose." Natarajasana affects various muscles such as the hamstrings, quadriceps muscles, and calves [12]. The various benefits of natarajasana are to stabilize the human body; improve quality of life [41]; and reduce the anxiety of performing music [38]. It is helpful for women to reduce menopausal symptoms and depression; increase positive energy; and prevent cortisol hormones increasing in post-menopausal women [46]. The king of the dance pose is shown in Figure 4.1(k).

4.2.10   Prasarita Padottanasana (Wide-Legged Forward Bend)

Prasarita padottanasana is an inverted symmetrical forward bend pose. It is a combination of five words "prasarita," "pada," "ut," "tan," and "asana" in which "prasarita" means "spread and expanded," "pada" means "foot or leg," "ut" means "intense," "tan" means "to stretch out," and "asana" means "pose," so prasarita padottanasana is known as the "wide-legged forward bend pose" or "wide-stance forward bend pose" [12]. The benefits of prasarita padottanasana are to stretch the hamstrings [5]; improve sleep and mood in women suffering from restless legs syndrome (RLS) [47]; help with muscular strength, endurance, and flexibility [48]. The wide-legged forward bend pose is shown in Figure 4.1(l).

4.2.11   Parsvakonasana (Side Angle Pose)

Parsvakonasana is a stretching standing pose. It is a combination of three words "parsva," "kona" and "asana" in which "parsva" means "side," "kona" means "angle," and "asana" means "pose," so parsvakonasana means the "side angle pose." The benefits of parsvakonasana help to strengthen and stretch the hips, stomach, knees, ankles, and legs; improve stamina; tone the legs; reduce back pain problems [12, 32]. There are two versions of parsvakonasana: (i) utthita parsvakonasana, and (ii) parivrtta baddha parsvakonasana.

4.2.11.1   Utthita Parsvakonasana (Extended Side Angle Pose)

Utthita parsvakonasana is a combination four words "utthita," "parsva," "kona" and "asana" in which "utthita" means "extended," "parsva" means "side," "kona" means "angle," and "asana" means "pose," so utthita parsvakonasana means the "extended side angle pose." The benefits of utthita parsvakonasana are to give muscle strength; improve flexibility in the body; strengthen and stretch the knees, feet, and ankles [12, 34, 37, 48]. The extended side angle pose is shown in Figure 4.1(m).

4.2.11.2   Parivrtta Baddha Parsvakonasana (Revolved Side Angle Pose)

Parivrtta baddha parsvakonasana is a rotated standing pose. It is a combination of five words "parivrtta," "baddha," "parsva," "kona" and "asana" in which "Parivrtta" means "revolve and twist," "baddha" means "bound," "parsva" means "side," "kona" means "angle," and "asana" means "pose," so parivrtta baddha parsvakonasana means the "revolved side angle pose." The benefits of parivrtta baddha parsvakonasana are to relax the mind; improve concentration; strengthen the physical and mental status of the body [5, 12]. The revolved side angle pose is shown in Figure 4.1(n).

4.2.12 Ardha Chandrasana (Half-Moon Pose)

Ardha chandrasana is a combination of three words "ardha," "chandr" and "asana" in which "Ardha" means "half," "chandr" means "moon" and "asana" means "pose," so ardha chandrasana means the "half-moon pose" [12]. The benefits of ardha chandrasana are to strengthen ankle muscles; tone buttocks and hips; stretch the hamstrings and thigh muscles; reduce levels of aggression [49]. It helps to prevent falls in older adults by strengthening the anterior tibialis and gastrointestinal muscles [5, 35]. It is also beneficial for breast cancer survivors by improving stability, flexibility, breathing, and energy; relieving pain; and reducing stress levels [33]. The half-moon pose is shown in Figure 4.1(o).

4.2.13 Plank Pose

The plank pose is based on arm balance. The benefits of the plank pose are to strengthen the core muscles, i.e., abdomen, hip, trunk, and lower back; strengthen the spine muscles [13, 16, 50]; improve physical fitness and balancing. It can be used as a transition pose in yoga to switch from one pose to another pose [32]. It also helps to improve the physical fitness of patients affected by scoliosis [51]. The plank pose is shown in Figure 4.1(p).

4.3 BENEFITS OF STANDING POSES

There are various benefits of standing poses for the human body, as given below:

a) To tone the legs; to stretch the shoulders; to strengthen the torso, thighs, and ankles in older adults [30].
b) To relieve back pain [32].
c) To improve the digestive system and body postures [33].
d) To help breast cancer survivors by improving stability, flexibility, breathing, and energy; relieving pain; reducing stress levels [33].
e) To stimulates the brain [34].
f) To strengthen the knee and thigh; to stretch hamstring [34].
g) To prevent falling in older adults [35].
h) To reduce depression level, especially in the younger generations [37].
i) To reduce MPA in musicians [38].
j) To balance metabolic risk [41].
k) To reduce hip fat and to improve physical balance, especially in adults [41].
l) To improve rheumatoid arthritis (chronic inflammatory disorder) [43].
m) To help diabetic patients [44].
n) To help pregnant women during the prenatal and postnatal periods [45].
o) To reduce menopausal symptoms and depression; increase positive energy; and prevent cortisol hormones increasing in post-menopausal women [46].
p) To reduce high blood pressure [47].
q) To strengthen the spine muscles [50].

4.4 INTRODUCTION TO ELECTROMYOGRAPHY

In 1849, Emile du Bois-Raymond [53] discovered that it was possible to record electrical movements during muscle contractions. In 1890, Étienne Jules Marey [54] made the first recording of this activity, as well as introducing the term electromyography. The first commercial electromyography system was introduced in 1950. The period from 1950 to 1973 was the era of analog EMG systems: EMG signals were recorded, and analyses were performed manually on film or paper [55]. In 1985, Basmaiijan and Luka [56] summarized existing muscle-related knowledge and research, according to electrical studies.

EMG is a technique for recording and evaluating neuromuscular activity generated during muscle relaxation and contraction in the skeletal system. The basic and smallest functional unit comprising of motor neurons and the muscle fibers that innervates during muscular contraction is called the motor unit. The activation of a motor neuron leads to a contraction of all muscle fibers in the unit [19]. Motor unit action potentials (MUAPs) are produced as soon as this muscle contraction is activated. These MUAPs play a very important role in EMG, with its features and firing rates providing information for diagnosing neuromuscular disorders [55]. In EMG, electromyography (an instrument) is used to produce electrical activity, and an electromyogram (a test) is used to record the electrical activity of the muscles.

Muscular electrical activity and artifacts are sources of EMG signals. There are three main steps in EMG: (i) recording, (ii) amplification, and (iii) displaying the signal. When recording and displaying an EMG signal, two significant problems may arise that affect the signal. The first problem is the signal-to-noise ratio, i.e., the ratio of an electrical or other signal carries unwanted interference energy information to the EMG signal. The second problem is signal distortion, i.e., the relationship between different frequency components or the basic waveform in the EMG signal should not be changed [55]. Initially, the EMG signal range is 0–10 mV (or 5 to –5 mV) when amplification has not occurred. EMG signals make noise while traveling through the various types of tissue around the muscles. However, it is important to understand the characteristics of electrical noises that can affect the EMG signal. The electrical noises can be classified into the following classes: i) inherent noise in the electrode, ii) ambient noise, iii) motion artifact and iv) Inherent instability of the signal. All electronic devices produce electrical noise known as "inherent noise." The frequency component in this noise ranges from 0 Hz to several thousand Hz. This type of electrical noise cannot be eliminated; it can be reduced only by using high-quality electronic components [22, 55]. Ambient noise is also called background noise. The main source of ambient noise is electromagnetic radiation. The range of amplitude of this noise can be one to three times higher than the EMG signal [57]. There are irregularities in the data due to motion artifacts. These noises also occur at the electrode–skin interface. There are two types of motion artifact: i) electrode cable and ii) electrode interfacing. Various factors can reduce the effect of motion artifacts by means of reducing the degree of motion, suppressing the signal in moving tissues, detecting motion, proper design of electronic circuitry, and set-up [22, 55].

EMG signals are influenced by various factors, such as body temperature, muscle blood flow, and type of muscle fibers. Therefore, different types of noise are present in the EMG signal as discussed above. Noise removal techniques can reduce various noises in EMG signals that are i) a low pass differential filter, ii) an adaptive noise cancellation, and iii) signal filtering based on wavelets. Low pass difference (LPD) filters are used in EMG signal processing on a wide scale. It is easy to access, which is implemented in the time domain that gives fast results for many real-time applications. Adaptive noise cancellation is used to reduce the effect of various types of noise present in the EMG signal. This technique is used to assess corrupted signals through additive noise or interference. Waves are used to remove noise from contaminated and unwanted signals in EMG signals. Deionizing a signal is characterized in three ways: signal decomposition, expansion coefficient thresholding, and signal reconstruction [57]

There are two types of EMG (i) surface EMG and (ii) intramuscular EMG. Surface EMG (sEMG) is a technique in which electrodes (non-invasive) are placed on the area of the skin where the muscle is more active to generate electrical signals for the measurement of muscle activity. The benefits of EMG are that it is painless to use; does not harm the skin; and does not need to be anesthetized. There are various applications where surface EMG can be used that are sports, clinical work, and superficial muscles. In intramuscular EMG, the term "intramuscular" is used to represent a needle or fine wire. The needle or wire (invasive) is inserted into the active muscle of the skin from which an electrical signal is generated. The benefits of intramuscular EMG are the ability to record the activity of the deep muscles or weak contractions, it has a low signal interference, and the ability to check individual muscles [55, 58, 59].

Various factors affect the performance of EMG signals: body temperature, type of muscle fibers, diameter of the muscle fiber involved in the electrical signal, number of muscle fibers involved in the electrical signal, distance from the electrode to the muscle fiber, types of muscle contraction, blood flow in the muscle, the area of the skin (detection surface), the firing (activation) rate of the motor unit, and the type of sensors [22, 55, 58].

EMG signals are used in the broad area of biomedical and clinical diagnosis. Its use consists of detecting, decomposition, processing, analyzing, and classification as per the field of application. With these general steps, one may develop a robust, flexible, and healthy application for various purposes. Various techniques are available in the literature to process EMG signals such as filtering, rectification, wavelet analysis, and modeling, as raw EMG signals do not give any useful information [22, 55].

There are various applications where EMG may be used, and these include neurological disorders, muscle disorders, training of muscle strength, sports physiology, bipolar disorders, and nerve disorders [58, 59]. EMG has already been applied in different fields as per the literature, such as: (i) it helps to prevent adults from falling and also helps with physiotherapy [35], (ii) spatial filtering (an image processing technique) [60], (iii) prosthesis control [61], (iv) to remove the unwanted part and interferences using EMG signal classification [62], (v) accurate measurement of lumbar multifidus muscle activity [63], and (vi) to improve arm function after stroke [64].

TABLE 4.1
Electrode Configuration of EMG [61]

Parameters	Range
Bandwidth of signal	20–450 Hz
The input range of signal	11 mV
Transmission range of EMG sensors	20 m
The sampling rate of EMG signal	2000 samples/sec
Baseline noise	<750 nVrms
Common mode rejection ratio (CMRR)	>80 db
EMG signal resolution	16 bit
Accelerometer outputs	$\pm 2g$ to $\pm 16g$
Bandwidth of accelerometer	24–470 Hz
Gyroscope outputs	± 250 dps to ± 2000 dps
Bandwidth of Gyroscope	24–360 Hz

There are three parts of the EMG instrument: (i) the electrodes, (ii) the leads, and (iii) the differential amplifiers. Electrodes are divided into two types: (i) the invasive electrode and (ii) the non-invasive electrode. These electrodes are used to acquire muscle signals. Leads are divided into three types: (i) the black lead for the active electrode (E1), (ii) the red lead for reference input (E2), and (iii) the green or white lead for ground. The leads are used to make a connection between the amplifier and the electrode. Differential amplifiers are used to magnify the potential difference between the active electrode (E1) and reference input (E2) [22, 55].

Characteristics of the EMG signal are divided into two parts: i) amplitude characteristics, and ii) frequency characteristics. Amplitude characteristics are in a time domain in which EMG amplitude parameters are studied based on their mean, maximum, minimum, root mean square (RMS), and peak time. Frequency characteristics are in the Fourier transform in which the amount of signal in the frequency bandwidth is known as the EMG frequency spectrum. The parameters most often used for these characteristics are their median, mean, and peak power frequency [52, 55, 57].

The processing of the EMG signal is divided into two types: (i) filter and (ii) rectifier. EMG activity is filtered in the range of 10 Hz (high pass filter) to 5 kHz (low pass filter). Rectifiers are divided into two types: (i) full-wave (integrates all modulating signals), and (ii) half-wave (accepts only positive values). The electrode configuration of an EMG instrument is shown in Table 4.1 [58, 59, 65].

4.5   ANALYSIS OF AN EMG SIGNAL FOR MUSCLE ACTIVITY DURING STANDING POSES

It is difficult to get any useful information from the raw EMG signal acquired from humans performing yoga until it is processed and quantified [66]. A system is required to use the characteristics of the acquired raw EMG signal, based on which it is classified for different postures. In Figure 4.2, the raw EMG signal is shown by the temporal signal. Several techniques are available in the literature to process the EMG signal such as the wavelet transform analysis [67], time-frequency approach [68], artificial intelligence (AI) [55], proprioceptive neuromuscular facilitation (PNF) [69], higher-order statistics [70], principal component analysis [70], and autoregressive model [71]. In the past, the processing of an EMG signal using the DWT has been used to evaluate muscle fatigue (neck and shoulder muscles) [67], identify four emotions, such as joy, anger, sadness, and pleasure [72], and detect muscle activity times under dynamic conditions [73].

For EMG signal acquisition from a muscle, first, shave the area of skin where muscle activity is to be measured and clean it properly with antiseptic alcohol to remove interference (noise) between the sensor and muscle. Thereafter, place the trigno wireless sensors on the muscle as per the requirements of the different poses. Figure 4.3 shows the general framework for measuring an EMG signal.

The following is a detailed description of the framework for measuring an EMG signal.

1. **Raw EMG data:** To record the EMG signal of a particular muscle, a wireless sensor is attached, which is connected to the base station. The base station receives raw EMG signals in terms of frequency, amplitude, and power, i.e., the function of time. Thereafter, raw EMG signals are then fed into the system for extracting features for pattern recognition and its analysis.
2. **EMG signal processing:** The signal is processed in six steps: i) raw signal amplification (differential mode), ii) analog band-pass filter, iii) analog to digital conversion (ADC), iv) rectify and digital low pass filter ("linear envelope"), or RMS filter, v) on/off time determination, and vi) time-frequency analysis [74].
   i) Raw signal amplification (differential mode): Record the EMG signals in a differential mode, i.e., measure the difference in voltage between two surface or needle electrodes [75].

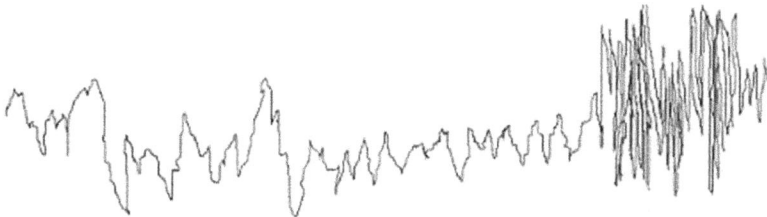

FIGURE 4.2   A raw EMG signal.

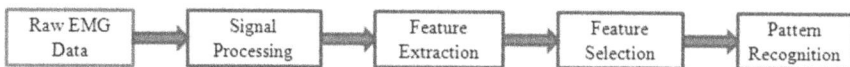

FIGURE 4.3   General framework for the analysis of EMG signals.

ii)   Analog band-pass filter: The analog filter is a band-pass filter that is applied to the raw EMG signal before it is digitized to remove low and high frequencies from the signal [74].

iii) Analog to digital conversion: During the digitization of the EMG surface, an analog to digital (A/D) converter is used to convert the signal to a specific sampling rate [76].

iv)   Rectify and digital low pass filter: Rectify and digital low pass filter is a very important part of signal processing because it performs multi-tasking, i.e., rectification and filtering. The combination of both rectification and low pass filtering is called the "linear envelope" of the signal [74].

v)   On/off times: The EMG is used to determine at what time the muscles "turn on" and "turn off" so that the muscle activity can be measured [74].

vi)   Time-frequency analysis: The analysis is carried out to measure the frequency of the EMG signal that changes continuously over time [77].

3. **Feature Extraction:** The feature extraction consists of identifying and structuring the useful characteristics from the raw signals and eliminating noise, interference, and the unusable portions of the EMG signal [62]. The features of the EMG signal are divided into three main groups that are time domain, frequency domain, and time-frequency domain. However, due to the need for rapid calculation of the changes in the feature values in the time-frequency domain, it cannot be used directly [78].

i)   **Time domain:** Time-domain features are very simple, easy to use, and quick to calculate. The time-domain features are often used in pattern recognition of the EMG signal. There are a large number of time-domain features in the EMG signal. Some of them are discussed below.

a)   Integrated EMG: The integrated EMG (IEMG) feature is widely used in the pre-activation index for muscle relaxation and contraction in the skeletal system [62]. It can be expressed as:

$$IEMG = \sum_{i=1}^{M} |e_i| \qquad (4.1)$$

where e_i is the EMG signal in a segment i and M is the length of the EMG signal.

b)   Mean absolute value: The mean absolute value (MAV) is a very popular feature in EMG signal analysis. It is used to detect and measure muscle contraction in the skeletal system [62, 78]. It can be expressed as:

$$MAV = \frac{1}{M} \sum_{i=1}^{M} |e_i| \qquad (4.2)$$

c) Mean absolute value slope: The mean absolute value slope (MAVSLP) is defined as the difference between the MAVs of the adjacent segments [78]. It can be expressed as:

$$MAVS_i = MAV_{i+1} - MAV \tag{4.3}$$

d) Simple square integral: The simple square integral (SSI) is defined as the energy of the EMG signal as a feature [62, 78]. It can be expressed as:

$$SSI = \sum_{i=1}^{M} e_i^2 \tag{4.4}$$

e) The variance of EMG: The variance of EMG (VAR) is defined as the power index. Generally, the variance is the average of the squared deviation of that random variable [62, 78]. It can be expressed as:

$$VAR = \frac{1}{M-1} \sum_{i=1}^{M} e_i^2 \tag{4.5}$$

f) Root mean square: The RMS is another popular feature in EMG signal analysis. The RMS is modeled as an amplitude modulated Gaussian random process in which it is related to constant force and the non-fatiguing of the muscle contractions in the skeletal system [62]. It can be expressed as:

$$RMS = \sqrt{\frac{1}{M} \sum_{i=1}^{M} e_i^2} \tag{4.6}$$

ii) Frequency domain: Frequency domain features are used to study muscle fatigue. They are based on the estimated power spectral density (PSD) of the signal. PSD is widely used in two variables: mean frequency (MNF) and median frequency (MDF). There are other variables such as mean power (MNP), peak frequency (PKF), and total power (TTP). Some of the frequency domain features are discussed below [62].

a) Mean frequency (MNF): The mean frequency (MNF) is the ratio of the sum of the product of the EMG power spectrum to the sum of the spectrum intensities [62]. It can be expressed as:

$$MNF = \sum_{j=1}^{K} f_j P_j / \sum_{j=1}^{K} P_j \tag{4.7}$$

where f_j represents the frequency of the spectrum at frequency j, P_j represents the EMG power spectrum at frequency j, and K denotes the length of the frequency bi.

b) Median frequency (MDF): The power spectrum is divided into two regions with equal amplitude known as median frequency (MDF), in other words, MDF is half of the total power (TTP) feature [62]. It can be expressed as:

$$\sum_{j=1}^{MDF} P_j = \sum_{j=MDF}^{K} P_j = \frac{1}{2}\sum_{j=1}^{K} P_j \tag{4.8}$$

c) Peak frequency: Peak frequency (PFK) is defined as the maximum power of the signal [62] and it can be expressed as:

$$PKF = \max\left(P_j\right), \quad j = 1,...K \tag{4.9}$$

d) Mean power: Mean power (MNP) is the average power at which some of the EMG power spectra is divided by the length of the frequency [62]. It can be expressed as:

$$MNP = \sum_{j=1}^{K} P_j / K \tag{4.10}$$

e) Total power: An aggregate of the EMG power spectrum is known as TTP. Which is also equal to zero spectral moments (SM0) [62]. It can be expressed as:

$$TTP = \sum_{j=1}^{K} P_j = SM0 \tag{4.11}$$

f) The first, second, and third spectral moment: The spectral moment (SM) is a process of statistical analysis to extract features from the EMG power spectrum. The first three spectral moments (SM1–SM3) are the most important [62], and it can be expressed as:

$$SM1 = \sum_{j=1}^{K} P_j f_j \tag{4.12}$$

$$SM2 = \sum_{j=1}^{K} P_j f_j^2 \tag{4.13}$$

$$SM3 = \sum_{j=1}^{K} P_j f_j^3 \tag{4.14}$$

4. **Feature Selection:** In recent years, many works have been done in the fields of emotion recognition [79], independent component analysis [80], digestive system [81], and rheumatoid arthritis [82] to further improve the accuracy of using feature selection techniques. Irrelevant features can reduce accuracy in pattern recognition and increase computational complexity. The extracted features often include a large amount of data which increases the training time of the classification system. Therefore, feature selection is an essential step in pattern recognition to improve response time, select the best combination of relevant features for dimension reduction, and improve classification accuracy [16]. A large number of feature selection algorithms are available in the literature, such as principal component analysis [83], ant colony optimization (ACO) [84], particle swarm optimization (PSO) [85], and artificial bee colony (ABC) [86]. ACO, PSO, and ABC are meta-heuristic algorithms which have shown prominent results in selecting the features and improving the accuracy of the model.

5. **Pattern Recognition:** Recognition of human/muscle activity using the EMG signals generated during muscular relaxation and contraction in the skeletal system is known as EMG pattern recognition. A large number of EMG pattern recognition applications are available in the literature, such as emotion recognition [87], hand prosthesis movements [88], and evaluation in clinical applications [89]. Analysis of EMG signals is more challenging than other biological signals (electrocardiogram (ECG), electrooculogram (EOG), and electroencephalogram (EEG)). The fast Fourier transform (FFT) and ANN algorithms are a very popular methods for pattern recognition in EMG signals [90].

Figure 4.4 shows the complete framework for measuring the muscle activity of the musculoskeletal system while doing various standing asanas. In this chapter, standing poses of yoga were taken into consideration for measuring muscle activity and its analysis. One may choose three different types of human to differentiate muscle activity while doing yoga: (i) yoga trainer (ii) practitioner, and (iii) non-practitioner. The participants have to perform specific standing poses for the analysis, which is carried out for the diagnosis of various problems, such as classification of whether the digestive system is in a good or bad class. To perform the EMG signal analysis during yoga, first, record the EMG signal of a particular muscle using a wireless sensor that is connected to the base station. The base station receives raw EMG signals in terms of frequency, amplitude, and power, i.e., the function of time. Thereafter raw EMG signals are fed into the system for the DWT analysis to extract the features using various functions [91]. The output signal of the DWT analysis is known as the preprocessed EMG signal.

Further, the preprocessed EMG signals can be quantified using two types of measures, i.e., statistical measures and signal processing measures. The mean, median, and the standard deviation can be taken as statistical measures whereas root means square, minimum, and maximum power spectral density can be taken as signal processing measures. The calculated statistical measures and signal processing

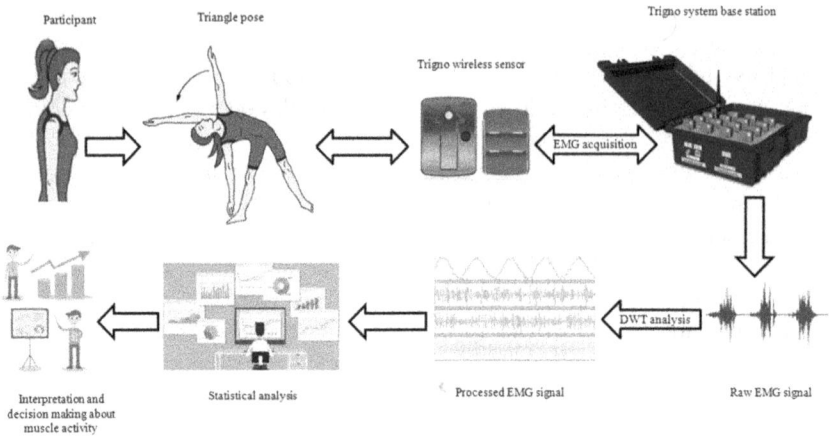

FIGURE 4.4   Framework for measuring muscle activity during yoga standing poses.

measures can be used for decision making. For example, one can say that yoga can be very effective to improve the musculoskeletal system if the muscle activity of the yoga trainer is better than the other two participants.

4.6   CONCLUSION AND FUTURE SCOPE

This chapter gave an overview of the EMG signal. A detailed description was given of the acquisition, preprocessing, and analysis of the EMG signal. The standing yoga poses were taken into consideration for acquiring, preprocessing, and analyzing the EMG signal to see the effect of muscle activity in the musculoskeletal system. The effect of standing poses can easily be differentiated by analyzing the EMG signals acquired from three different participants types. In the future, an extensive study and analysis can be carried out on a real dataset by involving other important poses such as sitting poses, kneeling poses, and arm support poses. Feature extraction and selection techniques for EMG signals may also be explored for improvement of the musculoskeletal system.

ACKNOWLEDGMENTS

Authors would like to thank NIT Raipur for providing necessary facilities to carry out the current work.

REFERENCES

1. M. Keegan. *2,100 Asanas: The Complete Yoga Poses*. New York: Reed Business Information, 2015.
2. B. Hauser. *Yoga Traveling: Bodily Practice in Transcultural Perspective*. Springer, Heidelberg, 2013.

3. J. A. Raub. "Psychophysiologic effects of Hatha Yoga on musculoskeletal and cardio-pulmonary function: a literature review." *The Journal of Alternative & Complementary Medicine*, vol. 8, pp. 797–812, 2002.
4. E. F. Crangle. *The origin and development of early Indian Contemplative Practices, Edward Fitzpatrick Crangle*, vol. 29. Otto Harrassowitz Verlag, Germany 1994.
5. N. Florek. *Stretch! An Illustrated Step-By-Step Guide To 90 Slimming Yoga Postures.* 2010.
6. S. P. Purohit, et al. "Effect of yoga program on minimum muscular fitness of orphan adolescents by using Kraus-weber test: A randomized wait-list controlled study." *Indian Journal of Positive Psychology*, vol. 6, p. 389, 2015.
7. H. M. Hill, et al. "Merging yoga and occupational therapy for Parkinson's disease improves fatigue management and activity and participation measures." *British Journal of Occupational Therapy*, p. 0308022620909086, 2020.
8. S. Ravi. "Physiotherapy and yoga for joint pain treatment: A review." *Journal of Yoga and Physical Therapy*, vol. 6, p. 2, 2016.
9. P. Dev, et al. "Effect of yoga on hemodynamic changes at prefrontal cortex during sustained attention task." in *2019* 5th International Conference on Advanced Computing & Communication Systems (ICACCS), 2019, pp. 728–731.
10. P. Posadzki, et al. "Yoga for hypertension: a systematic review of randomized clinical trials." *Complementary Therapies in Medicine*, vol. 22, pp. 511–522, 2014.
11. N. Sharma, et al. "Tele-Yoga therapy for patients with chronic pain during covid-19 lockdown: A prospective nonrandomized single arm clinical trial." *medRxiv*, 2020.
12. L. Kaminoff, et al. *Yoga anatomy.* Champaign, IL: Human Kinetics, 2007.
13. M. Ni, et al. "Core muscle function during specific yoga poses." *Complementary Therapies in Medicine*, vol. 22, pp. 235–243, 2014.
14. H. S. Longpré, et al. "Identifying yoga-based knee strengthening exercises using the knee adduction moment." *Clinical Biomechanics*, vol. 30, pp. 820–826, 2015.
15. D. Coulter. *Anatomy of Hatha Yoga: A manual for students, teachers, and practitioners.* Motilal Banarsidass Publ., Delhi, 2004.
16. M.-Y. Wang, et al. "The biomechanical demands of standing yoga poses in seniors: The Yoga empowers seniors study (YESS)." *BMC complementary and Alternative Medicine*, vol. 13, p. 8, 2013.
17. R. A. Ekstrom, et al. "Electromyographic analysis of core trunk, hip, and thigh muscles during 9 rehabilitation exercises." *Journal of Orthopaedic & Sports Physical Therapy*, vol. 37, pp. 754–762, 2007.
18. J. Yousefi and A. Hamilton-Wright. "Characterizing EMG data using machine-learning tools." *Computers in biology and medicine*, vol. 51, pp. 1–13, 2014.
19. A. L. Hof. "EMG and muscle force: an introduction." *Human Movement Science*, vol. 3, pp. 119–153, 1984.
20. C. J. De Luca. "The use of surface electromyography in biomechanics." *Journal of Applied Biomechanics*, vol. 13, pp. 135–163, 1997.
21. R. A. Ekstrom, et al. "Surface electromyographic analysis of exercises for the trapezius and serratus anterior muscles." *Journal of orthopaedic & Sports Physical Therapy*, vol. 33, pp. 247–258, 2003.
22. F. Sadikoglu, et al. "Electromyogram (EMG) signal detection, classification of EMG signals and diagnosis of neuropathy muscle disease." *Procedia Computer Science*, vol. 120, pp. 422–429, 2017.
23. R. Bose, et al. "Cross-correlation based feature extraction from EMG signals for classification of neuro-muscular diseases." in *2016* International Conference *on* Intelligent Control Power *and* Instrumentation (ICICPI), pp. 241–245, 2016.

24. F. Demir, et al. "Surface EMG signals and deep transfer learning-based physical action classification." *Neural Computing and Applications*, vol. 31, pp. 8455–8462, 2019.

25. V. K. Mishra, et al. "Analysis of ALS and normal EMG signals based on empirical mode decomposition." *IET Science, Measurement & Technology*, vol. 10, pp. 963–971, 2016.

26. S. Chada, et al. "An efficient approach for physical actions classification using surface EMG signals." *Health Information Science and Systems*, vol. 8, p. 3, 2020.

27. J. Wu, et al. "Real-time American sign language recognition using wrist-worn motion and surface EMG sensors." in *2015* IEEE 12th International Conference *on* Wearable *and* Implantable Body Sensor Networks (BSN), 2015, pp. 1–6.

28. M. Atzori, et al. "Electromyography data for non-invasive naturally-controlled robotic hand prostheses." *Scientific data*, vol. 1, pp. 1–13, 2014.

29. X. Xi, et al. "Feature-level fusion of surface electromyography for activity monitoring." *Sensors*, vol. 18, p. 614, 2018.

30. G. J. Salem, et al. "Physical demand profiles of hatha yoga postures performed by older adults." *Evidence-Based Complementary and Alternative Medicine*, vol. 2013, 2013.

31. T. Desikachar and R. H. Cravens. *Health, Healing, and Beyond: Yoga and the Living Tradition of T. Krishnamacharya*. North Point Press, New York, 2011.

32. E. Adamson and J. Budilovsky. *The Complete Idiot's Guide to Yoga*. Pearson Education, Indianapolis, United States, 2001.

33. M. L. Galantino, et al. "A qualitative exploration of the impact of yoga on breast cancer survivors with aromatase inhibitor-associated arthralgias." *Explore*, vol. 8, pp. 40–47, 2012.

34. K. L. Kappmeier and D. M. Ambrosini. *Instructing Hatha Yoga*. Human Kinetics, United States of America, 2006.

35. K. K. Kelley, et al. "A comparison of EMG output of four lower extremity muscles during selected yoga postures." *Journal of bodywork and Movement Therapies*, vol. 23, pp. 329–333, 2019.

36. K. A. Williams, et al. "Effect of Iyengar yoga therapy for chronic low back pain." *Pain*, vol. 115, pp. 107–117, 2005.

37. K. Williams, et al. "Evaluation of the effectiveness and efficacy of Iyengar yoga therapy on chronic low back pain." *Spine*, vol. 34, p. 2066, 2009.

38. S. Khalsa, et al. "Yoga reduces performance anxiety in adolescent musicians." *Alternative Therapies, Health and Medicine*, vol. 19, pp. 34–45, 2013.

39. N. P. Gothe and E. McAuley. "Yoga is as good as stretching–strengthening exercises in improving functional fitness outcomes: Results from a randomized controlled trial." *Journals of Gerontology Series A: Biomedical Sciences and Medical Sciences*, vol. 71, pp. 406–411, 2016.

40. A. Tiedemann, et al. "A 12-week Iyengar yoga program improved balance and mobility in older community-dwelling people: a pilot randomized controlled trial." *Journals of Gerontology Series A: Biomedical Sciences and Medical Sciences*, vol. 68, pp. 1068–1075, 2013.

41. C. Lau, et al. "Effects of a 12-week hatha yoga intervention on metabolic risk and quality of life in Hong Kong Chinese adults with and without metabolic syndrome." *PloS one*, vol. 10, pp. 1–18, 2015.

42. A. Kumar, et al. "Musculoskeletal modeling and analysis of trikonasana." *International journal of yoga*, vol. 11, p. 201, 2018.

43. H. Badsha, et al. "The benefits of yoga for rheumatoid arthritis: results of a preliminary, structured 8-week program." *Rheumatology International*, vol. 29, pp. 1417–1421, 2009.

44. Y. S. Kumar and J. Nishi. "Yoga and Diabetes Mellitus: Recommendations and Benefits-Systematic." *International Journal of Ayurvedic & Herbal Medicine*, vol. 7, pp. 2651–2655, 2017.

45. M. S. Dhapola and M. R. K. Prasad. "Role of different asanas during prenatal and postnatal pregnancy." *International Journal of Physical Education & Sports*, vol. 3, pp. 09–16.

46. M. P. Jorge, et al. "Hatha Yoga practice decreases menopause symptoms and improves quality of life: A randomized controlled trial." *Complementary therapies in medicine*, vol. 26, pp. 128–135, 2016.

47. K. E. Innes, et al. "Efficacy of an eight-week yoga intervention on symptoms of restless legs syndrome (RLS): a pilot study." *The journal of alternative and Complementary Medicine*, vol. 19, pp. 527–535, 2013.

48. C. Lau, et al. "Effects of a 12-week hatha yoga intervention on cardiorespiratory endurance, muscular strength and endurance, and flexibility in Hong Kong Chinese adults: a controlled clinical trial." *Evidence-Based Complementary and Alternative Medicine*, vol. 2015, pp. 1–12, 2015.

49. M. Tripathy. "Monitoring aggression in adolescents: Chandra Namaskara as a panacea." *Elysium Journal of Engineering Research and Management*, vol. 4, pp. 1–6, 2017.

50. D. Beazley, et al. "Trunk and hip muscle activation during yoga poses: Implications for physical therapy practice." *Complementary therapies in Clinical Practice*, vol. 29, pp. 130–135, 2017.

51. L. M. Fishman, et al. "Two isometric yoga poses reduce the curves in degenerative and adolescent idiopathic scoliosis." *Topics in Geriatric Rehabilitation*, vol. 33, pp. 231–237, 2017.

52. M. Kazamel and P. P. Warren. "History of electromyography and nerve conduction studies: A tribute to the founding fathers." *Journal of Clinical Neuroscience*, vol. 43, pp. 54–60, 2017.

53. N. Sikder, et al. "Heterogeneous hand guise classification based on surface electromyographic signals using multichannel convolutional neural network." in *2019 22nd International Conference on Computer and Information Technology (ICCIT)*, pp. 1–6, 2019.

54. M. B. I. Reaz, et al. "Techniques of EMG signal analysis: detection, processing, classification and applications." *Biological Procedures Online*, vol. 8, pp. 11–35, 2006.

55. J. N. Côté. "Electromyography (EMG): Intro to the muscle activity assessment technique." Montreal, Canada, p. 60, 2015.

56. N. Amrutha and V. Arul. "A review on noises in EMG signal and its removal." *International Journal of Scientific and Research Publication*, vol. 7, pp. 23–27, 2017.

57. B. Mokhlesabadifarahani and V. K. Gunjan, "Introduction to EMG technique and feature extraction," in *EMG Signals Characterization in Three States of Contraction by Fuzzy Network and Feature Extraction*, ed: Springer, Singapore, pp. 1–9, 2015.

58. P. Konrad. *"The ABC of EMG: A Practical Introduction to Kinesiological Electromyography*, vol. 1, pp. 30–35, 2005.

59. H. Reucher, et al. "Spatial filtering of non-invasive multielectrode EMG: Part I-Introduction to measuring technique and applications." *IEEE transactions on Biomedical Engineering*, vol. 2, pp. 98–105, 1987.

60. F. H. Chan, et al. "Fuzzy EMG classification for prosthesis control." *IEEE transactions on Rehabilitation Engineering*, vol. 8, pp. 305–311, 2000.

61. A. Phinyomark, et al. "Feature reduction and selection for EMG signal classification." *Expert systems with applications*, vol. 39, pp. 7420–7431, 2012.

62. I. A. Stokes, et al. "Surface EMG electrodes do not accurately record from lumbar multifidus muscles." *Clinical Biomechanics*, vol. 18, pp. 9–13, 2003.

63. J. Crow, et al. "The effectiveness of EMG biofeedback in the treatment of arm function after stroke." *International Disability Studies*, vol. 11, pp. 155–160, 1989.

64. Yogic way of life, Yoga Asanas, (27-04-2020). Available: https://www.yogicwayoflife.com/yoga_asana/

65. DELSYS, Trigno Wireless Biofeedback System, Galileo EMG Sensor User's Guide online, (28-04-2020). Available: https://www.delsys.com/downloads/USERSGUIDE/trigno/galileo-sensor.pdf

66. S. Sharma, et al. "Techniques for feature extraction from EMG signal." *International Journal of Advanced Research in Computer Science and Software Engineering*, vol. 2, pp. 1-4, 2012.

67. S. K. Chowdhury, et al. "Discrete wavelet transform analysis of surface electromyography for the fatigue assessment of neck and shoulder muscles." *Journal of Electromyography and Kinesiology*, vol. 23, pp. 995–1003, 2013.

68. A. Subasi and M. K. Kiymik. "Muscle fatigue detection in EMG using time–frequency methods, ICA and neural networks." *Journal of Medical Systems*, vol. 34, pp. 777–785, 2010.

69. Y. Miyahara, et al. "Effects of proprioceptive neuromuscular facilitation stretching and static stretching on maximal voluntary contraction." *The Journal of Strength & Conditioning Research*, vol. 27, pp. 195–201, 2013.

70. S. Jerritta, et al. "Emotion recognition from facial EMG signals using higher order statistics and principal component analysis." *Journal of the Chinese Institute of Engineers*, vol. 37, pp. 385–394, 2014.

71. D. Farina and R. Merletti. "Comparison of algorithms for estimation of EMG variables during voluntary isometric contractions." *Journal of Electromyography and Kinesiology*, vol. 10, pp. 337–349, 2000.

72. B. Cheng and G. Liu. "Emotion recognition from surface EMG signal using wavelet transform and neural network." in *Proceedings of the 2nd International Conference on Bioinformatics and Biomedical Engineering* (ICBBE), pp. 1363–1366, 2008.

73. G. Vannozzi, et al. "Automatic detection of surface EMG activation timing using a wavelet transform based method." *Journal of Electromyography and Kinesiology*, vol. 20, pp. 767–772, 2010.

74. W. Rose. *Electromyogram analysis*, (27-04-2020). Availanle: https://www1.udel.edu/biology/rosewc/kaap686/notes/EMG%20analysis.pdf .

75. W. Rose. "Electromyogram analysis." *Online Course Material.* University of Delaware. *Retrieved July*, vol. 5, p. 2016, 2011.

76. A. Sadhukhan, et al. "Effect of sampling frequency on EMG power spectral characteristics." *Electromyography and Clinical Neurophysiology*, vol. 34, pp. 159–163, 1994.

77. M. R. Davies and S. Reisman. "Time frequency analysis of the electromyogram during fatigue." in *Proceedings of 1994 20th Annual Northeast Bioengineering Conference*, pp. 93–95, 1994.

78. C. Spiewak, et al. "A comprehensive study on EMG feature extraction and classifiers." *Open Access Journal of Biomedical Engineering and Biosciences*, vol. 1, pp. 17–26, 2018.

79. X. Zhu. "Emotion recognition of EMG based on BP neural network." in *Proceedings of Int Symposium Network. Network Security*, Jinggangshan, China, pp. 227–229, 2010.

80. Y. Du, et al. "The effect of combining stationary wavelet transform and independent component analysis in the multichannel SEMGs hand motion identification system." *Journal of Medical and Biological Engineering*, vol. 26, p. 9, 2006.

81. P. Alagumariappan, et al. "Selection of surface electrodes for electrogastrography and analysis of normal and abnormal electrogastrograms using Hjorth information." *International Journal of Biomedical Engineering and Technology*, vol. 32, pp. 317–330, 2020.

82. R. Barn, et al. "Reliability study of tibialis posterior and selected leg muscle EMG and multi-segment foot kinematics in rheumatoid arthritis associated pes planovalgus." *Gait & posture*, vol. 36, pp. 567–571, 2012.

83. L. Rocchi, et al. "Feature selection of stabilometric parameters based on principal component analysis." *Medical and Biological Engineering and Computing*, vol. 42, pp. 71–79, 2004.

84. H. Huang, et al. "Ant colony optimization-based feature selection method for surface electromyography signals classification." *Computers in biology and medicine*, vol. 42, pp. 30–38, 2012.

85. J. Too, et al. "EMG feature selection and classification using a Pbest-guide binary particle swarm optimization." *Computation*, vol. 7, p. 12, 2019.

86. J. Too, et al. "A new competitive binary Grey Wolf Optimizer to solve the feature selection problem in EMG signals classification." *Computers*, vol. 7, p. 58, 2018.

87. A. Phinyomark and E. Scheme. "EMG pattern recognition in the era of big data and deep learning." *Big Data and Cognitive Computing*, vol. 2, p. 21, 2018.

88. M. Khezri and M. Jahed. "Real-time intelligent pattern recognition algorithm for surface EMG signals." *Biomedical Engineering Online*, vol. 6, pp. 1–12, 2007.

89. S.-H. Park and S.-P. Lee. "EMG pattern recognition based on artificial intelligence techniques." *IEEE transactions on Rehabilitation Engineering*, vol. 6, pp. 400–405, 1998.

90. J. C. Gonzalez-Ibarra, et al. "EMG pattern recognition system based on neural networks." in *2012* 11th Mexican International Conference on Artificial Intelligence, pp. 71–74, 2012.

91. M. Misiti, et al. *Wavelet Toolbox Getting Started Guide.* The Mathworks, 1997, Natick, MA 01760, United States.

5 Early Detection of Parkinson's Disease and SWEDD Using SMOTE and Ensemble Classifier

Emina Aličković and Abdulhamit Subasi

CONTENTS

5.1 INTRODUCTION

Physical oscillations in the human brain cause different neurological and psychiatric disorders, such as Alzheimer's disease (AD) and Parkinson's disease (PD). PD is a progressive, remediless neurodegenerative disorder, and it is the second most frequent neurodegenerative disease affecting the aging population (aged 65 or older) worldwide. At present, no absolute PD diagnostic test exists, and PD clinical diagnosis is established on the existence of basic diagnostic measures. However, accurate detection and diagnosis of PD using these measures is a challenging task, particularly in its early phase. Recently, machine learning tools have been used to solve this challenging task. It is crucial to have an automatic accurate model for early identification of PD. Besides PD, Scans without Evidence of Dopaminergic Deficit (SWEDD) conditions have medical symptoms that resemble symptoms of those found in PD. Accurate identification and diagnosis of SWEDD is extremely important, since misdiagnosis of SWEDD as PD might cause avoidable medical checkups and treatments with the related side-effects [1].

It is not unusual, especially in a medical dataset, to have datasets which are highly imbalanced. In binary classification, which is used in this study, an under-represented class, i.e., a class comprising only a few samples is called a minority or positive class, and the remaining class is called a majority or negative class. An imbalanced dataset is challenging for machine learning classifiers, since these classifiers result in a significant drop in the performance and classification of instances of a minority class, as these classifiers are biased toward majority class instances. To address this issue, the synthetic minority oversampling technique (SMOTE) [2] was suggested to randomly reproduce instances in a minority class in order to solve the biasing problem.

Although a lot of research was been conducted to build a computer-aided diagnostic (CAD) system for early identification of PD and SWEDD, the research has resulted in drawbacks: (1) As they work with a vast amount of features, they need efficient feature reduction tools and such tools may possibly result in data losses; these losses may negatively influence decision making; (2) the restricted size of the data sets used in the research; (3) not much of the research focused on the construction of predictive or prediction systems for estimation of the likelihood of PD and SWEDD in subjects, which may be valuable for the classification of subjects into various risk groups [1]; (4) there is a lac of research that distinguishes between different conditions similar to PD such as SWEDD; (5) the datasets used in different studies could be highly imbalanced, i.e., there were some classes with a very small number of instances and other class(s) with a more significant number of instances; (6) the unsatisfactory performances resulted from applying traditional machine learning techniques. Because of these drawbacks, additional research is necessary to create the diagnostic high-accuracy system to find the populations at high risk of PD, and to improve neuro-protective schemes [3].

In this chapter, we present a study that aimed to overcome these drawbacks and significantly improve up-to-date automatic PD predictions and SWEDD detection systems. To overcome the first drawback, four features were used in this study. These features are striatal binding ratio (SBR) values calculated for every one of the four striatal regions, i.e., left caudate (LC), right caudate (RC), left putamen (LP), and right putamen (RP), by employing either semi-automatic or fully automatic machine learning (ML) tools [4], taken from the Parkinson's progression markers initiative (PPMI) [5] database. These four features were significantly important with $p < 0.05$. It was observed that the performance of ensemble machine learning methods resulted in higher performances compared to results obtained by applying the traditional machine learning methods reported in previous similar research [1, 6–8]. To overcome the second drawback, the sample size of the data set employed in this research was best when weighed against other similar research [1, 6–9]. To overcome the third drawback, different statistical measures were used to prove the effectiveness of the model in the classification of subjects into various risk groups, such as PD and SWEDD. To overcome the fourth drawback, an automated model for differentiating early PD from SWEDD was employed in this research. Besides differentiating early PD from healthy individuals, which was the case in previous studies, the model presented in this work is used to distinguish SWEDD subjects from early PD. To overcome the fifth drawback—an imbalanced dataset—A SMOTE oversampling method is employed. To overcome the sixth drawback, a RFBoostRotF ensemble classifier is employed.

The present work achieved the best possible prediction and detection performances with respect to the following metrics: sensitivity, specificity, and overall accuracy. In the literature, different approaches were proposed to address the issue of the unsatisfactory performances obtained by using traditional machine learning techniques [10–12]. In order to solve this problem, four different ensemble models were proposed and evaluated, and it was revealed that ensemble classifiers resulted in very high performances. In this chapter, a novel classification model, called the RFBoostRotF, for automatic detection of early PDs and SWEDDs was employed. The model employed in this study was the improvement of a rotation forest proposed by Rodriguez, Kuncheva, and Alonso [13], and it used a principal component analysis (PCA) for rotation and an AdaBoost ensemble learning method, which used a random forest for classification. Experimental results obtained in the present work showed that RFBoostRotF reduces bias and offers improved performance when compared to traditional and ensemble learning techniques.

5.2   RELATED WORK

As PD is an increasingly pressing issue worldwide, on-time treatment and disease avoidance is imperative and the only way to realize this is a precocious PD diagnosis. It is of vital importance to have early PD detection due to several reasons: timely supervision, unnecessary therapeutic tests, treatment prevention supplementary expenses, side-effects, and security threats prevention. It is important to have an accurate PD diagnostic model for subjects being registered for experimental research such as the Parkinson's progression markers initiative [1, 14]. It is doubtless that early PD detection is going to open the gate for the most important improvements in PD treatments. For this reason, much research is being conducted to design the best-performing systems for early PD identification. From the research being conducted presently, it can be seen that medical laboratory, imaging, and molecular genetic data is going to influence upcoming PD detection.

Different imaging technologies, among which are magnetic resonance imaging (MRI), positron emission tomography (PET), and single-photon emission computed tomography (SPECT), have been employed to measure different biomarkers and have helped diagnostic applications. The creation of an appropriate automatic system to study PD progress and the finding of discriminatory molecular biomarkers for early PD detection is the main aim of many studies being conducted in recent years. In most cases, the acquired DATSPECT images are visually inspected and analyzed by medical experts. Such processes typically consist of a predefined rating [15, 16], or a region of interest (ROIs) investigation [17]. Such processes are potentially subjective and inclined to errors, as they depend on unpredictable deviations in transporter density all through the ROI. On the other hand, several automated techniques were suggested in other research [6, 9, 18], which form semi-quantitative parameters to indicate the absolute differences among identifiable/unidentifiable significances in the tomographic studies. Currently, ML-based approaches are employed for the classification of different patterns in image analysis, resulting in the building of CAD systems for several neurodegenerative disorders, and particularly for PD [1, 7, 19–21]. Such CAD systems are employed for the analysis of MRI, PET, or SPECT images, with the purpose of complex high-dimensional features extraction and classification, i.e., the differentiation of "normal" images (recorded from healthy individuals) from "pathological" images (recorded from sick individuals), accomplishing automatic disorder identification and diagnosis [18]. Hirschauer et al. [22] suggested a model for PD identification based on the motor, non-motor, and neuroimaging features employing the enhanced probabilistic neural network (EPNN). Ozcift [23] proposed a new model that uses a support vector machine (SVM) to select features and rotation forest classifiers for PD identification.

In an imbalanced classification problem, the number of samples for different classes is different, i.e., one class comprises a large number of samples while the other classes comprise only a few samples. When attempting to train machine learning algorithms on an imbalanced dataset, there will be over-fitting in the majority classes and bias in the results. In an imbalanced classification problem, the minority classes are generally the more important classes. For example, in the binary classification of an unbalanced dataset, access to healthy individuals is easier which in turn allows access to a significantly higher number of the negative samples, and the generated algorithm is likely to have detrimental effects on the positive samples obtained from sick individuals [24].

Imbalanced data classification has been extensively studied [25–31]. One of the solutions for resolving the imbalanced data classification problem is a resampling approach, which is used to balance an imbalanced dataset. Recent papers [32, 33] have suggested that balanced datasets give improved classification results when compared to imbalanced datasets. In contrast, other works [34, 35] have suggested that imbalanced datasets may not be accountable for the marginal classification performances. Resampling methods are often imperative under these circumstances, since resampling fine-tunes the training dataset, instead of adapting the learning algorithm. Such a method [33, 36] offers an efficient tool to work with classification problems, including imbalanced data using traditional machine learning classifiers. The SMOTE [2] is a widely known oversampling method in which the positive class is oversampled by creating synthetic instances in the feature space generated by the positive instances and their k-nearest neighbors [24]. In the present work, we combined SMOTE and ensemble classifiers to solve the imbalanced dataset problem and to create a well-performing tool for the early detection of PD and SWEDD.

5.3 MATERIALS AND METHODS

5.3.1 Database and Study Cohort Details

Data obtained from the Parkinson's progression markers initiative database (http://www.ppmi-info.org/data) were used in this study. For the latest information related to PPMI please go to www.ppmi-info.org. The PPMI [5] is the first extensive, broad, experimental, transnational clinical study, taking place in the United States, Europe, Israel, and Australia, to detect progression biomarkers in PD subjects to enhance the comprehension of PD causes and the efficiency of PD transforming therapeutic experimental studies. PPMI constructed as a dynamic, broad framework, which is commonly accepted as a significant pattern for PD biomarker research. In this research, four different features were considered. These four different features are the SBR values of the four striatal sections, namely the left and right caudate, and left and right putamen, and they were calculated from DaTSCAN SPECT images obtained from the PPMI database. Database used in the present work was taken from the PPMI website on September 18, 2014. This database contains SBR values for 195 healthy, 422 early PD, and 63 SWEDD subjects. More information about this dataset can be found in [37].

The PPMI study generally includes early untreated PD and healthy individuals, with both groups having the same age range. Conditions for PPMI study cohort and for selecting PPMI subject, protocols followed in DaTSCAN image acquisition and the steps employed by the Imaging Core (http://www.ppmi-info.org/about-ppmi/who-we-are/study-cores/) of the PPMI for SBR calculation. Regarding PPMI study selection criteria, PD subjects are enrolled based on the disease threshold. Prerequisites for PD subjects are: (1) to have an asymmetric resting tremor or asymmetric bradykinesia, (2) resting tremor and rigidity with a diagnosis within two years, and (3) not to be cured of PD. All PPMI subjects were submitted to dopamine transporter (DAT) imaging, and for any subject to be classified as having PD, a DAT deficiency was requisite. Healthy subjects did not have any recorded important neurologic dysfunction, and they did not have first-degree family members having identified PD. The PPMI Central SPECT Core Lab reconstructed data obtained from 123-I Ioflupane SPECT scans [15], corrected attenuation and corrected data with a standardized ROI template to extract regional count densities in the right and left putamen and caudate. Striatal binding ratio values were computed for the right caudate, left caudate, right putamen, and left putamen, where the reference was the occipital lobe region [1, 37].

Detailed information about the protocols followed in DaTSCAN image acquisition is explained in [38]. PD identified subjects underwent DaTSCAN imaging at the initial screening and at months 12, 24, and 48 (or visits 4, 6, 10, respectively) or at premature withdrawal. Healthy control subjects underwent DaTSCAN imaging at the screening. Before DaTSCAN injection, all subjects were pretreated with saturated iodine solution (ten drops in water) or perchlorate (1000 mg). The objective

Brain of HC subject   **Brain of PD subject**

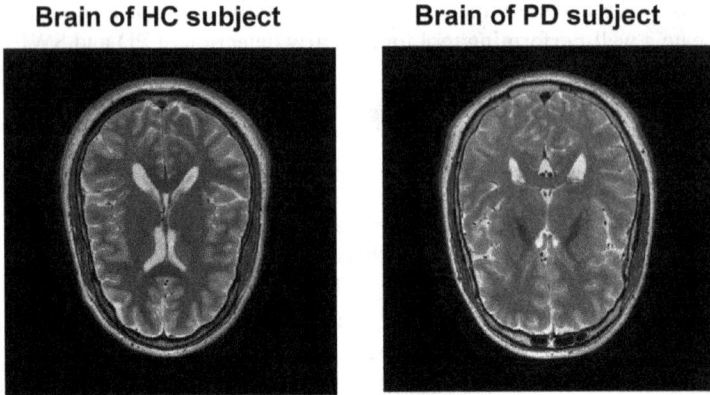

FIGURE 5.1   Brain images for healthy control subject and subject with Parkinson's disease.

dosage for all subjects was 185 MBq or 5.0 mCi of DaTSCAN. The injection dosage range for injection was between 111 and 185 MBq. Subject image acquisition was performed 4 hours later (+/– 30 minutes). Brain images for healthy control subjects and PD subjects are given in Figure 5.1.

The Imaging Central SPECT lab computes SBR values by employing the occipital lobe reference region, and SBR values are the main results that are used to quantify dopamine transporters in assumed PD cases. The SBR calculation procedure is explained in detail in [1, 37]. In this study, 181 observations obtained from 179 different normal (healthy) subjects (HCs), 822 observations obtained from 391 subjects with PDs, and 76 observations taken from 58 different SWEDD subjects were included. Moreover, two different datasets were created. The first dataset, S1, consisted of SBR values for healthy controls (HCs) and PD subjects, and it was used to evaluate performances of the system for early PD diagnosis proposed in this study. So the first study has 181 HC scans and 822 PD scans. The second dataset, S2, consisted of SBR values for PD subjects and SWEDD subjects and it was used to evaluate the performance of the system for detection of SWEDD subjects to distinguish them from PD subjects proposed in this study. This dataset contained 822 PD scans and 76 SWEDD scans. Since each of these two datasets contained only two different classes, in this study, we had a binary classification.

5.3.2   SYNTHETIC MINORITY OVERSAMPLING TECHNIQUE

In an imbalanced dataset, there is a noticeably more significant number of samples of one class when compared to the number of samples belonging to the remaining class(es). As a consequence, high accuracy is observed for the majority class, but this would not be true for the minority class. To resolve this problem, SMOTE [2] can be used. In SMOTE, the minority class is oversampled, i.e., new synthetic samples are generated in the feature space from the samples comprising the minority class and their k-nearest neighbors. Synthetic data is generated in the following way. First, every minority class instance is joined with a line to its k-nearest neighbors. Dependent on the required amount of oversampling, say λ, a portion of these neighbors or all of them, can be chosen, and one instance is generated in all directions of the chosen neighbors. To illustrate this, five nearest neighbors can be used, but if only 200% oversampling is needed for the application, only two out of these five will be chosen, and new instances in these two directions will be generated. For any instance s_0 from the minority class, new synthetic instances will be generated as given in $s_{SMOTE} = s_0 + \omega \cdot (s_0^k - s_0)$ where s_{SMOTE} is a new synthetic instance, s_0^k is the kth nearest neighbor, and ω is a random number between zero and one, and this parameter causes a random point selection on the line between any instance of the minority class and its kth neighbor.

5.3.3 CLASSIFICATION METHODS

5.3.3.1 Naïve Bayes (NB)

The naïve Bayes technique offers a simple methodology and transparent interpretation to apply to probabilistic learning. Specifically, naïve Bayes assumes that attributes are conditionally independent of each other, given that the class can express conditional dependencies using simplified formulas. For every discrete attribute and class, the probability that the attribute will handle all values in its sphere of influence, specified by the class, must be calculated. For every continuous attribute and every class, the attribute's mean and standard deviation (SD), specified by the class, is calculated. The maximum likelihood estimate (MLE) of such parameters is simple. The assessed probability that a nominal random variable acquires a particular value is the same as its sample frequency f, where f can be computed as the number of times the value was divided by the whole number of notes. The MLE of the mean is equal to the sample average, and SD of the normal distribution is equal to the sample SD [39].

5.3.3.2 Logistic Regression (LR)

LR is a broadly applied tool in clinical research [40, 41]. The simple linear regression can only be used on one independent variable, but in many cases LR can be seen as a multivariate method that forms a functional relationship between at least two independent parameters (predictors) and one dependent parameter (result). The resulting parameter is continuous, but it can be discretized into two or more classes. In such a case, LR may also be a suitable methodology for classification tasks. In this study, binary logistic regression was used as it only predicts the relationship between two categorical results, i.e., normal and PD [42]. In this study, a multinomial LR with a ridge estimator was used.

5.3.3.3 Support Vector Machines

A SVM was suggested for the first time in [43] and [44]. Since then, SVMs have been employed in numerous different applications where the task is classification, regression, prediction, estimation, and forecasting. It is based on the Vapnik-Chervonenkis

(VC) theory and Structural Risk Minimization (SRM) principle. The main principle of SVMs is to discover the minimum training set error through maximizing the margin by splitting the hyperplane and provided data. One of the key advantages of SVMs is the usage of convex quadratic programming (QP), because the outcome of complex QP is only one global minima [44–46]. Different kernels, such as the RBF kernel function, polynomial kernel function, and normalized polynomial kernel function, can be used, and appropriate kernel selection depends on the application. In the literature, there are different ways to solve QP. To deal with QP, in this study Sequential Minimal Optimization (SMO) algorithm first proposed by Platt [47] was employed in the SVM training. The principle of SMO is the splitting of the QP optimization problem into a chain of low-scale QPs. Any low-scale QP operates with only two Lagrange multipliers concurrently for every iteration step. Additional information about SMO is provided in [48].

5.3.3.4   Artificial Neural Networks (ANN)

An ANN is a computational representation of a nervous system composed of an immense number of uncomplicated, tremendously interrelated processing modules, identified as neurons. Therefore, an ANN may be viewed as a parallel distributed processing system. One of the most famous types of ANN is Multilayer Perceptron (MLP). A MLP has one input layer, at least one hidden layer where all processing is performed, and the output layer. The number of hidden layers depends on the application. An improper number of neurons in the hidden layer results in poor generalization potentials and over-fitting [49], [50]. A MLP is a valuable classifier for linearly inseparable data, such as SPECT imaging data, as long as there is a sufficient amount of reliable data to be used in the training process.

5.3.3.5   Decision Tree (C4.5)

A C4.5 classifier is a decision tree (DT) classifier established on knowledge approach used to represent several classification rules. A decision tree generated from a C4.5 algorithm comprises a single root, numerous branches, numerous decision nodes, and numerous leaves. The path from the root to the leaf denotes one branch. Each of the nodes contains one attribute. Leaves in the tree represent classes, i.e., class labels. Tests to be done on one attribute, with one branch and one sub-tree for every test result, are identified by decision nodes. Splitting is performed by applying an information gain ratio measure. When this process finally (and compulsorily) ends at the leaf, this leaf shows the class of that instance [51].

5.3.3.6   Random Forest (RF)

A random forest [52] is an ensemble machine learning classifier which consists of a committee of tree-structured classifiers for solving prediction, detection, and classification problems. In a random forest, a classification and regression tree (CART) is used as a decision tree method. CART is a binary tree similar to C4.5 with the main difference being that it uses a Gini index measure as splitting criteria at every node in a decision tree with the lowest impurity. In a random forest, features are randomly selected at every node, and there is no rule for pruning or stopping. Every tree in a random forest is grown by employing bootstrapped instance λ' from original instance λ. Bootstrapped instance λ' contains around two-thirds of the observations (independent variables) from the original instance λ. The rest of the observations from λ not found in λ' are out-of-bag (OOB) instances. But, at every tree node, k independent variables are randomly selected from the K available independent variable, and from this selection a split is selected. Typically, the initial k is selected to be the first integer less than $\log_2 K + 1$, i.e., $k = \log_2 K + 1$ and then k is adjusted (decreased or increased) so that at the minimum error rate is achieved for an OOB dataset. At every node, only one independent variable with the best split is employed from the k chosen independent variables. By randomly choosing k independent variables, the correlation among trees in the forest is decreased, and therefore the error rate is decreased as well. In classification tasks, such as tasks in this study, the RF decision is made based on the majority of class votes. Trees are grown to the maximum depth in order to obtain low bias. After every tree in the forest is constructed using bootstrapped data, OOB may be employed to test a constructed tree.

5.3.3.7   Rotation Forest (RotF)

A RotF [13] is another ensemble machine learning classifier that has two different components. The first component is a heuristic component based on principal component analysis, which is applied to rotate initial independent variable axes. PCA is employed to construct a feature subset for the base classifier. The second component in a RotF is a base classifier, which is employed to construct the classification system [45]. In this study, three different rotation forest approaches were suggested, and these are the PCA-based C4.5, PCA-based random forest, and PCA-based AdaBoost. We name these tree rotation forest approaches: DTRotF, RFRotF, and RFBoostRotF. AdaBoost [53, 54] is a serial technique where every new classifier is constructed by considering performances received in the earlier created classifiers. In an AdaBoost algorithm, there is a set of weights $Wj(k)$, $(k = 1, 2, ..., K)$, where K is the number of independent variables for jth classifier kept over training set T created after applying PCA and in the beginning these weights are adjusted to be equal. In the subsequent iterations, weights are modified so that the weights of wrongly classified samples are augmented, while the weights of the correctly classified samples are reduced. There are two approaches to build the training set T_j for classifier A_j, where A_j is base classifier: (1) resampling of T [53] and (2) reweighting of T based on modified probability distribution W_j kept over T [54]. The second approach was selected in this study because of its simplicity and better performances. Weight is assigned to every classifier A_j during training and the final decision on AdaBoost is completed by weighted voting of every classifier A_j.

5.4   RESULTS AND DISCUSSION

In this chapter, we defined how classifiers can be constructed from imbalanced datasets. If the classes are not equally included, then the dataset is said to be imbalanced, and it contains a considerable number of consistent instances with only a small percentage of unbalanced examples. Most of the time, the classification of the imbalanced dataset is challenging. Machine learning algorithms are not efficient at discriminating between minority classes in imbalanced datasets. Consequently, a data balancing task is required in the data preprocessing phase [55]. In this chapter, SMOTE is used as a data balancing method by using different classifiers in order to construct a system, which will automatically discriminate PD patients from healthy controls. SWEDD is a condition different than PD, but it has similar medical symptoms. Accurate SWEDD diagnosis is crucial since misdiagnosis of PD may result in redundant medical checkups, treatments, and accompanying side-effects. It was demonstrated in different dopaminergic research that nearly 10% of suspected PD subjects have normal dopaminergic imaging scans [56]. Such subjects are known as SWEDDs. There is concrete proof that SWEDDs, most probably, don't have PD since long term checkups of SWEDDs showed a reduced in the reaction to PD treatments (levodopa) [5]. Mean SWEDD subjects' SBR values (1.66) are similar to the SBR values of healthy controls (1.69), and cross-sectional data across an age range demonstrates an age-related decline in SWEDD, like in healthy controls [57]. Due to these reasons, it is important to have a system which will accurately classify PD subjects as PDs and SWEDDs, as many SWEDDs are exposed to unnecessary treatments and medication for years [56]. In this study, a unique automated classification system is proposed to perform two different classification tasks. The first task is the detection of early PD subjects. The second task is to distinguish PD subjects from early PD subjects.

5.4.1 Experimental Setup

It is a challenging task to obtain a high classification accuracy by using machine learning techniques. For any classification problem, it is necessary to make a decision on how to create the training and testing sets from the original dataset so that that classifier will result in high-performance rates. In order to construct a classification model, the training set is employed. To validate the constructed classification model, the testing set is used. When the number of samples in the training set is limited, cross-validation (CV) is typically employed. In order to confirm the high performance of the proposed system based on combining ensemble classifiers in this study, a ten-fold CV was employed. In a ten-fold CV, the dataset was divided into ten equal subsets, where one of these ten subsets were employed for testing, and nine subsets were employed for training. SMOTE was run inside of each CV loop. The benefit of a CV is that the test set on which the model is to be evaluated is independent. For smaller datasets, as the datasets used in this study, it is more efficient to use a CV in which more samples are found in the training set [58]. Since the used data set is so small, we also used different a CV, such as a two-fold CV, three-fold CV, four-fold CV, and five-fold CV, to see the difference between different cross-validation techniques.

In this study, two real datasets S1 and S2, explained in section "Database and study cohort details" were employed. S1, consisted of SBR values for 181 HCs and 822 PD subjects; S2 consisted of SBR values for 822 PD subjects and 76 SWEDD subjects. Both S1 and S2 datasets were highly imbalanced datasets; hence, the SMOTE was applied to make the dataset balanced. Every minority class in S1 and S2 was over-sampled. As a result, a quantity of new synthetic instances was generated depending on λ and k. For both datasets, k was selected to be 5, i.e., $k = 5$, while for S1, λ was set to be 400% and for S2, λ was set to be 1000%, i.e., $\lambda(S1) = 400\%$, $\lambda(S2) = 1000\%$.

Different machine learning techniques, namely a naïve Bayes, LR, SVM, ANN, and C4.5, were tested in the first step. Since these methods did not result in a satisfactory performance, there was a need to find another approach which could be used for the realization of the automated system for early PD detection and SWEDD detection with better performance. Therefore, attention was paid to ensemble classifiers. Two different ensemble techniques were selected in this study based on their high performances when compared to other ensemble classifiers. These two ensemble classifiers were a random forest (RF) and a rotation forest (RotF). Since a rotation forest consists of two components, a PCA and a classifier, in this study three different classifiers were tested in combination with the PCA, namely C4.5, RF, and AdaBoost with RF as the base classifier. Since all these approaches calculate a split based on information gain or the Gini measure, and these measures are not sensitive to monotonous alteration in the data, this solves the problem if the data does not have a zero-mean. The three cases in this study were called DTRotF, RFRotF, and RFBoostRotF, respectively. The whole procedure is presented with a flowchart given in Figure 5.2, describing how the proposed system works in a graphical way.

Five different statistical measures, namely an area under ROC (receiver operating characteristic) curve (AUC) value, F-measure, sensitivity, specificity, and overall (CV) accuracy, were employed to evaluate and compare the different machine learning techniques suggested in this study. For two-class datasets (early PD vs healthy individuals and early PD vs SWEDDs), we used the specificity and sensitivity metrics to assess the performances of the proposed models. Sensitivity and specificity metrics define how well the suggested machine learning classifiers can distinguish between individuals with and without PD. Specificity indicates how well the suggested model(s) performs when the task is distinguishing between healthy and SWEDD individuals, and it indicates false predictions/detections. In contrast, sensitivity indicates how accurate the suggested model(s) are in identifying early PD. The AUC shows the classification results, when there was no concern for the distribution of the classes or error costs. AUC values ranged between 0 and 1 where 0 was the worst case, 1 was the best case, and 0.5 was random classification.

F-measure, sensitivity, specificity, and overall accuracy were defined by Equations 5.1–5.4. Sensitivity gives the percentage of correctly predicted PD patients in both datasets; specificity gives the number of correctly classified HCs in S1 and SWEDDs in S2.

$$F - \text{measure} = 2 \times \frac{\text{Precision} \times \text{Recall}}{\text{Precision} + \text{Recall}},$$

$$\left(\text{Precision} = \frac{\text{TP}}{\text{TP} + \text{FP}}, \text{Recall} = \frac{\text{TP}}{\text{TP} + \text{FN}} \right) \qquad (5.1)$$

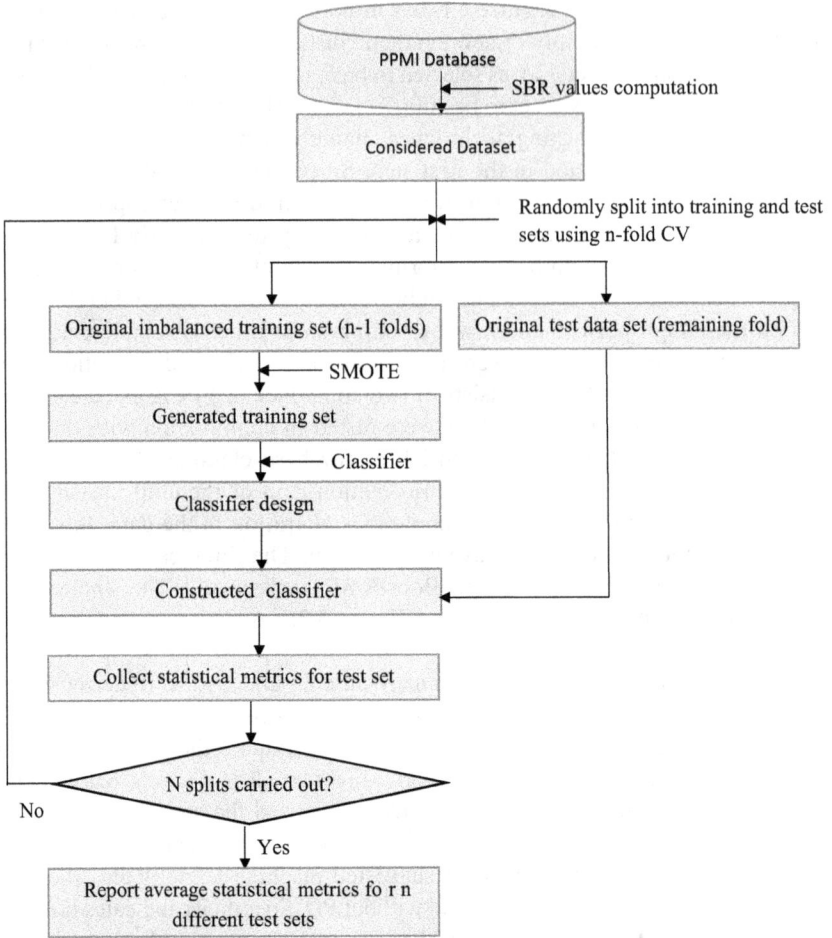

FIGURE 5.2 Flowchart for the proposed approach for early detection of PD and SWEDD subjects.

$$\text{Sensitivity} = \frac{TP}{TP + FN} \times 100\% \qquad (5.2)$$

$$\text{Specificity} = \frac{TN}{TN + FP} \times 100\% \qquad (5.3)$$

$$\text{Accuracy} = \frac{TP + TN}{TP + FN + TN + FP} \times 100\% \qquad (5.4)$$

where TP, FP, TN, and FN are the number of true positives, false positives, true negatives, and false negatives, respectively.

Tables 5.1, 5.2, and Figures 5.3, 5.4 show the overall results for the different classifier models that were used in this study in order to construct a system that will correctly perform two different tasks, i.e., the detection of early PDs in the first task and discriminate SWEDDs from PDs in the second task. To perform these two tasks, two different experiments were conducted in this study. The first experiment was to build an automated system that would discriminate early PDs from healthy controls, i.e., that would predict early PDs, and the second experiment was to build a model that would discriminate PDs from SWEDDs. The experimental results obtained in the present work gave preliminary evidence that the same system could perform both of these tasks with a high performance. After setting up the parameters in these two experiments, the obtained results were compared in order to find the most accurate technique.

5.4.2 Experiment 1: Prediction of Early PD

Classification results obtained for the first experiment, which consisted of discriminating between early PD and healthy controls, are presented in this section. To assess how well-proposed these classifiers can perform, an extensive comparison between the obtained results was performed. Table 5.1 shows the results of the performances of nine different approaches (five traditional learning methods and four ensemble approaches) which are assessed in terms of F-measure, AUC, sensitivity, specificity, and accuracy. This table shows the results for both original dataset and dataset S1 (after SMOTE was applied). From this table, it can be seen that sensitivity increases with SMOTE, while maintaining high specificity in some classifiers but lower specificity in others. Because the imbalance degree of PD classes to HCs was approximately 4:1, oversampled HC instances were balanced at $\lambda = 400\%$. Compared to the five traditional methods, it can be seen that SMOTE + RF, SMOTE + DTRotF, SMOTE + RFRotF, and SMOTE + RFBoostRotF show better performances when compared to a naïve Bayes, LR, SVM, ANN, and C4.5. In this study, different kernels were tested for SVM classifiers, and the highest performances were obtained with a PUK kernel and $C = 100$. In an ANN, the learning rate was set to 0.3, and momentum was set to 0.2 with one hidden layer.

With reference to Table 5.1 and Figure 5.3, it can be observed that

(i) NB, ANN, and SVM perform better without SMOTE in terms of F-measure, AUC value, specificity, and accuracy; even after the SMOTE was applied, the obtained performance results of these methods were still unsatisfactory.

(ii) When SMOTE was applied, LR and C4.5 gave a better overall accuracy than when compared to without it. SMOTE was applied to the original dataset as compared to a NB, ANN or SVM in terms of five performance measures.

(iii) Since RF is an ensemble of CARTs, and the best number of trees in the forest depends on the application, experimental results showed that the highest performance was obtained when the forest was composed of 13 CARTs.

(iv) Because in a RFRotF, a random forest performs classification on a PCA-rotated feature set, the number of trees in a RF also depends on the

TABLE 5.1
Classification Results for Experiment 1

HC/PD

Classifier	F-Measure		AUC		Sensitivity		Specificity		Accuracy	
	Original	SMOTE	Original	SMOTE	Original	SMOTE	Original	SMOTE	Original	SMOTE
Naïve Bayes	0.950	0.879	0.980	0.876	95.01	100.00	93.92	75.10	94.82	88.16
LR	0.975	0.988	0.991	0.998	98.54	99.12	92.82	98.48	97.51	98.82
SVM	0.977	0.879	0.964	0.876	98.42	100.00	94.48	75.10	97.71	88.16
ANN	0.976	0.878	0.995	0.880	98.30	100.00	94.48	74.97	97.61	88.09
C4.5	0.979	0.993	0.961	0.991	98.54	99.50	95.03	99.03	97.91	99.28
Random Forest	0.975	0.993	0.985	0.998	98.42	99.37	93.37	99.17	97.51	99.28
DTRotF	0.975	0.993	0.985	0.996	98.30	99.62	93.92	98.89	97.51	99.28
RFRotF	0.978	0.991	0.988	0.997	98.66	99.37	93.92	98.89	98.81	99.14
Proposed RFBoostRotF	0.976	0.994	0.991	0.998	98.54	99.75	93.37	99.03	97.61	99.41

TABLE 5.2

Classification Results for Experiment 2

PD/SWEDD

Classifier	F-Measure		AUC		Sensitivity		Specificity		Accuracy	
	Original	SMOTE	Original	SMOTE	Original	SMOTE	Original	SMOTE	Original	SMOTE
Naïve Bayes	0.947	0.951	0.976	0.963	95.01	100.00	86.84	90.00	94.32	95.11
LR	0.972	0.989	0.991	0.998	98.91	99.50	78.95	98.29	97.22	98.91
SVM	0.98	0.951	0.923	0.95	99.15	100.00	100.00	90.00	99.21	95.11
ANN	0.98	0.95	0.991	0.954	98.66	100.00	90.79	89.87	98.00	95.05
C4.5	0.975	0.994	0.926	0.996	98.78	99.62	84.21	99.08	97.55	99.36
Random Forest	0.976	0.994	0.958	0.998	99.03	99.75	82.89	98.95	97.66	99.36
DTRotF	0.978	0.992	0.96	0.999	98.91	99.50	85.53	98.95	97.77	99.23
RFRotF	0.981	0.994	0.981	0.999	99.15	99.62	86.84	99.21	98.11	99.42
Proposed RFBoostRotF	0.979	0.994	0.979	0.999	99.15	99.87	84.21	98.95	97.88	99.42

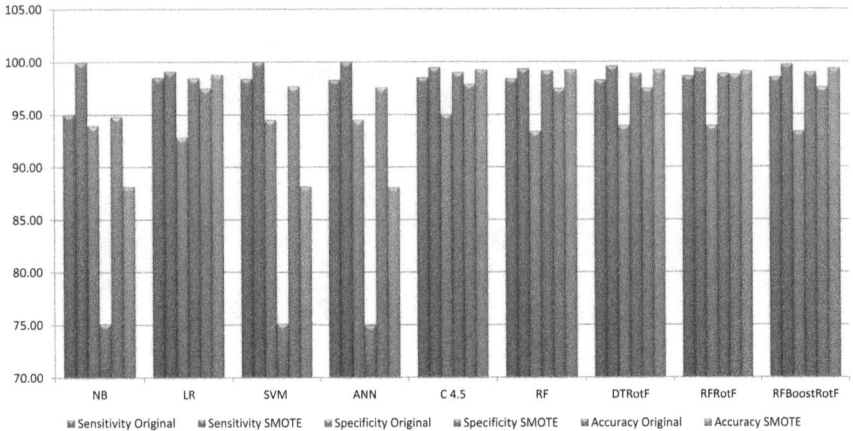

FIGURE 5.3 Comparison of classification results for Experiment 1 (HC-PD classification).

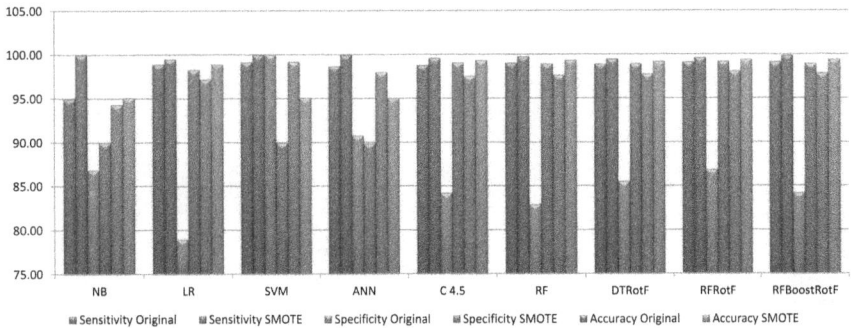

FIGURE 5.4 Comparison of classification results for Experiment 2 (PD-SWEDD classification).

application, and experimental results showed the best performance when 13 CARTs were used in the forest.

(v) Since the RFBoostRotF used an AdaBoost for classification and the AdaBoost was also an ensemble learning technique, experimental results showed that the highest results were obtained when a RF was used as a base classifier in an AdaBoost, and when 13 CARTs were found in the RF classifier.

(vi) The highest performance measures obtained for SMOTE combined with a RFBoostRotF model, which was proposed in this study, in terms of F-measure, AUC value, sensitivity, specificity, and overall accuracy.

(vii) F-measure, AUC value, sensitivity, specificity, and overall accuracy for SMOTE + RFBoostRotF are 0.994, 0.998, 99.75%, 99.03%, and 99.41%, respectively, meaning that this algorithm could detect early PD with very high guarantees.

With reference to Table 5.1 and Figure 5.3, it can be seen that SMOTE + C4.5 used either as a stand-alone traditional learning classifier or as a base classifier for a rotation forest, namely a DTRotF model, proposed in this study, results in a very high performance. Since C4.5 can be understood as a set of if-then rules, the following decisions can be concluded:

(i) If the value of the left putamen is less than or equal to 1.25, with an accuracy of 99.48%, the subjects can be classified as early PD.

(ii) If the value of the left putamen is more significant than 1.25 and the value of the right putamen is more significant than 1.18, the patient can be classified as HC with an accuracy of 99.31%.

(iii) All other cases can be considered as PD with an accuracy of 96.97%.

Furthermore, we compared our results with the literature. In the literature, there is only one study [1], which considered SBR features for early PD detection. When the obtained results were compared to the results obtained in this study, it could be seen that our proposed model SMOTE+ RFBoostRotF outperformed the results obtained in [1]. The overall sensitivity, specificity and classification accuracies were 96.55 %, 95.03 %, and 96.14 % for SVM with a RBF kernel, 95.33 %, 83.98 %, and 92.28 % for SVM with a linear kernel, and 97.74%, 77.35 %, and 90.8 % for LR, respectively [1].

5.4.3 EXPERIMENT 2: DISCRIMINATION OF SWEDDs FROM EARLY PDs

Classification results obtained for the second experiment, which consisted of discriminating between SWEDDs from PDs, are presented in this section. To assess how well the proposed classifiers performed, an extensive comparison between the obtained results was performed. Table 5.2 shows the results of the performances of nine different approaches (five traditional learning methods and four ensemble approaches) which are evaluated in terms of F-measure, AUC, sensitivity, specificity, and accuracy. Table 5.2 shows the results for both the original dataset and dataset S2 (after SMOTE was applied). From this table, it can be seen that sensitivity increases with SMOTE, while maintaining high specificity, except for with an ANN and SVM. Because the imbalance degree of the PD classes to SWEDDs was approximately 10:1, oversampled SWEDD instances were balanced at $\lambda = 1000\%$. Compared to the five traditional methods, it can be seen that two modifications of the rotation forest algorithm, namely a RFRotF and RFBoostRotF when combined with SMOTE, result in the highest performances when compared to other models. Similar to Experiment 1, different kernels were tried with a SVM classifier, and the highest performances were obtained with a PUK kernel and $C = 100$. In an ANN, the learning rate was set to 0.3, and momentum was set to 0.2 with one hidden layer.

With reference to Table 5.2 and Figure 5.4, it can also be seen that SMOTE + RFBoostRotF outperformed the other models evaluated in this study. Therefore,

similar to Experiment 1, SMOTE + RFBoostRotF could be used to build an automated diagnosis system for detection of SWEDDs. Besides:

(i) A SVM and an ANN perform better without using SMOTE in terms of
    F-measure, specificity, and accuracy.
(ii) A NB, LR, and C4.5 give better overall accuracy with SMOTE when com-
    pared to NB, LR, and C4.5 without applying SMOTE on the original data-
    set in terms of all performance measures.
(iii) All ensemble classifiers result in higher performance in terms of five per-
    formance measures.
(iv) Since RF, RFRotF, and RFBoostRotF use a CART as a base classifier, and
    the best number of trees in the forest depends on the application, our experi-
    mental results show that the highest performance is obtained when the for-
    est is composed of 13 CARTs.
(v) The highest performance results were obtained when applying SMOTE
    with a RFRotF and RFBoostRotF model, which was proposed in this work.
(vi) SMOTE + RFBoostRotF outperforms all other approaches in terms of
    F-measure, AUC value, sensitivity, specificity, and overall accuracy because
    it resulted in the highest performance in both experiments.
(vii) *F*-measure, AUC value, sensitivity, specificity, and overall accuracy for
    SMOTE + RFBoostRotF were 0.994, 0.999, 99.87%, 97.88%, and 99.42%,
    respectively, meaning that the algorithm could detect SWEDD with very
    high guarantees.

A novel model for the automated classification of PDs and SWEDDs using machine
learning techniques was presented. Furthermore, when this study was compared
to previous studies [1], the discrimination of conditions of SWEDD is different
from PD, but shows similar characteristics. Correct diagnosis is very important
since many medications and therapies can be avoided if SWEDDs are correctly
classified [56].

5.4.4 DISCUSSION

In this study, the performances of different models were compared. The efficiency
of the suggested SMOTE + RFBoostRotF model was examined by employing two
different real imbalanced datasets. First, it was verified that there was a significant
difference among the RFBoostRotF and the SMOTE + RFBoostRotF when using a
t-test of the form $T = \Delta / t$ where $\Delta = \sqrt{k \cdot \bar{A} / \sigma}, t = \tilde{\sigma} / \sigma$ and $\bar{A}$ is data sample mean,
σ is population standard deviation, $\tilde{\sigma}$ is sample standard deviation, and k repre-
sents the sample size. The obtained results show that there is a difference between
RFBoostRotF and SMOTE + RFBoostRotF under *t*-test for both experiments.
Secondly, it was verified that there was a significant difference between the models
with or without SMOTE.

Furthermore, Table 5.3 gives a comparison of the overall accuracy results when
the different fold of CV was applied. With reference to this table, it can be seen that

TABLE 5.3

Comparative Accuracy Results When Different CV Were Used to Evaluate Classifier Performances

Classifier	Two-fold CV		Three-fold CV		Four-fold CV		Five-fold CV	
	HC/PD	PD/SWEDD	HC/PD	PD/SWEDD	HC/PD	PD/SWEDD	HC/PD	PD/SWEDD
Naïve Bayes	88.16	95.11	88.16	95.11	88.16	95.11	88.16	95.11
LR	88.03	95.05	88.16	94.89	88.09	94.89	87.89	94.89
SVM	88.09	95.11	88.16	95.05	88.16	95.05	88.16	95.05
ANN	88.09	95.05	88.09	95.05	88.09	95.05	88.09	95.05
C4.5	98.42	98.65	98.01	98.26	98.82	98.26	98.08	99.29
Random Forest	98.42	99.10	98.21	99.16	98.49	99.16	98.28	99.55
DTRotF	98.95	98.26	98.01	99.42	98.21	99.42	98.72	99.49
RFRotF	98.88	98.71	98.21	99.36	98.27	99.36	98.08	99.42
Proposed RFBoostRotF	99.21	99.16	99.21	99.42	99.28	99.42	99.08	99.55

the SMOTE + RFBoostRotF outperformed the other methods which prove the high performances in this study.

In addition to the fact that the model proposed in this study, the SMOTE and RFBoostRotF, results in better performances, the following are observed as well:

(i) By comparing the performance results obtained for the DTRotF and two other models, namely the RFRotF and RFBoostRotF, it is easy to see that the models that used SMOTE for oversampling achieved a better performance than the DTRotF, RFRotF, and RFBoostRotF without SMOTE oversampling.

(ii) All models except the ANN and SVM using SMOTE for oversampling achieved a better performance than without using SMOTE for oversampling.

(iii) There was an improvement in terms of performance metrics in almost all evaluated models that SMOTE increases the performance of the evaluated models when on-time detection of early PDs and discrimination between PDs and SWEDDs is considered.

(iv) Among the four different ensemble classifiers, the RFBoostRotF performed the best in both experiments.

5.5 CONCLUSION

In this study, a robust and competent model was suggested for detection of early PDs and SWEDDs, by combining the SMOTE, rotation forest, AdaBoost, and random forest ensemble classifier to create the RFBoostRotF. Furthermore, it was demonstrated that the RFBoostRotF results in lower classification errors and higher performances when compared to other ensemble learning techniques, such as the random forest and rotation forest models (DTRotF and RFRotF). Early detection of PD is significant for better treatment, since PD only shows when there is a loss of dopaminergic neurons of more than 60%. Discrimination between conditions similar to PD is also essential. Up to now, there is no experimental test for early PD detection and its discrimination from conditions similar to PD, such as SWEDD. As a consequence, there are vast numbers of misdiagnoses, particularly when PD is at an early stage. Therefore, finding an accurate diagnostic model is of crucial importance since it can help medical workers while diagnosing PD. The SBR values of four different striatal regions taken from the PPMI database were used in this study. The received experimental results showed that the suggested SMOTE + RFBoostRotF provides a promising answer to current advanced techniques for solving classification tasks with imbalanced data. Experimental performance results obtained in this study showed that the system built on the SMOTE and RFBoostRotF model outperformed other machine learning models for two different tasks: (1) detection of early PDs and discriminating them from HC in terms of F-measure, AUC value, sensitivity, specificity, and overall accuracy, (2) discrimination of SWEDDS from early PDs. Therefore, it can be seen that the SMOTE + RFBoostRotF can help the clinician in making a correct diagnosis in the early detection of PDs.

ACKNOWLEDGMENT

PPMI—a public-private partnership—is funded by the Michael J. Fox Foundation for Parkinson's Research and funding partners, including [list of all of the PPMI funding partners found at www.ppmi-info.org/fundingpartners].

REFERENCES

1. Prashanth R, Roy SD, Mandal PK and Ghosh S. 2014. Automatic classification and prediction models for early Parkinson's disease diagnosis from SPECT imaging. *Expert Syst. Appl.* **41**, 3333–3342.
2. Chawla NV, Bowyer KW, Hall LO and Kegelmeyer WP. 2002. SMOTE: synthetic minority over-sampling technique. *J. Artif. Intell. Res.* **16**, 321–357.
3. Berg D. 2008. Biomarkers for the early detection of Parkinson's and Alzheimer's disease. *Neurodegener. Dis.* **5**, 133–136.
4. Zubal G, Wisniewski G, Marek K and Seibyl J. 2011. Automated program for analyzing striatal uptake of DaTSCAN SPECT images in humans suspected of Parkinson's disease. *J. Nucl. Med.* **52**, 2098–2098.
5. Marek K, Jennings D, Lasch S, Siderowf A, Tanner C, Simuni T et al. 2011. The Parkinson progression marker initiative (PPMI). *Prog. Neurobiol.* **95**, 629–635.
6. Illán I, Górriz J, Ramírez J, Segovia F, Jiménez-Hoyuela J and Ortega Lozano S. 2012. Automatic assistance to Parkinson's disease diagnosis in DaTSCAN SPECT imaging. *Med. Phys.* **39**, 5971–5980.
7. Rojas A, Górriz J, Ramírez J, Illán I, Martínez-Murcia FJ, Ortiz A et al. 2013. Application of empirical mode decomposition (EMD) on DaTSCAN SPECT images to explore Parkinson disease. *Expert Syst. Appl.* **40**, 2756–2766.
8. Towey DJ, Bain PG and Nijran KS. 2011. Automatic classification of 123I-FP-CIT (DaTSCAN) SPECT images. *Nucl. Med. Commun.* **32**, 699–707.
9. Segovia F, Górriz J, Ramírez J, Salas-Gonzalez D and Álvarez I. 2013. Early diagnosis of Alzheimer's disease based on partial least squares and support vector machine. *Expert Syst. Appl.* **40**, 677–683.

10. Woźniak M, Graña M and Corchado E. 2014. A survey of multiple classifier systems as hybrid systems. *Inf. Fusion* **16**, 3–17.

11. Sesmero MP, Alonso-Weber JM, Gutierrez G, Ledezma A and Sanchis A. 2015. An ensemble approach of dual base learners for multi-class classification problems. *Inf. Fusion* **24**, 122–136.

12. Yin XC, Huang K, Yang C and Hao HW. 2014. Convex ensemble learning with sparsity and diversity. *Inf. Fusion* **20**, 49–59.

13. Rodriguez JJ, Kuncheva LI and Alonso CJ. 2006. Rotation forest: A new classifier ensemble method. *IEEE Trans. Pattern Anal. Mach. Intell.* **28**, 1619–1630.

14. Cummings JL, Henchcliffe C, Schaier S, Simuni T, Waxman A and Kemp P. 2011. The role of dopaminergic imaging in patients with symptoms of dopaminergic system neurodegeneration. *Brain* **134**, 3146–3166.

15. Benamer HT, Patterson J, Grosset DG, Booij J, De Bruin K, Van Royen E et al. 2000. Accurate differentiation of parkinsonism and essential tremor using visual assessment of [123I]-FP-CIT SPECT imaging: The [123I]-FP-CIT study group, *Mov. Disord.* **15**, 503–510.

16. Tolosa E, Borght TV and Moreno E. 2007. Accuracy of DaTSCAN (123I-ioflupane) SPECT in diagnosis of patients with clinically uncertain parkinsonism: 2-Year follow-up of an open-label study. *Mov. Disord.* **22**, 2346–2351.

17. Lozano SO, del Valle Torres MM, Moreno ER, Viedma SS, Raissouni TA and Jiménez-Hoyuela J. 2010. Quantitative evaluation of SPECT with FP-CIT. Importance of the reference area. *Rev. Esp. Med. Nucl. Engl. Ed.* **29**, 246–250.

18. Martínez-Murcia FJ, Górriz JM, Ramírez J, Illán I, Ortiz A and Parkinson's Progression Markers Initiative. 2014. Automatic detection of Parkinsonism using significance measures and component analysis in DaTSCAN imaging. *Neurocomputing* **126**, 58–70.

19. Morales DA, Vives-Gilabert Y, Gómez-Ansón B, Bengoetxea E, Larrañaga P, Bielza C et al. 2013. Predicting dementia development in Parkinson's disease using Bayesian network classifiers. *Psychiatry Res. NeuroImaging* **213**, 92–98.

20. Ornelas-Vences C, Sanchez-Fernandez LP, Sanchez-Perez LA, Garza-Rodriguez A and Villegas-Bastida A. 2017. Fuzzy inference model evaluating turn for Parkinson's disease patients. *Comput. Biol. Med.* **89**, 379–388.

21. Exarchos TP, Tzallas AT, Baga D, Chaloglou D, Fotiadis DI, Tsouli S et al. 2012. Using partial decision trees to predict Parkinson's symptoms: A new approach for diagnosis and therapy in patients suffering from Parkinson's disease. *Comput. Biol. Med.* **42**, 195–204.

22. Hirschauer TJ, Adeli H and Buford JA. 2015. Computer-aided diagnosis of Parkinson's disease using enhanced probabilistic neural network. *J. Med. Syst.* **39**, 179.

23. Ozcift A. 2012. SVM feature selection based rotation forest ensemble classifiers to improve computer-aided diagnosis of Parkinson disease. *J. Med. Syst.* **36**, 2141–2147.

24. Gao M, Hong X, Chen S and Harris CJ. 2011. A combined SMOTE and PSO based RBF classifier for two-class imbalanced problems. *Neurocomputing* **74**, 3456–3466.

25. Provost F. 2000. Machine learning from imbalanced data sets. *Proceedings of the AAAI'2000 workshop on Imbalanced Data Sets*, 1–3.

26. Barandela R, Sánchez JS, Garcìa V and Rangel E. 2003. Strategies for learning in class imbalance problems. *Pattern Recognit.* **36**, 849–851.

27. Chawla NV, Cieslak DA, Hall LO and Joshi A. 2008. Automatically countering imbalance and its empirical relationship to cost. *Data Min. Knowl. Discov.* **17**, 225–252.

28. Weiss GM, McCarthy K and Zabar B. 2007. Cost-sensitive learning vs. sampling: Which is best for handling unbalanced classes with unequal error costs? *DMIN* **7**, 35–41.

29. Thammasiri D, Delen D, Meesad P and Kasap N. 2014. A critical assessment of imbalanced class distribution problem: The case of predicting freshmen student attrition. *Expert Syst. Appl.* **41**, 321–330.

30. Cheng F, Zhang J, Wen C, Liu Z and Li Z. 2017. Large cost-sensitive margin distribution machine for imbalanced data classification. *Neurocomputing* **224**, 45–57.

31. Karabulut EM and Ibrikci T. 2014. Effective automated prediction of vertebral column pathologies based on logistic model tree with SMOTE preprocessing. *J. Med. Syst.* **38**, 50.

32. Weiss GM and Provost F. 2003. Learning when training data are costly: the effect of class distribution on tree induction. *J. Artif. Intell. Res.* **19**, 315–354.

33. Estabrooks A, Jo T and Japkowicz N. 2004. A multiple resampling method for learning from imbalanced data sets. *Comput. Intell.* **20**, 18–36.

34. Japkowicz N and Stephen S. 2002. The class imbalance problem: A systematic study. *Intell. Data Anal.* **6**, 429–449.

35. Batista GE, Prati RC and Monard MC. 2004. A study of the behavior of several methods for balancing machine learning training data. *ACM Sigkdd Explor. Newsl.* **6**, 20–29.

36. Drummond C and Holte RC. 2003. C4. 5, class imbalance, and cost sensitivity: why under-sampling beats over-sampling. Workshop *on* Learning *from* Imbalanced Datasets II 11.

37. Seibyl J, Jennings D, Grachev I, Coffey C and Marek K. 2013. 123-I Ioflupane SPECT measures of Parkinson disease progression in the Parkinson Progression Marker Initiative (PPMI) trial. *J. Nucl. Med.* **54**, 190.

38. Initiative TPPM. 2010. *PPMI: Imaging Technical Operations Manual.*

39. Schalkoff RJ. 1992. *Pattern Recognition.* Wiley Online Library.

40. Bagley SC, White H and Golomb BA. 2001. Logistic regression in the medical literature: Standards for use and reporting, with particular attention to one medical domain. *J. Clin. Epidemiol.* **54**, 979–985.

41. Mazzocco T and Hussain A. 2012. Novel logistic regression models to aid the diagnosis of dementia. *Expert Syst. Appl.* **39**, 3356–3361.

42. Le Cessie S and Van Houwelingen JC. 1992. Ridge estimators in logistic regression. *Appl. Stat.* **41**(1), 191–201.

43. Boser BE, Guyon IM and Vapnik VN. 1992. A training algorithm for optimal margin classifiers. *Proceedings of the Fifth Annual Workshop on Computational learning theory*, 144–152.

44. Vapnik V. 2013. *The Nature of Statistical Learning Theory.* Springer Science & Business Media, 2013.

45. Aličković E and Subasi A. 2017. Breast cancer diagnosis using GA feature selection and rotation forest. *Neural Comput. Appl.* **28**, 753–763.

46. Subasi A. 2012. Medical decision support system for diagnosis of neuromuscular disorders using DWT and fuzzy support vector machines. *Comput. Biol. Med.* **42**, 806–815.

47. Platt J. 1998. Sequential minimal optimization: A fast algorithm for training support vector machines. Microsoft.com. https://www.microsoft.com/en-us/research/wp-content/uploads/2016/02/tr-98-14.pdf

48. Song X, Chen W, Chen Y-PP and Jiang B. 2009. Candidate working set strategy based SMO algorithm in support vector machine. *Inf. Process. Manag.* **45**, 584–592.

49. Alickovic E and Subasi A. 2015. Effect of multiscale PCA de-noising in ECG beat classification for diagnosis of cardiovascular diseases. *Circuits Syst. Signal Process.* **34**, 513–533.

50. Haykin S. 1994. *Neural Networks: A Comprehensive Foundation.* Prentice Hall PTR.

51. Quinlan JR. 1993. *C4. 5: Programming for Machine Learning.* Morgan Kauffmann 38.

52. Breiman L. 2001. Random forests. *Mach. Learn.* **45**, 5–32.

53. Freund Y and Schapire RE. 1996. Experiments with a new boosting algorithm. *ICML'96: Proceedings of the Thirteenth International Conference on International Conference on Machine Learning*, **96**, 148–156.

54. Freund Y and Schapire RE. 1995. A decision-theoretic generalization of on-line learning and an application to boosting. *Computational Learning Theory*, 23–37.
55. Thammasiri D, Delen D, Meesad P and Kasap N. 2014. A critical assessment of imbalanced class distribution problem: The case of predicting freshmen student attrition. *Expert Syst. Appl.* **41**, 321–330.
56. Schwingenschuh P, Ruge D, Edwards MJ, Terranova C, Katschnig P, Carrillo F et al. 2010. Distinguishing SWEDDs patients with asymmetric resting tremor from Parkinson's disease: a clinical and electrophysiological study. *Mov. Disord.* **25**, 560–569.
57. Seibyl J. 2012. Baseline neuroimaging characteristics of the Parkinson's progression marker initiative (PPMI) Parkinson's and healthy cohorts. *Mov. Disord.* **27**, S255–S256.
58. Salzberg SL. 1997. On comparing classifiers: Pitfalls to avoid and a recommended approach. *Data Min. Knowl. Discov.* **1**, 317–328.

6 Computer-aided Design and Diagnosis Method for Cancer Detection

Johra Khan and Yousef Rasmi

CONTENTS

6.1 INTRODUCTION

Computer-aided design and diagnosis (CAD) is a modern diagnostic technique to improve traditional histopathology image analysis, which in turn helps to improve the treatment that every patient receives [1]. A CAD system uses image-based information along with other diagnostic data to provide a qualitative and quantitative result for identification of different abnormalities, such as lung nodule detection [2], cancer cell tracking, mitosis in cell detection, cerebral micro bleed analysis, nuclei related abnormalities, and lesions in the tissue or organ detection [3]. In this chapter, we discuss some of the approaches of CAD in early cancer detection and diagnosis.

6.2 CAD-BASED TECHNIQUES FOR EARLY TUMOR DETECTION IN BREAST CANCER

Early detection of a cancerous tumor in the breast gives the chance of an earlier cure. A medical image examination depends on a radiologists experience and mental status, any deficiency of these may cause tumors to get overlooked. CAD makes detection easy by highlighting the suspicious regions, and as a result increases accuracy [4]. CAD based on extreme learning machines (ELM) is gaining more attention as it helps to increase performance standards with rapid training with less human intervention [5, 6]. An algorithm-based ELM classifier provides the highest accuracy even with small breast tumor images [7]. Some researchers used ELM to detect microcalcification in mammograms using texture features, and discovered that wavelet texture features result in the best classification accuracy [8, 9]. There are several commercially available technologies based on computer-aided design and diagnosis such as "imageChecker" [10] and "SecondLook" [11], and because of which around 12,860 mammograms were screened in a period of one year, cancer tumor detection increased by 19.5%, and the early-stage detection of malignancies increased from 73% to 78% [12]. A study by Burhenn et al., which used a CAD system for detection of mass calcification in a mammography, resulted in a 75% sensitivity in mass and architectural disproportion detection with only one false-positive image analysis result [13].

For detection of breast tumors using a CAD system to achieve a high-efficiency image, four stages are followed as, shown in Figure 6.1:

Step 1. Image pre-processing: In this step ultrasound is used to enhance the image quality with a reduction in noise without image distortion. This is a recent development in CAD, and it is not currently available in many CAD systems [14].

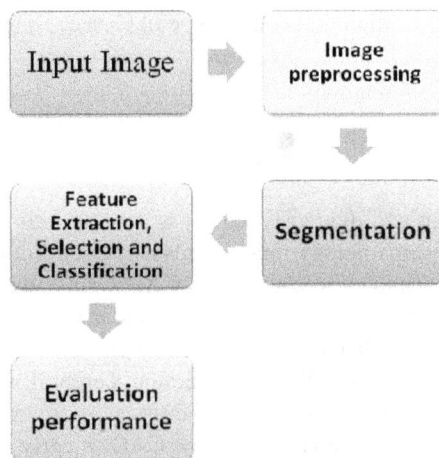

FIGURE 6.1 Overview of a CAD system for breast tumor diagnosis.

Step 2. Image segmentation: Image segmentation is an important step in effective CAD system development, which is to separate the region of interest (ROI) in correspondence with the desired properties [15]. MRI, 3D ultrasound, CT (computed tomography), and many other modalities are able to produce a 3D image. This 3D image segmentation provides more accurate volumetric imagery for tumor detection [16].

Step 3: Selection, extraction, and classification: In this step features of a lesion are extracted from the image to distinguish whether a lesion is benign or malignant. These features are used to classify a large set, and only the most effective features are taken for evaluation [17].

6.2.1 TECHNIQUES FOR BREAST CANCER DETECTION USING CAD

There is regular development in the algorithms for CAD tumor detection in breast cancer research, including techniques for the analysis of features like bilateral asymmetry [18], curvilinear structure [19], and breast density [20]. Some similar techniques used for detection and diagnosis of breast cancer using CAD are described in detail below:

6.2.1.1 Image Enhancement Technique

The characteristic features, such as calcification and mass, that a radiologist finds difficult to identify in mammograms can be improved using a contrast enhancement technique, and the detection of breast cancer can be increased up to 50% [21, 22]. Image segmentation is one methods of image enhancement, as it helps to highlight various structures by selecting significant regions [23]. It also helps to detect tumors or organ with suspicious structures [24]. Image segmentation can be broadly classified into two groups: the fully automated and semi-automated method, but an automated algorithm method is more desirable in medicine for detection and localization of abnormalities [25]. On the basis of image properties, the segmentation method is broadly classified into: i) the similarity based approach, and ii) the discontinuity-based approach [26]. The similarity based approach is further divided into threshold-based, region-based, and clustering-based methods on the basis of pre-determined criteria [27], whereas the discontinuity-based approach divides an image on the basis of sudden changes in intensity [28].

6.2.1.2 Edge-Based Segmentation Technique

This technique deals with the detection of the edge or pixels between different regions of an image with a sudden change in intensity. This method works well with images that have high contrast and no noise. The edge-based segmentation method is mainly used in human organ detection. Yu-qian et al. [29] presented a mathematical morphological edge identifying algorithm to detect a lung CT image with salt and pepper noise. Haris et al. [30] proposed an integrated 2D/3D segmentation method for magnetic resonance images on edge-based and region growing watershed transforms [31].

6.2.1.3   Threshold-Based Segmentation Technique

This method is used to develop a CAD system for the extraction of a significant area for an image analysis [32]. A CAD system based on a histogram threshold was developed to study automated nuclear segmentation in a breast histopathology image and was able to achieve 97% accuracy in nucleus detection [33, 34].

6.2.1.4   Clustering-Based Segmentation Technique

This technique is useful to study clinical image segmentation. The 3D image segmentation method is preferred for accurate volumetric imagery. A 3D image can be obtained using different image modalities, like MRI [35], CT [36], CTLM [37], and 3D ultrasound [38]. Due to various applications of the 3D reconstruction algorithm in surgery, anatomy, diagnosis, and teaching, it is gaining attention in the field of breast cancer screening [39].

6.2.2   Feature Extraction Technique

Computational feature extraction descriptors help to specify the data to be extracted from an image, resulting in less labor for the radiologist and reduced chances of error [40]. Proper feature selection of the whole image or particular region of interest has a significant influence on memory size, cost of classification using the computational method [40], and accuracy of classification and robustness. Image descriptors are generally classified into three basic forms: color-based, shape-based, and texture-based from which shape-based features are more important for human tissue as colors and textures are not so distinct in humans [41]. In Table 6.1, we summarize different classification techniques and results obtained for breast cancer detection.

6.3   PROSTATE CANCER DIAGNOSIS: IMAGE-BASED CAD

Prostate cancer is one of the non-skin cancers, and the second main cause of cancer deaths around the world. Incidences of prostate cancer increase with age [59, 60]. More than 60 % of prostate cancer cases get detected in men aged more than 65 years [61]. For the diagnosis of prostate cancer, pathologists depend on histopathology image grading, according to Gleason's grading system as endorsed by the WHO in 2003 [62]. Due to rapid development in CAD-based multiparametric MRI, specifically the PI-RADS scoring system, deep-learning and diagnosis of prostate cancer is possible with more accuracy [63].

6.3.1   Histopathology Image-Based CAD

6.3.1.1   Input Modalities

The use of prostate cancer imaging in diagnosis plays a very important role, and the choice of image modalities, which can be CAD input, depend on many factors, like accuracy, cost, availability, and clinical practice [64]. In the case of MRI, mapping is done either by the spin-lattice technique (T1W) [65] or by the spin-spin (T2W) [65]

TABLE 6.1
Techniques of Classification Used to Study Breast Cancer Using CAD

Method	Use	Accuracy and Sensitivity	References
Segmentation techniques	Automated, and Modified Seeded Region Grown as per PSO technique	Relative overlap (RO) (mean) = 0.75 True negative fraction (TNF) (mean) = 0.90 True positive fraction (TPF) (mean) = 0.82	[3, 42, 43]
	Adjusted Otsu's threshold	Sensitivity /Segmentation: Benign = 69.4 % Malignant = 90.2 % Sensitivity/Classification = 79.39%	[44, 45]
	Particle Swarm Optimized Wavelet Neural Network (PSOWNN)	Sensitivity = 83.58% Specificity = 93.43%	[46, 47]
	Multilayer perceptron neural networks (MLPNN)	Mean absolute error (MAE) = 0.0339 Root mean squared error (RMSE) = error 0.1433 Relative absolute error (Tiedeu et al., 2012) = 7.535 %	[48, 49]
Classification techniques	Comparison between the performance of SVM and MLPNN for diagnosis breast tumor on ultrasound images	SVM: Sensitivity = 99.37 % Specificity = 99.64% MLPNN: Sensitivity = 91.19 % Specificity = 91.92%	[50, 51]
	Comparison of accuracies of different classifiers include MLPNN, CNN, PNN, RNN and SVM	SVM; Sensitivity = 90 % Specificity = 75% RNN: Sensitivity = 99.37 % Specificity PNN: Sensitivity = 100 % Specificity = 85% CNN: Sensitivity = 99.37 % Specificity MLPNN: Sensitivity = 95 % Specificity = 82%	[52, 53]
	Genetically Optimized Neural Network (GONN) algorithm used for classification of breast cancer	BPNN: Sensitivity = 94.46 % Specificity = 77.21% GONN: Sensitivity = 98.77 % Specificity = 100%	[54–56]
	CAD system for breast cancer developed based on a deep belief network	EM-PCA-CART = 93.2% ART-1 = 93.2% LP-SVM = 97.3% SVM-RBF = 93.47%	[57, 58]

technique. These techniques have limitations as T1W images have issues like low zonal anatomy perception [66] and in T2W images prostate cancer tissues appear ill delineated and as homogeneous lesions due to the low water content [67]. The diagnosis of benign tumor tissues is difficult using these techniques due to their similar nature to healthy tissues. Some studies show only a 50–70% sensitivity [68] and 70–80% specificity [69] using the T2W technique, which are not very promising results.

6.3.1.2   Prostate Image Segmentation

Prostate image segmentation of some of the histological structures are needed for a system that requires the morphometric information of the tissue for it to be used as a feature vector [70]. For the detection of prostate cancer, shape, size, extent, and other morphological features of the glands and cancer tissue provide important information. In the grading of a prostatic tumor, the density and distribution of cancer cells in a particular area of the tissue are used as indicators [71]. In this step, more emphasis is given to zonal segmentation, i.e., peripheral, transition, or central zone of the tumor [72, 73]. Zone segmentation helps to identify both healthy and malignant tissues as the difference is clearly visible in different zones, which can be observed using PI-RADS [74, 75]. The nuclear segmentation method has been used by many researchers to separate the nuclei from other structures in histopathology images. The different techniques applied are clustering algorithms, thresholding, k-means, fuzzy c means [76], Bayesian classifications [77], color region mapping, watershed transforms, and the novel edge detection methods [78, 79]. In a histopathology image, the nucleus of a normal gland appears as a uniform dark blue spot without nucleoli spots, whereas in a cancerous gland it appears in a light blue color with clear nucleoli. In the segmentation process, they are first separated from the background by thresholding blue channels in the image. Then the circularity and area of each blob is used as a vector feature to refine the segmentation [80].

6.3.2   CAD-BASED FEATURE EXTRACTION

This tool mostly depends on texture feature extraction, model-based feature extraction, and the extraction of subtle patterns which are typical for malignancy conditions [81]. Some researchers like Huang et al. [82] have established systems which analyzed texture complexities in histological images by using fractal measurements such as a fractal dimension calculation through a differential box-counting method, and entropy-based fractal dimension measurements. A DCE-MRI (dynamic contrast-enhanced MRI) [83, 84] frequently has a temporal component, and feature extraction, which is important to reduce the dimensionality of the input. A computer computer-aided classification system developed by Yoon et al. [85] extracted texture features from the CMRT (cardinal multi-ridgelet transform) to distinguish images of Gleason 3 from Gleason 4. Another researcher, Almuntashri et al. [86] developed an automated classification method for combining features from the wavelet transform and wavelet-based fractals analysis to capture cancerous image complexity for classification of prostate cancer biopsy images. The results of this study showed an average support vector machine (SVM) classification accuracy of 95% in Gleason grades 3, 4, and 5 for a set of 45 images [87, 88].

As a routine procedure lesion classification of prostate cancer is validated using biopsy results, and sometimes target biopsies are required, which is based on 10–12 biopsy core template and serves as a recommended guideline approach to prostate cancer diagnosis [89, 90]. But the use of biopsy results for a whole prostate map leads to under diagnosis of malignancies and an average of 30% negative results when a repeat biopsy found positive results. Studies show that patients with suspected prostate cancer even after more than eight core biopsies have a more than 40% chance to be detected as positive. Radical prostatectomy (RP) is a more suitable method to validate biopsy results [91, 92].

6.3.3   CAD-Based Classification System

The detection and grading of prostate cancer applying automated system depend on pattern recognition technique as a pattern is a vector for the feature description of an object [70]. Different techniques developed for classification include: K-NN (k-nearest neighbors) [93], Bayesian classifier [94], SVM [95], neural networks, Markov random field (MRF), Gaussian classifier, and CLD (classical linear discrimination) [92, 96]. Although each classification system has its advantages and disadvantages, but several studies compared these methods to find the most suitable classification method for prostate cancer diagnosis [97]. A similar study was done by Alexandratou et al., using 16 supervised MLA (machine learning algorithm) [98] to be compared based on their performance. Classification problems included were tumor vs nontumor, low vs high-grade recognition and multi-class problem Gleason. Thirteen different Haralick texture features were calculated on the basis of gray level co-occurrence matrix of prostate cancer tissues. For each performance level of accuracy recorded was 97.9% for cancer detection [99], 77.75% for detection and Gleason grading, and 81% for low-high grad differentiation [100]. Results obtained in these studies shows Logistic regression and sequential minimal optimizations were the top-scoring algorithms for classification of prostate cancer [101].

6.3.3.1   System Accuracy Evaluation

To diagnose prostate cancer with accuracy it's important to evaluate classification. Cross-validation is one of the most known methods to estimate classification performance [102]. The three methods used for cross-validation are; k-fold, leave-one-out, and hold-out [103, 104].

  1) K-fold validation technique: In this technique, the data is divided randomly into k in equal folds. K-iteration [105] of the training and test are to be carried out such that with each iteration different segment will be held for validation and k-1 folds are used to fit the model [106].
  2) Leave-one out technique: In the special case of k-fold validation, this technique is used for one sample at a time, and the result is considered most accurate [107].

3) Hold on validation method: In this method, the data sheet is divided into two non-overlapped segments; one for training and other for testing [108].

Some of the performance indicators used are:

a) **Correct classification:** To calculate correct classification, the ratio of correct classification to the total number of classified cases [109].

$$CCR = \frac{TP + TN}{Total\ samples}$$

TP: True positive, TN: True negative

b) **Sensitivity:** It is the proportion of case that belong to one class, calculated as a true positive rate (TPR) [110].

$$TPR = \frac{TP}{TP + FN}$$

FN = False negative

c) **Specificity:** It is the proportion of non-cases of a specific class, which is calculated as a true negative rate (TNR) [111].

$$TNR = \frac{TN}{TN + FP}$$

FP; False-positive

6.4 CAD AND ITS APPLICATION IS ANOTHER TYPE OF CANCER DIAGNOSIS

CAD for prostate cancer developed in past years with efforts to develop a more accurate and automated system for detection of prostate cancer using digital biopsy images [65, 112]. The biggest challenge for automated prostate cancer diagnosis is the large amount of data which require fast and accurate image analysis [113, 114]. 3D tissue evaluation is one such method which still needs more research and development will allow pathologists to get more information about the structure of prostate cancer tissues [115, 116].

Due to lung cancer diagnosis at an advanced stage causes large number death around the world, however, with the development of CT a significant reduction in lung cancer mortality and 96% positive screening result was noted [117]. Many studies show that computer-aided design and diagnosis help radiologists to detect false-positive CT in lung cancer screening. Some of the common non-invasive imaging modulation for detection and diagnosis of lung nodules using CAD are; PET (Position emission tomography) [118], CT, LDCT (low-dose computer tomography) [119, 120], and CE-CT (contrast-enhanced computed tomography) [119, 121] (Figure 6.2).

The efficiency of a lung cancer diagnosis can be increased by reducing search space for lung nodules in CT scan, and also by accurate segmentation of lung field

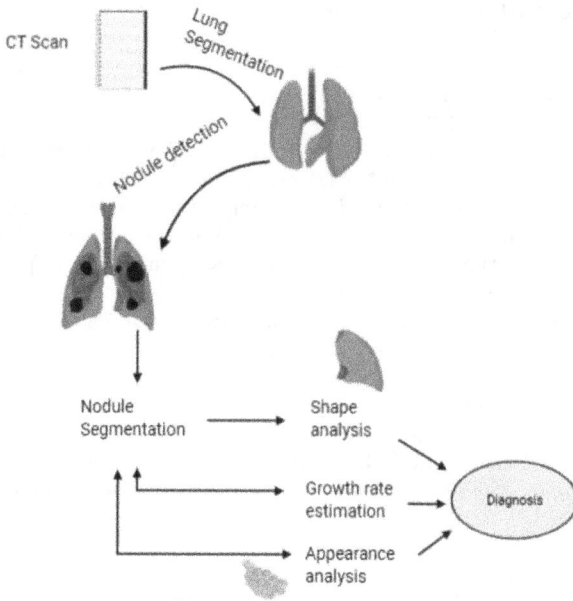

FIGURE 6.2   A CAD-based diagnostic system for lung cancer detection.

[122]. CAD-based diagnostic techniques still have some limitations like sensitivity for scanning parameters, the efficiency of the algorithm and their ability for proper lung segmentation with severe pathogenesis and inhomogeneities in pathological lung condition [121, 123].

6.5   CONCLUSION

In recent years development of Computer-aided design and diagnosis helped not only in reduction of errors, but it also increased accuracy and speed of diagnosis, which in many types of cancer play crucial role for diagnosis and treatment. CAD system improved the performance of radiologists in finding and distinguishing between malignant and benign tumor. There are many limitations and disadvantages in the CAD system, and many techniques need validation and further research for more efficient diagnosis. The computer-aided design and diagnosis system is typically taken as a one-size-fits-all tool, which deals with all users in a similar manner but the condition of all users can't be same so, a personalized CAD system can be more useful to radiologists in a flexible manner to evaluate different disease condition.

REFERENCES

1. Borchartt, T.B., et al., Breast thermography from an image processing viewpoint: A survey. *Signal Processing*, 2013. 93(10): pp. 2785–2803.
2. Awai, K., et al., Pulmonary nodules at chest CT: Effect of computer-aided diagnosis on radiologists' detection performance. *Radiology*, 2004. 230(2): pp. 347–352.

3. Lee, H. and Y.-P.P.P. Chen, Image based computer aided diagnosis system for cancer detection. *Expert Systems with Applications*, 2015. 42(12): pp. 5356–5365.

4. Rangayyan, R.M., F.J. Ayres, and J.L. Desautels, A review of computer-aided diagnosis of breast cancer: Toward the detection of subtle signs. *Journal of the Franklin Institute*, 2007. 344(3–4): pp. 312–348.

5. Cao, J., Z. Lin, and G.-B. Huang, Self-adaptive evolutionary extreme learning machine. *Neural Processing Letters*, 2012. 36(3): pp. 285–305.

6. Han, F., H.-F. Yao, and Q.-H. Ling, An improved evolutionary extreme learning machine based on particle swarm optimization. *Neurocomputing*, 2013. 116: pp. 87–93.

7. Huang, G.-B., X. Ding, and H. Zhou, Optimization method based extreme learning machine for classification. *Neurocomputing*, 2010. 74(1–3): pp. 155–163.

8. Huang, G.-B., D.H. Wang, and Y. Lan, Extreme learning machines: A survey. *International Journal of Machine Learning and Cybernetics*, 2011. 2(2): pp. 107–122.

9. Malar, E., et al., A novel approach for detection and classification of mammographic microcalcifications using wavelet analysis and extreme learning machine. *Computers in Biology and Medicine*, 2012. 42(9): pp. 898–905.

10. Schneider, M., and M. Yaffe. Better detection: Improving our chances. In *Digital Mammography: 5th International Workshop on Digital Mammography IWDM*. 2000.

11. Wangpipatwong, S., W. Chutimaskul, and B. Papasratorn, Understanding citizen's continuance intention to use e-government website: A composite view of technology acceptance model and computer self-efficacy. *Electronic Journal of e-Government*, 2008. 6(1). pp 55-64

12. Ciatto, S., et al., Comparison of standard reading and computer aided detection (CAD) on a national proficiency test of screening mammography. *European Journal of Radiology*, 2003. 45(2): pp. 135–138.

13. Warren Burhenne, L.J., et al., Potential contribution of computer-aided detection to the sensitivity of screening mammography. *Radiology*, 2000. 215(2): pp. 554–562.

14. Nie, K., *Development of Breast MRI Computer-Aided Diagnosis System*. University of California, Irvine, 2009.

15. Engan, K., M.R. Lillo, and T.O. Gulsrud. Compression of digital mammograms with region-of-interest coding evaluated on a CAD system. In *Medical Imaging 2005: Image Processing*. International Society for Optics and Photonics, 2005.

16. Gustafson, G.A., *Mammography Information System*. Google Patents, 2014.

17. Jalalian, A., et al., Foundation and methodologies in computer-aided diagnosis systems for breast cancer detection. *EXCLI Journal*, 2017. 16: pp. 113.

18. Lee, L.K., S.C. Liew, and W.J. Thong, A review of image segmentation methodologies in medical image. In *Advanced Computer and Communication Engineering Technology*. Springer, 2015, pp. 1069–1080.

19. Obara, B., et al., Contrast-independent curvilinear structure detection in biomedical images. *IEEE Transactions on Image Processing*, 2012. 21(5): pp. 2572–2581.

20. Kyaw, M.M., Pre-segmentation for the computer aided diagnosis system. *International Journal of Computer Science & Information Technology*, 2013. 5(1): p. 79.

21. Tang, J., et al., Computer-aided detection and diagnosis of breast cancer with mammography: Recent advances. *IEEE Transactions on Information Technology in Biomedicine*, 2009. 13(2): pp. 236–251.

22. Rahman, M.M., et al., Interactive cross and multimodal biomedical image retrieval based on automatic region-of-interest (ROI) identification and classification. *International Journal of Multimedia Information Retrieval*, 2014. 3(3): pp. 131–146.

23. Liu, M., et al., *Matching of Regions of Interest across Multiple Views*. Google Patents, 2014.

24. Cheng, H.-D., et al., Automated breast cancer detection and classification using ultrasound images: A survey. *Pattern Recognition*, 2010. 43(1): pp. 299–317.

25. Jochelson, M.S., et al., Bilateral contrast-enhanced dual-energy digital mammography: Feasibility and comparison with conventional digital mammography and MR imaging in women with known breast carcinoma. *Radiology*, 2013. 266(3): pp. 743–751.

26. Al-Amri, S.S., N. Kalyankar, and S. Khamitkar, Image segmentation by using edge detection. *International Journal on Computer Science and Engineering*, 2010. 2(3): pp. 804–807.

27. Saini, S. and K. Arora, A study analysis on the different image segmentation techniques. *International Journal of Information & Computation Technology*, 2014. 4(14): pp. 1445–1452.

28. Muthukrishnan, R. and M. Radha, Edge detection techniques for image segmentation. *International Journal of Computer Science & Information Technology*, 2011. 3(6): pp. 259.

29. Yu-Qian, Z., et al. Medical images edge detection based on mathematical morphology. In *2005 IEEE Engineering in Medicine and Biology 27th Annual Conference*. IEEE, 2006.

30. Haris, K., et al., Hybrid image segmentation using watersheds and fast region merging. *IEEE Transactions on Image Processing*, 1998. 7(12): pp. 1684–1699.

31. Zhang, Y., et al. A hybrid image segmentation approach using watershed transform and FCM. In *Fourth International Conference on Fuzzy Systems and Knowledge Discovery (FSKD 2007)*. IEEE, 2007.

32. Al-Bayati, M. and A. El-Zaart, Mammogram images thresholding for breast cancer detection using different thresholding methods, 2013. 2(3): pp. 72–77.

33. Ganesan, K., et al., Computer-aided breast cancer detection using mammograms: A review. *IEEE Reviews in Biomedical Engineering*, 2012. 6: pp. 77–98.

34. Saha, M., et al., Histogram based thresholding for automated nucleus segmentation using breast imprint cytology. In *Advancements of Medical Electronics*. Springer, 2015, pp. 49–57.

35. Song, S.E., et al., Computer-aided detection (CAD) system for breast MRI in assessment of local tumor extent, nodal status, and multifocality of invasive breast cancers: Preliminary study. *Cancer Imaging*, 2015. 15(1): pp. 1–9.

36. Li, Y., et al., Three-dimensional volume reconstruction from slice data using phase-field models. *Computer Vision and Image Understanding*, 2015. 137: pp. 115–124.

37. Moon, W.K., et al., Computer-aided diagnosis for the classification of breast masses in automated whole breast ultrasound images. *Ultrasound in Medicine & Biology*, 2011. 37(4): pp. 539–548.

38. Tan, T., et al., Computer-aided detection of cancer in automated 3-D breast ultrasound. *IEEE Transactions on Medical Imaging*, 2013. 32(9): pp. 1698–1706.

39. Gnonnou, C. and N. Smaoui. Segmentation and 3D reconstruction of MRI images for breast cancer detection. In *International Image Processing, Applications and Systems Conference*. IEEE, 2014.

40. Kowal, M., et al., Computer-aided diagnosis of breast cancer based on fine needle biopsy microscopic images. *Computers in Biology and Medicine*, 2013. 43(10): pp. 1563–1572.

41. Al-Antari, M.A., et al., A fully integrated computer-aided diagnosis system for digital X-ray mammograms via deep learning detection, segmentation, and classification. *International Journal of Medical Informatics*, 2018. 117: pp. 44–54.

42. Al-Faris, A.Q., et al., Computer-aided segmentation system for breast MRI tumour using modified automatic seeded region growing (BMRI-MASRG). *Journal of Digital Imaging*, 2014. 27(1): pp. 133–144.

43. Ball, J.E. and L.M. Bruce. Digital mammographic computer aided diagnosis (cad) using adaptive level set segmentation. In *2007 29th Annual International Conference of the IEEE Engineering in Medicine and Biology Society*. IEEE, 2007.

44. Mohammed, M.A., et al., Neural network and multi-fractal dimension features for breast cancer classification from ultrasound images. *Computers & Electrical Engineering*, 2018. 70: pp. 871–882.

45. Qayyum, A. and A. Basit. Automatic breast segmentation and cancer detection via SVM in mammograms. In *2016 International Conference on Emerging Technologies (ICET)*. IEEE, 2016.

46. Saini, S. and R. Vijay, Performance analysis of artificial neural network based breast cancer detection system. *International Journal of Soft Computing and Engineering (IJSCE)*, 2014. 4: pp. 70–72.

47. Dheeba, J., N.A. Singh, and S.T. Selvi, Computer-aided detection of breast cancer on mammograms: A swarm intelligence optimized wavelet neural network approach. *Journal of Biomedical Informatics*, 2014. 49: pp. 45–52.

48. Hassanien, A.E., et al., MRI breast cancer diagnosis hybrid approach using adaptive ant-based segmentation and multilayer perceptron neural networks classifier. *Applied Soft Computing*, 2014. 14: pp. 62–71.

49. Gupta, N., et al., A survey on crop prediction using back propagation neural network, 2014.

50. Übeyli, E.D., Implementing automated diagnostic systems for breast cancer detection. *Expert Systems with Applications*, 2007. 33(4): pp. 1054–1062.

51. Razia, S. and M.N. Rao, Machine learning techniques for thyroid disease diagnosis-a review. *Indian Journal of Science and Technology*, 2016. 9(28): pp. 1–9.

52. Keyvanfard, F., M. Shoorehdeli, and M. Teshnehlab, Feature selection and classification of breast cancer on dynamic magnetic resonance imaging using ANN and SVM. *American Journal of Biomedical Engineering*, 2011. 1(1): pp. 20–25.

53. Richard, M.D. and R.PP. Lippmann, Neural network classifiers estimate Bayesian a posteriori probabilities. *Neural Computation*, 1991. 3(4): pp. 461–483.

54. Bhardwaj, A., et al., A genetically optimized neural network model for multi-class classification. *Expert Systems with Applications*, 2016. 60: pp. 211–221.

55. Devarriya, D., et al., Unbalanced breast cancer data classification using novel fitness functions in genetic programming. *Expert Systems with Applications*, 2020. 140: pp. 112866.

56. Leema, N., H.K. Nehemiah, and A. Kannan, Neural network classifier optimization using differential evolution with global information and back propagation algorithm for clinical datasets. *Applied Soft Computing*, 2016. 49: pp. 834–844.

57. Karthik, S., R.S. Perumal, and P.C. Mouli, Breast cancer classification using deep neural networks. In *Knowledge Computing and Its Applications*. Springer, 2018, pp. 227–241.

58. Jafari-Marandi, R., et al., An optimum ANN-based breast cancer diagnosis: Bridging gaps between ANN learning and decision-making goals. *Applied Soft Computing*, 2018. 72: pp. 108–120.

59. Ho, T.K., Random decision forests. In *Proceedings of 3rd International Conference on Document Analysis and Recognition*. IEEE, 1995.

60. Kasivisvanathan, V., et al., MRI-targeted or standard biopsy for prostate-cancer diagnosis. *New England Journal of Medicine*, 2018. 378(19): pp. 1767–1777.

61. Welch, H.G. and P.C. Albertsen, Prostate cancer diagnosis and treatment after the introduction of prostate-specific antigen screening: 1986–2005. *Journal of the National Cancer Institute*, 2009. 101(19): pp. 1325–1329.

62. Epstein, J.I., An update of the Gleason grading system. *The Journal of Urology*, 2010. 183(2): pp. 433–440.
63. Mosquera-Lopez, C., et al., Computer-aided prostate cancer diagnosis from digitized histopathology: A review on texture-based systems. *IEEE Reviews in Biomedical Engineering*, 2014. 8: pp. 98–113.
64. Wildeboer, R.R., et al., Artificial intelligence in multiparametric prostate cancer imaging with focus on deep-learning methods. *Computer Methods and Programs in Biomedicine*, 2020. 189: pp. 105316.
65. Peng, Y., et al., Quantitative analysis of multiparametric prostate MR images: Differentiation between prostate cancer and normal tissue and correlation with Gleason score—a computer-aided diagnosis development study. *Radiology*, 2013. 267(3): pp. 787–796.
66. Grover, V.P., et al., Magnetic resonance imaging: Principles and techniques: Lessons for clinicians. *Journal of Clinical and Experimental Hepatology*, 2015. 5(3): pp. 246–255.
67. Carroll, P.R., F.V. Coakley, and J. Kurhanewicz, Magnetic resonance imaging and spectroscopy of prostate cancer. *Reviews in Urology*, 2006. 8(Suppl 1): p. S4.
68. Akin, O., et al., Transition zone prostate cancers: Features, detection, localization, and staging at endorectal MR imaging. *Radiology*, 2006. 239(3): pp. 784–792.
69. Mazaheri, Y., et al., Prostate cancer: Identification with combined diffusion-weighted MR imaging and 3D 1H MR spectroscopic imaging—correlation with pathologic findings. *Radiology*, 2008. 246(2): pp. 480–488.
70. Ghose, S., et al., A survey of prostate segmentation methodologies in ultrasound, magnetic resonance and computed tomography images. *Computer Methods and Programs in Biomedicine*, 2012. 108(1): pp. 262–287.
71. Singh, R.P., S. Gupta, and U.R. Acharya, Segmentation of prostate contours for automated diagnosis using ultrasound images: A survey. *Journal of Computational Science*, 2017. 21: pp. 223–231.
72. Zhu, Q., B. Du, and PP. Yan, Boundary-weighted domain adaptive neural network for prostate MR image segmentation. *IEEE Transactions on Medical Imaging*, 2019. 39(3): pp. 753–763.
73. Mazonakis, M., et al., Image segmentation in treatment planning for prostate cancer using the region growing technique. *The British Journal of Radiology*, 2001. 74(879): pp. 243–249.
74. Weinreb, J.C., et al., PI-RADS prostate imaging–reporting and data system: 2015, version 2. *European Urology*, 2016. 69(1): pp. 16–40.
75. Futterer, J.J., et al., Prostate cancer localization with dynamic contrast-enhanced MR imaging and proton MR spectroscopic imaging. *Radiology*, 2006. 241(2): pp. 449–458.
76. Yan, P., et al., Discrete deformable model guided by partial active shape model for TRUS image segmentation. *IEEE Transactions on Bio-medical Engineering*, 2010. 57(5): pp. 1158–1166.
77. Jin, J., et al., Detection of prostate cancer with multiparametric MRI utilizing the anatomic structure of the prostate. *Statistics in Medicine*, 2018. 37(22): pp. 3214–3229.
78. Durmus, T., A. Baur, and B. Hamm. Multiparametric magnetic resonance imaging in the detection of prostate cancer. In *RöFo Fortschritte auf dem Gebiet der Röntgenstrahlen und der Bildgebenden Verfahren*. ©Georg Thieme Verlag KG, 2014.
79. Hricak, H., et al., Imaging prostate cancer: A multidisciplinary perspective. *Radiology*, 2007. 243(1): pp. 28–53.
80. Rampun, A., et al., Computer aided diagnosis of prostate cancer: A texton based approach. *Medical Physics*, 2016. 43(10): pp. 5412–5425.
81. Rampun, A., et al., Computer-aided detection of prostate cancer in T2-weighted MRI within the peripheral zone. *Physics in Medicine & Biology*, 2016. 61(13): p. 4796.

82. Ma, W., L. Huang, and C. Liu. Crowd density analysis using co-occurrence texture features. In *5th International Conference on Computer Sciences and Convergence Information Technology*. IEEE, 2010.
83. Tiwari, P., A. Madabhushi, and M. Rosen. A hierarchical unsupervised spectral clustering scheme for detection of prostate cancer from magnetic resonance spectroscopy (MRS). In *International Conference on Medical Image Computing and Computer-Assisted Intervention*. Springer, 2007.
84. Choyke, P.L., A.J. Dwyer, and M.V. Knopp, Functional tumor imaging with dynamic contrast-enhanced magnetic resonance imaging. *Journal of Magnetic Resonance Imaging: An Official Journal of the International Society for Magnetic Resonance in Medicine*, 2003. 17(5): pp. 509–520.
85. Yoon, H.-J., et al. Cardinal multiridgelet-based prostate cancer histological image classification for Gleason grading. In *2011 IEEE International Conference on Bioinformatics and Biomedicine*. IEEE, 2011.
86. Almuntashri, A., et al. Gleason grade-based automatic classification of prostate cancer pathological images. In *2011 IEEE International Conference on Systems, Man, and Cybernetics*. IEEE, 2011.
87. Tiwari, P., M. Rosen, and A. Madabhushi. Consensus-locally linear embedding (C-LLE): Application to prostate cancer detection on magnetic resonance spectroscopy. In *International Conference on Medical Image Computing and Computer-Assisted Intervention*. Springer, 2008.
88. Doyle, S., et al. Automated grading of prostate cancer using architectural and textural image features. In *2007 4th IEEE International Symposium on Biomedical Imaging: From Nano to Macro*. IEEE, 2007.
89. Tiwari, P., M. Rosen, and A. Madabhushi, A hierarchical spectral clustering and non-linear dimensionality reduction scheme for detection of prostate cancer from magnetic resonance spectroscopy (MRS). *Medical Physics*, 2009. 36(9Part1): pp. 3927–3939.
90. Lopez, C.M. and S. Again. A new set of wavelet-and fractals-based features for Gleason grading of prostate cancer histopathology images. In *Image Processing: Algorithms and Systems XI*. International Society for Optics and Photonics, 2013.
91. Tiwari, P., et al., Multimodal wavelet embedding representation for data combination (MaWERiC): Integrating magnetic resonance imaging and spectroscopy for prostate cancer detection. *NMR in Biomedicine*, 2012. 25(4): pp. 607–619.
92. Zincke, H., et al., Radical prostatectomy for clinically localized prostate cancer: Long-term results of 1,143 patients from a single institution. *Journal of Clinical Oncology*, 1994. 12(11): pp. 2254–2263.
93. Loizidou, K., et al., An automated breast micro-calcification detection and classification technique using temporal subtraction of mammograms. *IEEE Access*, 2020. 8: pp. 52785–52795.
94. Rashed, E.A., Neural networks approach for mammography diagnosis using wavelets features. *arXiv Preprint arXiv:2003.03000*, 2020.
95. Mavroforakis, M.E. and S. Theodoridis, A geometric approach to support vector machine (SVM) classification. *IEEE Transactions on Neural Networks*, 2006. 17(3): pp. 671–682.
96. Hamed, G., et al. Deep learning in breast cancer detection and classification. In *Joint European-US Workshop on Applications of Invariance in Computer Vision*. Springer, 2020.
97. Shrivastava, N. and J. Bharti, Breast tumor detection and classification based on density. *Multimedia Tools and Applications*, 2020. 79(35) 26467–87.
98. Alexandratou, E., et al., Evaluation of machine learning techniques for prostate cancer diagnosis and Gleason grading. *International Journal of Computational Intelligence in Bioinformatics and Systems Biology*, 2010. 1(3): pp. 297–315.

99. Hosseini, R., et al., An automatic approach for learning and tuning Gaussian interval type-2 fuzzy membership functions applied to lung CAD classification system. *IEEE Transactions on Fuzzy Systems*, 2011. 20(2): pp. 224–234.

100. Bellotti, R., et al., A completely automated CAD system for mass detection in a large mammographic database. *Medical Physics*, 2006. 33(8): pp. 3066–3075.

101. Doyle, S., et al. Cascaded multi-class pairwise classifier (CASCAMPA) for normal, cancerous, and cancer confounder classes in prostate histology. In *2011 IEEE International Symposium on Biomedical Imaging: From Nano to Macro*. IEEE, 2011.

102. Lu, D. and Q. Weng, A survey of image classification methods and techniques for improving classification performance. *International Journal of Remote Sensing*, 2007. 28(5): pp. 823–870.

103. Pal, M. and P.M. Mather, An assessment of the effectiveness of decision tree methods for land cover classification. *Remote Sensing of the Environment*, 2003. 86(4): pp. 554–565.

104. Liu, L. and M.T. Özsu, *Encyclopedia of Database Systems*. Vol. 6. Springer, 2009.

105. Refaeilzadeh, P., L. Tang, and H. Liu, *Cross Validation, Encyclopedia of Database Systems (EDBS)*. Arizona State University, Springer, 2009, p. 6.

106. Rodriguez, J.D., A. Perez, and J.A. Lozano, Sensitivity analysis of k-fold cross validation in prediction error estimation. *IEEE Transactions on Pattern Analysis and Machine Intelligence*, 2009. 32(3): pp. 569–575.

107. Wong, T.-T., Performance evaluation of classification algorithms by k-fold and leave-one-out cross validation. *Pattern Recognition*, 2015. 48(9): pp. 2839–2846.

108. Moreno-Torres, J.G., J.A. Sáez, and F. Herrera, Study on the impact of partition-induced dataset shift on $ k $-fold cross-validation. *IEEE Transactions on Neural Networks and Learning Systems*, 2012. 23(8): pp. 1304–1312.

109. Triba, M.N., et al., PLS/OPLS models in metabolomics: The impact of permutation of dataset rows on the K-fold cross-validation quality parameters. *Molecular BioSystems*, 2015. 11(1): pp. 13–19.

110. Jung, Y., Multiple predicting K-fold cross-validation for model selection. *Journal of Nonparametric Statistics*, 2018. 30(1): pp. 197–215.

111. Lalkhen, A.G. and A. McCluskey, Clinical tests: Sensitivity and specificity. *Continuing Education in Anaesthesia Critical Care & Pain*, 2008. 8(6): pp. 221–223.

112. Doyle, S., et al. Detecting prostatic adenocarcinoma from digitized histology using a multi-scale hierarchical classification approach. In *2006 International Conference of the IEEE Engineering in Medicine and Biology Society*. IEEE, 2006.

113. Goo, J.M., A computer-aided diagnosis for evaluating lung nodules on chest CT: The current status and perspective. *Korean Journal of Radiology*, 2011. 12(2): pp. 145–155.

114. Wang, S., et al., Computer aided-diagnosis of prostate cancer on multiparametric MRI: A technical review of current research. *BioMed Research International*, 2014. 2014.

115. Hand, D.J. and R.J. Till, A simple generalisation of the area under the ROC curve for multiple class classification problems. *Machine Learning*, 2001. 45(2): pp. 171–186.

116. Bouatmane, S., et al., Round-Robin sequential forward selection algorithm for prostate cancer classification and diagnosis using multispectral imagery. *Machine Vision and Applications*, 2011. 22(5): pp. 865–878.

117. Armato III, S.G., et al., Lung image database consortium: Developing a resource for the medical imaging research community. *Radiology*, 2004. 232(3): pp. 739–748.

118. Brown, M., et al., Automated method for detecting lung micronodules on thin section CT images. *Radiology*, 2003. 226: pp. 256–262.

119. Henschke, C.I., et al., CT screening for lung cancer: Suspiciousness of nodules according to size on baseline scans. *Radiology*, 2004. 231(1): pp. 164–168.

120. Patz, E.F., et al., Overdiagnosis in low-dose computed tomography screening for lung cancer. *JAMA Internal Medicine*, 2014. 174(2): pp. 269–274.
121. Brown, M.S., et al., Toward clinically usable CAD for lung cancer screening with computed tomography. *European Radiology*, 2014. 24(11): pp. 2719–2728.
122. Chan, H.-P., et al., Computer-aided diagnosis of lung cancer and pulmonary embolism in computed tomography—a review. *Academic Radiology*, 2008. 15(5): pp. 535–555.
123. Sun, T., et al., Comparative evaluation of support vector machines for computer aided diagnosis of lung cancer in CT based on a multi-dimensional data set. *Computer Methods and Programs in Biomedicine*, 2013. 111(2): pp. 519–524.

7 Automated COVID-19 Detection from CT Images Using Deep Learning

Abdulhamit Subasi, Arka Mitra,
Fatih Ozyurt, and Turker Tuncer

CONTENTS

7.1 INTRODUCTION

The severe acute respiratory syndrome coronavirus-2 (SARS-CoV-2) has caused havoc all around the globe. The virus was first contracted in China, and it dates back to November 17, 2019 [1]. In the initial stages of the spread, it followed an exponential growth, and as of September 1, 2020, there were over 25 million cases worldwide, with around 850,000 deaths and the numbers rising each day [2]. It is the biggest pandemic that the modern world has faced, and it has changed the way human beings spend their day to day life and also how they interact with one another. It was classified as a public health emergency of international concern (PHEIC) by the World Health Organization (WHO) on January 30, 2020 [3]. SARS-CoV-2 belongs to the family of coronaviruses which cause illnesses such as respiratory and gastrointestinal diseases. As of now, there has been no clinical treatment or vaccine against the virus. Thereby early detection and isolation of the affected person to prevent the spread of the disease is of utmost importance. This increases not only the chances of survival of the infected person but also decreases the chances of the spread of the disease to a new set of susceptible populations. For early detection, a reverse transcription-polymerase chain reaction (RT-PCR) or gene sequencing in blood and respiratory samples was introduced as the main COVID-19 screening method [4]. The mean accuracy of this method was relatively low, with about a 30% to 60% ratio [5]. This is not good and could lead to an explosion in the number of cases due to the incorrect screening of patients.

The literature on the epidemiology of COVID-19 shows that there is a peak in the outbreak. During this, the hospitals get filled, and the number of available intensive care units (ICUs) decreases [6]. Patients in critical condition are brought into ICUs. In general, the ICUs are divided into COVID-19 and non-COVID-19 ICUs. However, if any of the patients are mistakenly sent to the wrong ward, it will be dangerous not only for the patient but also for the other patients on the ward. Thereby health systems can fail during this peak surge of cases, mostly due to the unavailability of ICUs, and this can happen in even the most developed countries. The identification of SARS-CoV-2 was made in early January, and the genetic sequence was shared publicly on January 11–12, 2020. The virus belongs to the class of coronaviruses, and thereby the symptoms range from cold and fever, gastrointestinal problems to loss of taste. Some of the sister viruses belonging to the same family are severe acute respiratory syndrome (SARS) and the Middle East respiratory syndrome (MERS). The novel coronavirus also damages the kidney and liver, unlike previous generations of the virus [7]. Due to the severity of the disease and the fast rate of transmission, early detection of the symptoms is essential and crucial. In accordance with the Chinese government reports [8], it was found that the gene sequencing of blood and spleen samples was a crucial part of the reverse transcription-polymerase chain reaction (RT-PCR). This was one of the most critical identification points for the novel coronavirus. Still, it does involve a considerable amount of time to be correctly placed (about four to six hours), which is rather sluggish compared to the speedy rate of the contagion of the virus. Another downside to this technique is that the kits were in very low supply [9]. The report also shows that the incubation period of the virus can be as long as 14 days, with the average incubation period being five days. This incubation period makes it increasingly hard to identify the people who have been infected, and therefore this increases the spread of the virus. Due to the low number of test kits, the fast rate of transmission, and the difficulty of detecting who has been infected during the non-symptomatic period, RT-PCR proves to be an inefficient method for the detection of the virus. Therefore, it is necessary to investigate other safe, accurate, and fast methods of detecting infected people. Using such a new technique will prevent the spread of the disease and also save people who have been infected, as it has been shown that early detection of SARS-CoV2 can drastically increase survival rate [10]. This has led to many studies on the possible sources from which the virus could be detected. Doctors, especially radiologists, can detect various diseases using computed tomography (CT) scans and X-rays. Investigations have also shown that CT scans can provide a better way for classifying the disease via observing ground-glass opacities in the CT scans [11]. Thus, the researchers have stated that a method based on CT scans and X-rays could be effective in detecting the disease [12]. CT scans are also better than RT-PCR, as shown by the report [13]. Since a computer system is used for the detection of the disease, an automated app can also be developed that does not even require the patient to visit the doctor. This will further help to decrease the spread of the disease by reducing the conglomeration of people in places like hospitals.

Artificial Intelligence has transformed the world and specifically the medical field in recent times. Advancements in computer vision, natural language processing, as well as reinforcement learning, has helped to give more insight into how humans do things. This has in turn helped to automate overall workflow in most areas of life and industry. When the number of doctors and expert radiologists is not enough, an automated image analysis becomes crucial. It could help with detecting anomalous CT scans or X-rays, thereby assisting in the detection of which people have the disease, and helping others by isolating them and increasing their chances of survival by helping with early screening. Another problem that is faced in the field of artificial intelligence or machine learning, in general, is the lack of labeled data. Even if it contains the images of patients infected by the notorious virus, it will be useless unless we know which images correspond to which disease. Although methods exist which can utilize the underlying features of the data and classify them in an unsupervised way [14], it is always better to have supervised data. But most importantly, with the fast rate of transmission, we need to work with whatever data we have and make the most of it. A small dataset was obtained from COVID-19-related papers containing the abnormalities and then checked by a radiologist to ensure that the labels are correct. Overall, they contained 349 CT images from 216 patients [15]. The various methods, like X-rays and CT scans, are relatively easy to conduct and do not pose a significant risk to the body. There have been considerable advancements to chest X-ray classification, and [16] contains a large number of images containing chest X-rays of 14 diseases. However, they do not contain the COVID-19 disease. Kanne et al. [17] identified important aspects for radiologists to identify COVID-19 , and pointed out the small irregular patterns that are present in coronavirus patients. The topic was further complicated in the X-rays as the characteristics of pneumonia and COVID-19 were similar in the images. However, as many studies have already been conducted with X-rays, we look into the field of CT scans. Ai et al. [8] showed that there was a correlation between CT Scans and the RT-PCR methods.

Extensive experiments have been carried out to predict COVID-19 from CT scan images, and these have been tested with both single-stage and dual-stage detection methods. In single-stage methods, the detection is performed using artificial algorithms applying an end to end process, while in the dual stage, transfer learning was used to obtain the features of the images and then classified according to these features. Deep learning is another subfield of machine learning which has helped to achieve the state-of-the-art in most of the problems that are faced. It has left a significant impact in all fields ranging from image classification, object detection, machine translation, face recognition, and super-resolution problems [18–21]. In some cases, such as the Imagenet challenge, these methods have also surpassed human-level performances (5.1%), and this is increasing almost every year [22].

The SARS-CoV-2 spreads through human to human contact and mainly enters the human body via the eyes, nose, and mouth. The virus can stay in the throat for some time before entering the lungs. After it enters the lungs, it starts to mutate. The immune system goes to fight the external virus, and this can sometimes come in contact with the alveoli and allow the blood from the vessels near the alveoli to get inside the lungs. This fills up the lungs with blood and this can lead to fatal consequences. X-rays and CT scan techniques make use of the difference in the density of the different substances in the lungs to give a detailed analysis of what could be present there. Due to the density of the blood being different from that of the rest of the lung, this can be seen as a different color on the scan as compared to the rest; this is called ground-glass opacity [23]. It was stated earlier that COVID-19 is quite similar to another disease known as pneumonia. Pneumonia is also fatal for people and so can quite easily be confused with the novel coronavirus. Apart from being contagious, like the coronavirus, pneumonia also shares many similar symptoms. Severe acute respiratory syndrome coronavirus-2 additionally had SARS, which makes it even dangerous; rather, it can be considered the deadliest form of pneumonia. There has been a lot of different research to find a vaccine for the disease and methods by which to screen patients for it. Screening using simple questions can help reduce fear among the general population and thereby help them decide whether they should visit the doctor or not [24]. This restrains the people from panicking and thereby reduces the number of people visiting the hospitals, which would put a strain on the resources. Due to the new disease, a large number of doctors are also succumbing to the deadly virus, thereby reducing the number of doctors per patient even more. This creates even more stress on the doctors, and they can misclassify due to extreme stress and fatigue. For this, the automated screening of the patients is a must as the resources should be allocated to those people at a greater risk. At the present moment, three techniques are being deployed to check whether a person is infected or not, and these are RT-PCR, Chest X-rays, and CT scans. The first is the lengthiest and can take about four to six hours. X-rays not only help to detect whether the person has the disease or not, but also helps to detect which regions of the image are responsible for coming to such a conclusion. However, these images do not contain the 3D information from the longitudinal axis of the body and cannot tell how deep inside the infection has occurred. For that, CT scans have been used that give a more accurate location of the air sacs [25]. We have thereby moved forward with CT scans for classifying the images.

Machines in hospitals and other institutions produce a large amount of data each day, and it is not possible to go through all the data and reports by hand for so many hospitals. The speed of data generation has almost exceeded the speed at which humans can process the data. So, it is essential to have automated systems that can either store the data more effectively or retrieve the essential parts of the data. Automation is most needed while detecting the diseases in the patients. The RT-PCR is very time consuming, and even if automation is performed to collect the data from the patients safely, the time taken will increase the number of cases overall. Thereby we look more into deep learning techniques for classifying the diseases. The best solution, as has been stated before, is the use of deep learning or a mix of deep learning and classical method learning techniques. The images are of a large size, which can take up a lot of space. However, deep learning methods can help to extract the 224 x 224 size images into a 1024 sized vector, which corresponds to a 50x times decrease in size. Thus, 50 Tbs of data can now effectively be stored in 1 Tb for further preprocessing. This can also help even more patients by enabling the databases to store the data of more patients. Machine learning is a subpart of artificial intelligence, and deep learning (DL) is a subpart of machine learning. By utilizing the recent advancements in DL, we can help the radiologists by reducing their workload. Different methods can also help to point out which region of the image the model is looking at to give the corresponding information [25]. Thereby the radiologists can just skim through the images to check whether the predictions are correct or not and select the images with incorrect predictions that can then be fed back into the model to improve the performance of the model even more. Deep learning helps by extracting the most important information from the images.

Since COVID-19 is a lung infection, we need to investigate chest X-rays and CT scans for identifying and becoming informed about the disease. Due to the several models currently available, it is also possible to deploy these models on the web or an app so that the patients can get even faster results. Care also needs to be taken that the model does not get confused between COVID-19 and CT scans. If the dataset contains all three classes, we can divide the dataset in two ways. First, the three classes are divided equally so that each image can either be COVID-19, pneumonia, or normal. Another way is that COVID-19 is kept as a separate class, and the other two are merged to form a non-COVID class. The latter is helpful when we come across a dataset that does not mention pneumonia. By considering that the second type of dividing the data is also possible, it removes the need for manually labeling the class of pneumonia as we are only interested in COVID-19.

COVID-19 has impacted people differently based on their age, sex, and other factors like humidity, climate, and temperature [26, 27], and even the degree of social distancing maintained. It has led most countries to go into lockdown and thereby affected how humans interact with one another. It has thereby stalled many industries and led to overall chaos in general with many people losing their jobs, and most companies incurring losses due to the extensive length of the lockdown. Thereby it is very important to detect this disease in the early stages and prevent its spread even further. The classification needs to be as accurate as possible.

In this chapter, deep learning is utilized for COVID-19 detection using CT scans. Moreover, we tried to see how transfer learning combined with other classical machine learning algorithms can help classify the CT images. We were also working

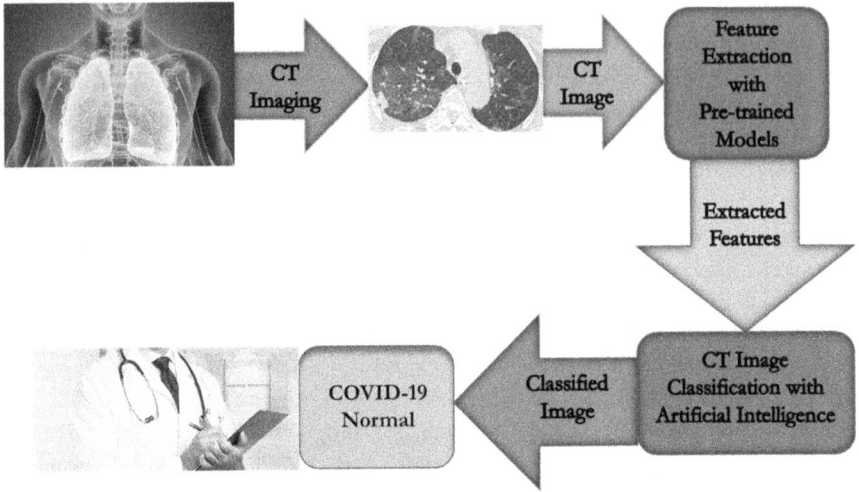

FIGURE 7.1   Pipeline of the architecture used.

with a small dataset as in the initial stages only a small amount of data was available, and this is the most important period for the detecting various diseases and classifying them. To deal with the difference in validation scores due to the small amount of data, we also used various techniques to ensure that the final metrics obtained were not due to the model memorizing the overall images. The pipeline, which gave the best results overall, was described hereby. In this pipeline, a CT scan was passed through a pre-trained model to obtain the features. A feature selection algorithm was used to obtain the important features, and classical techniques were implemented on them to get the final classification (Figure 7.1).

7.2   LITERATURE SURVEY IN CT SCAN PROGNOSIS

COVID-19 emerged in the city of Wuhan, a district in China, and later spread throughout the whole world. Thereby it has so far caused over 25 million cases and about a million deaths. The disease is infectious, and since it cannot be seen by the naked eye, it cannot be avoided unless we take precautionary measures. The worst part is that in most people the symptoms are the same as that of normal flu, while in others the symptoms are not even visible. Also, there is an incumbency period of a maximum of 14 days. All this make it very difficult to detect who has already been infected and thereby understand whom to be wary of. Thereby social distancing rules have been imposed to try and reduce the number of cases. The curve does not always follow the simple curve that is portrayed by the susceptible infectious removed (SIR) model [28], and there are regions where the disease has had two peaks. This makes the modeling of the disease even more difficult. Factors like comorbidity in the various patients make the disease even more dangerous as there is an even greater risk of death when infected [29]. Sun et al. [30] showed that early detection and intervention of COVID-19 leads to a lower mortality rate. All this points to the fact that early detection is necessary for keeping the disease under control. The three main methods of screening in people who are suspected of having the virus are RT-PCR, chest X-rays, and CT scans. RT-PCR takes a great deal of time to process (four to six hours), and this can further increase the spread of the disease. Also, they do not have high sensitivity, which is very much required as we need to lower the number of false negatives. Even if we have false positives, it is okay as just one person would have to go through the extra hassle and be admitted to the hospital and be treated. But in the other case of a false negative, the person with the disease would be allowed to roam throughout the city and thereby infect even more people. This would even increase the rate of spread of the already highly infectious disease. Thus, we need to look even more closely into the other two methods, which have a higher sensitivity. Also, the sensitivity in these types of models can be changed by changing the parameters of the model, and thus we have it under our control. CT scans give a better representation of the regions which lead to the model predicting if COVID-19 is present, and thus it has been selected as part of our study. We now talk about some of the other methods that have been performed with respect to the detection of the disease from CT scans.

Li et al. [31] used a COVID-19 detection neural network to obtain visual features from volumetric chest CT scans for COVID-19 detection. They added CT scans of community-acquired pneumonia and other non-pneumonia irregularities to make the model even more robust. The final dataset consisted of 4352 CT scans from 3322 patients, and it had an area under the receiver operating characteristic curve (AUC) of 0.96. Zhang et al. [32] extracted the features from the CT scans and compared them against the clinical evidence of disease severity based on other organ systems' parameters. They found a strong correlation between lung lesions as compared to other organs, and highlighted the importance of lung damage on the overall prognostic implications. Also, they were able to show that there is a strong association between age and a low mortality rate due to COVID-19. It also showed that there was a correlation among other organs as well that proved that multi-organ failure occurs in COVID-19.

Kang et al. [33] used the V-Net model [34] to extract different segments from the image. Thus, they extract features from each CT image and divided the features into radioomic features and handcrafted features. They then transformed the features into a structured latent representation and used added latent-representation-based classifiers to classify the image. They were able to obtain 95.5%, 96.6%, and 93.2% accuracy, sensitivity, and specificity, respectively, on the dataset from 3322 patients. Bai et al. [35] had a total of 521 patients with positive RT-PCR and abnormal chest findings, along with 665 patients with non-COVID CT scans. They used EfficientNet B4 [36] to classify the images after lung segmentation had been performed. The final model had an accuracy of 96%, sensitivity of 95%, and an AUROC of 0.95.

Fan et al. [37] introduced a novel architecture (IRFNet) that outperformed most cutting-edge segmentation models. They used a parallel partial decoder to concatenate the features and create a global map. Then, they used implicit reverse attention and explicit edge-attention to model the boundaries and enhance the boundaries.

Ardakani et al. [38] had their data on high-resolution CT scan images of 512 x 512 pixels. They tried to transfer learning with ten popular convolutional networks. From their studies, it was found that ResNet101 and Xception had the best results with an AUC of 0.994. Voulodimos et al. [39] showed the effectiveness of the unit [40] and fully convolutional neural networks (FCNN) [41] for segmentation of the regions containing COVID-19. They obtained data of high resolution, 630 x 630, which was segmented and verified by radiologists. They showed that their model performed well even when there was a class imbalance and when the annotations had some human-made errors near the boundaries.

Hu et al. [42] used various data-augmentation techniques on a total of 1042 images to see their effects on the overall performance of the system. According to them, the technique of data augmentation increased the performance of the model with an AUC of 0.97 and was also robust when there was a 10% noise in the training data, which changes the ROC to 0.92. Thus, it showed how to increase the model's performance even with low data. Yousefzadeh et al. [43] showed the effects of a CNN based model on the overall performance of the model. Their dataset consisted of 2121 total axial spiral chest CT scan images. Their data were divided into three classes of COVID-19: COVID-19 abnormal, non-COVID abnormal, and normal, and they demonstrated the AUC when one class was trained against the others. For this study, we were only interested in COVID vs others, which give an AUROC of 0.989. The paper also showcased that RT-PCR had a lower AUC than all the cases that had been mentioned.

Song et al. [44] used data obtained from the CT scans of 88 COVID-19 infected people, 101 patients with bacterial pneumonia, and 86 healthy patients. Their model had a high AUC of 0.99 and a sensitivity of 0.93. It also performed exceptionally well when differentiating COVID-19 from pneumonia, with an AUC of 0.95 and a recall of 0.96. Wang et al. [45] used a Densenet121-Feature Pyramid Network (FPN) [46, 47] pre-trained on 1.4 million natural images for automatic lung segmentation. The part that was not segmented correctly was cropped using the cubic bounding box of the segmented lung areas. Then they used a non-lung suppression to reduce the intensities of the non-lung regions inside the lungs, like the spine, and heart, they used the DenseNet model for training the network. They also used a 64-dimensional output vector combined with the clinical features to select the prognostic features. Thereafter they created a Cox proportional hazard (CPH) [48] model to determine how long a patient needed to stay at the hospital to reduce their chances of death. It also had an overall god AUC score of 0.87.

Mei et al. [49] published a paper in *Nature Medicine* about why RT-PCR was bad for classifying patients. They considered the non-image features of the patient, along with the features of the image to compare the results. The joint model had an AUC of 0.86, while a CNN for the image features gave a score of 0.86 and an MLP composed of non-image features gave a score of 0.80 AUC. Most of these models performed better than a senior thoracic radiologist who had an AUC of 0.84. The joint one had a sensitivity of 0.84. Zhou et al. [50] developed a novel method for finding a small object in a large 3D scene by decomposing the 3D segmentation problem into 2D segmentations. This reduced the complexity of the model, whereas it improved the segmentation accuracy. The authors stacked the CT scan images into a 3D tensor and then normalized the image and intensified the signal. This helped to embed a CT scan into a machine agnostic standard space. Han et al. [51] used a multi-instance learning approach for the screening of COVID-19 patients. The stack of CT scans was treated as a 3D image, and it is treated as a bag of instances. They proposed an attention-based 3D MIL model that learned the Bernoulli distribution of the bag-level labels. They had a high accuracy of 97 % and an AUC of 99.

Thus, most of the papers use a deep learning model for classifying the images. The other main aim of recent research is to use a multi-modal learning approach where both the image data and the features of the patients is used to get more information. There is also interest in lung segmentation to make processing easier in the later stages. Some authors have also used a 3D object segmentation task to mask out where the infected regions are. These all highlight the importance of CT scans over RT-PCR in the detection of the disease. However, the models have either used a large amount of publicly available data or obtained high-resolution data from hospitals. We shall demonstrate the performance of the models when fed with a small number of examples but show that they also achieve a relatively high AUC after following the proposed pipeline.

7.3   MACHINE LEARNING TECHNIQUES

7.3.1   MULTILAYER PERCEPTRON

A multi-layered perceptron (MLP) is referred to as an artificial neural network as it provides a close representation of the neurons in human brains. They follow a similar pattern where some of the neurons light up and propagate information to the neurons ahead of them, just like a human brain. They are used to map a given set of inputs onto their corresponding output. The universal approximation theorem states that any function can be approximated using a neural network. Leshno et al. [52] showed that a two-layered multi-layered perceptron is enough to represent any function. In a perceptron learning algorithm, the inputs are multiplied with weights, and the result is added to a bias term to get the output. When the idea is extended to multiple layers, it becomes a MLP. The layer which contains the input is called the input layer, and the one which includes the output is called the output layer. All intermediate layers are called the hidden layers. In the MLP, the whole network is viewed as a graph where the inputs are multiplied by various weight vectors and added to a bias term to get the output for a node in the next layer [53]. This is done for all the nodes in the first hidden layer. A similar process follows for all the layers to finally give the output. However, if we are multiplying with an affine transform in each of the layers, it can be shown that the final relationship can also be written as a linear relationship between the input and the output, and the MLP transforms back into a single-layer perceptron. Thereby it is essential to add non-linearity after each of the affine functions, which not only adds complexity to the whole model but all prevents the problem stated. In the new scenario, after the inputs to the last layers are multiplied by the affine transform to get the output at a node, the value of that node is passed through a non-linear function. In more general terms, the weights of each of the nodes can be seen as the strengths of the connections among the neurons. When an input is given, only one set of neurons get activated, which in turn helps to activate another set in the next layer. The bias term is added so that the linear transformation can cover more hypothesis space; it also makes the model more flexible and gives a better generalization to the model. One of the most popular methods to update the weights of the networks is called the backpropagation algorithm [54]. The weights of the neural network do not have a bound on themselves and can extend from negative infinity to positive infinity. However, during backpropagation, we take the differential with respect to the weights, and during that time a very high value of the gradient can lead to all the weights in the consequent layers of the backward propagation to have a very high value. This is known as the exploding gradient problem, and it causes the network to give an absurd value or a value of "not a number" (Nan). Another similar problem that is associated with the backpropagation algorithm is that of the vanishing gradients. If the gradient at any point becomes very low, the gradients at the consequent points are multiplied by the same very small gradient using the chain rule. Thereby all the gradients become very small, and there are almost no changes to the weights.

7.3.2   *k*-Nearest Neighbor (*k*-NN)

A *k*-Nearest Neighbor [55] is one of the most straightforward algorithms that is applied in the field of machine learning. It can be used both for classification and clustering of the different groups. When the different labels of the underlying distribution are not known, the *k*-NN can be used to cluster the data. In a supervised setting, as in our case, the clustering was already done, and thus the *k*-NN was used to classify new points into one of the given classes. It measures the similarity of the new class against all the available cases using a distance metric and then classifies it as the class that is the most voted for among the top *k* most-similar class. The *k*-NN is a simple learning algorithm as it stores all the data points and uses them all for classification. One of the ways to remove this downside is to use a modified type known as Condensed Nearest Neighbors [56], which only uses a subset of the total training data, thereby reducing the cost of classifying a single new data point. The *k*-NN algorithm has almost no learnable parameters and does not assume anything about the underlying data. However, the *k*-NN is not suitable for data with a substantial dimension. As the number of dimension increases, the dimension of the search space increases exponentially. For the *k*-NN to succeed, the data point needs to be close to the other points with respect to all the given axes; that is, the data needs to be dense. However, as the number of dimensions increases, it is not possible to store the data in a dense form. Another large problem that is faced by the algorithm is the value of *k* that should be chosen along with the ideal distance metric. Since we have thousands of distance metrics, it is not possible to be able to guess which one is the best. The problems introduced by these are solved in an algorithm known as Neighborhood Component Analysis (NCA) [57]. There is also some preprocessing that needs to be done to ensure that the algorithm does not favor any single feature just because it is close by a specific distance. For example, if we consider the Manhattan distance, data points which are almost equal to each other except one feature which has a high difference will be considered dissimilar, and data points where all the features are somewhat matching may be considered to be a better match.

7.3.3 SUPPORT VECTOR MACHINE (SVM)

Support vector machines [58] constitute one of the recent discoveries in the field of artificial intelligence. They are a set of supervised learning algorithms that can be used for classification, regression, as well as anomaly detection. They are mathematically intensive and can theoretically be shown how the best line is found out. SVMs are thus different from MLPs as they explain how they obtained the best fit line. They also show the contribution of each of the data points toward the best fit. Apart from that, they have a parameter near the margin by which the nearest data points from the line should be. They thus maximize the distance between the two nearest data points. The decision boundary created with this method is called the maximum margin hyperplane, and the points get separated into different classes using this hyperplane. In the case of a simple linear classifier, the SVM would give the best line such that the facts on one side belonged to one class and those on the other belonged to the other class. However, the data might not always be separable using a linear line. To tackle this, the SVM makes use of a kernel trick, which transforms the data into a higher-dimensional space in which the SVM can classify them using a single hyperplane. The data points that are the furthest from the hyperplane are usually the outliers or anomalous data, and thereby it can also help to improve the dataset. The values for SVMs are typically found using Lagrange Optimizers, which can be transformed into quadratic programming problems. These do not usually face problems like vanishing and exploding gradients like MLPs do.

7.3.4 RANDOM FOREST

A random forest [59] is created to deal with the defects of decision trees. Decision trees are very deterministic. So when given data, they always provide the same structure as in the decision trees. On the feature set, the lines along which they differentiate the features are thereby too abrupt. But we usually look for the fuzziness in the decision making as decisions are never yes or no based on the features. Decision trees are also very interpretable, and one can see the features and the decisions that the model has taken to come to the given output. They use grouping along with features that always maximize the Gini index. So, if the number of features present remains the same, it will keep on giving the same results. To introduce fuzziness into the decision making and to incorporate randomness, random forests were introduced. The forests in the name are used to refer to the fact that there are many different trees. Random forests are an ensemble of several decision trees. This type of ensemble is built on the bagging method. The main theory behind bagging is that the conglomeration of many learning algorithms increases the overall accuracy of the model. This is because each of the individual models might be overfitting onto a particular feature or have considerable depth, and in another case it might be too biased due to being very shallow. The errors tend to cancel each other when the models are combined. The trees thus support each other and help to remove the individual errors to make the overall forest more robust. An error might be present in a small percentage of trees, so when a large number of trees are considered, the negative effect of that error reduces even further. In random forests, the whole dataset is not considered. Instead, some of the rows and some of the features are randomly chosen to create individual decision trees. This diversity helps to make the decision boundary fuzzier and also helps to increase the overall performance. Random forests can also be used to interpret the importance of the individual features in the performance of the model. The random forests are initially trained on the whole dataset, and we get a corresponding score. After that, one of the features is randomly shuffled. This removes the correlation between the data samples and that corresponding feature, which is equivalent to removing the feature from the whole dataset. The scores for the random forest on the new shuffled dataset are now observed. The decrease in score corresponds to the importance of the feature. If the score decreases a lot, it is an essential feature, whereas if there is not much decrease, it is not that important of a feature. Thus, random forests help to improve the model and also make it more interpretable by giving an importance number to each of the features.

7.3.5 XGBoost

XGBoost [60] stands for extreme gradient boosting which is built onto decision trees, and relies on the method of boosting. Here short and straightforward decision trees are built during the first stage. Then similar trees are built on top of it iteratively. Each of the small decision trees that are built is a "weak learner" as it is very simple and has a high bias. XGboost then starts to build the learners iteratively until the stopping criteria are reached. The algorithm is very fast as there is parallelization to make use of all the cores of the CPU while training. It also has distributed computing for training extensive clusters models, and it has some parameters which use a cluster of machines. It includes gradient boosting to improve the overall performance of the model, and stochastic gradient descent to sub-sample the different rows, columns, and columns per split levels. The XGBoost also allows for the control of the overfitting data by having parameters for L1 and L2 regularization.

7.3.6 Convolutional Neural Network (CNN)

CNNs [61] are one of the networks that are used in deep learning. CNNs are now being used in all the fields of computer vision, speech recognition, natural language processing, and even reinforcement learning [62, 63]. They are composed of a matrix that acts as a filter. Filters are important in image processing as the images, on passing through the filters, can give useful information about the image. For example, edge detection filters are present, which, when passed through the image gives the edges of the images. When the filter is passed through the image, the matrix which makes up the filter is convoluted through the different parts of the image. The output of each of the filters is an activation map that denotes which parts of the image have been activated. For detecting whether a feature is present in a feature, it is enough to look at the activation map after the corresponding filter to that feature has been passed through the image. If the activation map is activated, it would mean that the image has that required feature present inside it. However, generating the feature by hand is very cumbersome and hard. That is why the CNNs filters are kept as trainable parameters that learn to generate the best filters which can find the most useful features present in the image. There are many ways in which the convolution can be done. After each convolution, the width and height of the image are reduced. To prevent this, the image can be padded with the required number of pixels. Apart from that, the filter can also skip some pixels instead of the convolution of one pixel after another. The number of pixels it will skip is called the stride of the filter. The filter size can also vary. Usually, at the start, the size of the filter or the kernel is seven, while later on it becomes three or five. There is also a dilation factor that is used in the CNN filters. After passing through the CNNs, the activation maps are then passed through an activation function. One of the most popular functions that is usually used together with a CNN is the rectified linear unit (ReLu). The ReLu keeps the maximum of the inputs and zero, that is, if the number is less than zero, the output becomes zero; otherwise, it remains unchanged. Pooling is utilized to reduce the height and width of the image by choosing the max among particular filters or average over filters with a stride of 2. Each of the filters generates a feature map. At each level of the architecture, several filters are used, and it generates a stack of feature maps. Each of these stacks constitutes a channel, and thus at each level we get a certain number of channels. The initial input image has three channels, which are known as "RGB" channels. They are also better at feature detection irrespective of where in the image they are located. If we wanted to create a dog detection algorithm, we would have to use extensive data augmentation techniques to put the image of the dog in all parts of the image and thereby train a MLP on it. But with a CNN, it is not necessary to do so. Due to the nature of the filters, the feature maps will be activated irrespective of where the dog is located, thereby saving a lot of data augmentation. Also, their size is relatively low as compared to an MLP. This is because the same filter is used to traverse through the whole image. Thus, the weights are shared, and it also helps in reducing the size of the model. In a normal pipeline, the images are passed through a CNN layer, passed through ReLu, and then after some iterations of this pooling is done, and the whole process repeats. After we have reached the required number of levels, we do a global pooling and flatten the layers to get the final features. Finally, we put a fully connected layer at the end to classify the images. We put raw images and labels into the CNNs with no information on the underlying data, and it almost gives us the state-of-the-art.

7.3.7 Neighborhood Component Analysis

We have already discussed the problems faced by k-NN and how NCA [64] is a possible solution for the same. The authors tried to learn a quadratic distance metric and a transformation such that the k-NN performed well in the newly transformed space.

Also, they used a soft classifier, which assigned a probability to each point. The authors then maximized the total number of points correctly classified and thereby obtained the necessary transformation. Also, the authors showed that by restricting the transformation matrix, they could make a low data rank and thereby decrease the overall dimension of the transformed space as well. This reduced the dimension of the data and also helped in clustering the different classes in a better way. This is the algorithm we have used on top of our features to see and observe if they result in any improvement of the performance of the model.

7.3.8 TRANSFER LEARNING

In this section, we talk about some of the architectures we used in our dataset, as well as why transfer learning provided such good results. Transfer learning is the process where the knowledge obtained by training the model on a dataset and task can be used to help with a task on another dataset. In deep learning architectures, this technique is quite widely used. As we have discussed in the subsection on CNNs, the different features of the CNNs learn different features from the images [65, 66]. The filters at the top learn very basic level features like edges and contours, while the feature maps at the bottom correspond to high-level features like the face of a person or the structure of a cat. It is obvious that the low-level features and some of the mid-level features will be reused in the new dataset as well, so the weights of those layers do not need to be changed that much. Transfer learning is so common that the pre-trained weights of most of the state-of-the-art architectures are available in most of the present common frameworks. The architectures of the VGG-16,19 [67], ResNet152 [68], DenseNet121 [46], InceptionV3 [69], MobileNetV2 [70], and Xception [71] were used as transfer learning (Figure 7.2).

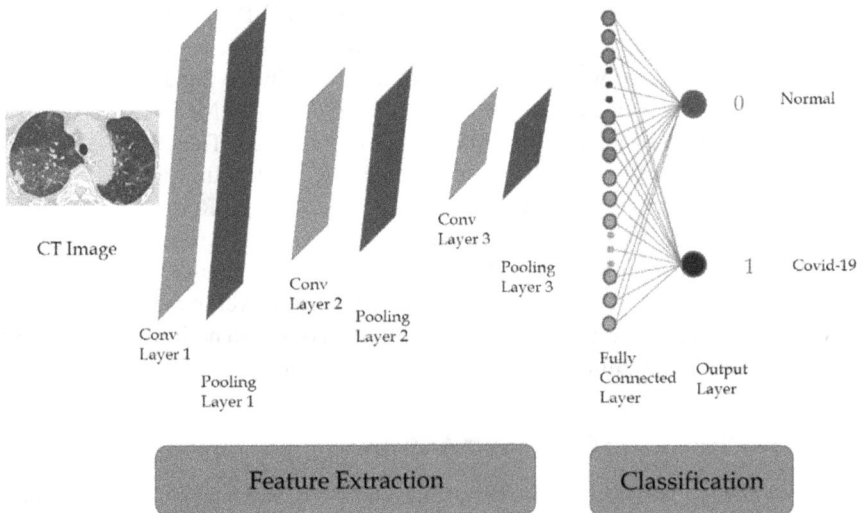

FIGURE 7.2 Deep learning classification model.

7.4 RESULTS AND DISCUSSIONS

7.4.1 DATASET

The dataset was taken from a publicly available dataset [15]. The correctness of the dataset was checked by a senior radiologist in Tonji Hospital, Wuhan, China, who had already diagnosed many COVID-19 patients. The dataset contained 349 CT images containing clinical findings of the novel virus from 216 patients. The dataset also had 397 CT scans of patients who didn't have COVID-19. The CT scan images from 760 preprints were collected from medrxiv and biorxiv. The height and width of the images were also not consistent and varied quite a lot. The minimum, average, and maximum width of the scans were 124, 383, and 1485, respectively; the heights were 153, 491, and 1853, respectively. The images were taken from preprints; thus, the image quality was a little degraded as the Hounsfield unit (HU) values were lost, the number of bits per pixel was reduced, and the resolution of the images was reduced. But they consulted a radiologist, and he confirmed that the factors did not significantly change the accuracy of the diagnosis. The dataset had already been divided into the training and testing split.

7.4.2 EXPERIMENTAL SETUP

To ensure that the models worked across mobile devices, we decreased the resolution of the images to 224 x 224. The images were both trained on the data using transfer learning and the classical methods. In the case of transfer learning, the feature maps at the output of the various networks were passed through a global average pooling, and a classifier was put on top of it. Later, the features of the images were extracted, and further experiments were done on them. NCA was performed to reduce their dimension and to check if that improved the performance of the model. The different architectures that were used were VGG16, VGG19, ResNet152V2, DenseNet121, InceptionV3, MobileNetV2, and Xception. NCA with different output feature dimensions was used to check how the number of dimensions would affect the overall performance of the models. In deep-learning-based models, reducing the dimension would create a bottleneck and reduce performance. However, in other methods where higher dimensions lead to problems, a decrease in the dimensions would help to get rid of the curse of dimensionality, and lead to an overall increase in performance. Also, whenever the model was trained on a deep-learning-based architecture, the training was done for 100 epochs, and the model with the best validation metric was saved.

7.4.3 PERFORMANCE METRICS

The model has approximately the same number of COVID-19 positive and negative CT images. So, the problem of class imbalance did not arise. However, since the data was very low, whenever the data was split into training and validation, the splitting was done such that there were more positive cases in the validation set. Now, if the model was biased toward recognizing more positive cases, it would give a very good score on the validation set, which was not correct. Thus, the model was trained on a five-fold cross-validation set to ensure that such a case did not arise. A five-fold cross-validation was used as it divided the non-testing dataset into an 80 to 20 ratio, which was expected from the machine learning pipeline. The model was run on four of the folds and validated on the remaining fold. The experiments were conducted five times so that each of the folds was used as a validation set once. The average results of the experiments were then taken and recorded. Since this was a classification problem in the medical domain, the area under the ROC curve (AUC) was selected. The overall accuracy, the cross-validation accuracy, the F1 scores, and the kappa scores were reported. The confusion matrix was also saved to take useful information from the data later.

7.4.4 HYPER-PARAMETERS SELECTION

A grid search was initially performed to get the best learning rate for the different algorithms, and they were continued later. The learning rate was 1×10^{-4}, and an Adam optimizer was used. During the selection of the model architecture, batch-normalization [72] and drop-out [73] were not added together as they lead to a degradation in performance [74]. Also, the effects of batch size on the overall performance of the model was seen. It was observed that a batch size of eight gave the best performance, and this was used in the consequent experiments.

7.4.5 EXPERIMENTAL RESULTS

Different architectures were trained with pre-trained weights. Apart from the popular architectures, we wished to see how the performance would change as the CNNs of higher levels were being used. We conducted experiments on the data for the same and came up with the following tables.

From Table 7.1, it can be observed that the DenseNet had the best performance while VGG19 had the worst performance among all the models. This might be

TABLE 7.1
The Calculated Results Using Different CNN Models

Model	Accuracy	F1 Measure	Kappa	ROC Area
CNN 1 Layer	0.73	0.7542	0.5067	0.83
CNN 3 Layer	0.704	0.7448	0.4482	0.7955
CNN 4 Layer	0.6836	0.7822	0.5684	0.8435
VGG16	0.681	0.817	0.633	0.891
VGG19	0.5	0.3	0	0.5
ResNet152V2	0.81	0.8248	0.6572	0.906
DenseNet121	0.8595	0.9199	0.8377	0.9839
Inception_v3	0.8512	0.8867	0.7619	0.9625
MobileNetV2	0.65	0.6353	0.3217	0.92519
Xception	0.8445	0.9128	0.8255	0.975

because, in DenseNet, the information from the front reaches the end almost directly due to the presence of dense blocks and dense layers. The same reason might be why the VGG-19 performed the worst. Due to the high number of layers with no residual skip in between the layers, the information could not get propagated throughout the model, and thus it gave a very poor performance.

Now we compare the methods of when the features from the pre-trained models were taken, and various machine learning algorithms were applied on top of these algorithms.

In Table 7.2, the data shows that the DenseNet 121 still gave the best performance in all of the metrics. The results were obtained after the NCA of different dimensions was performed. The best results were obtained when the NCA did not reduce the dimension. This can be explained by the fact that the reduction of the information through the NCA layer led to a decrease in the overall performance in this case.

In Table 7.3, it is seen that there is no consistency among the NCA dimension and the overall accuracy in the results shown by the RF. The best performance obtained by ResNet152 was when it had an NCA dimension of 50, while the best performance

TABLE 7.2

The Results of the Deep Features by Deploying ANN Classifier

Model	Accuracy	F1 Measure	Kappa	ROC Area
VGG16	0.8016	0.8273	0.6528	0.8944
VGG19	0.78	0.8121	0.6241	0.8495
ResNet152V2	0.7909	0.8381	0.6788	0.8902
DenseNet121	0.8468	0.8724	0.7449	0.9468
Inception_v3	0.7479	0.8187	0.6377	0.8786
MobileNetV2	0.768	0.8187	0.636	0.84981
Xception	0.7962	0.825	0.6529	0.888

TABLE 7.3

The Results of the Deep Features by Deploying RF Classifier

Model	Accuracy	F1 Measure	Kappa	ROC Area
VGG16	0.7976	0.7854	0.5703	0.8974
VGG19	0.788	0.8325	0.6616	0.9049
ResNet152V2	0.7708	0.8185	0.6301	0.9188
DenseNet121	0.84	0.8595	0.7107	0.9103
Inception_v3	0.7359	0.783	0.562	0.867
MobileNetV2	0.771	0.7978	9.5939	0.865
Xception	0.7855	0.8258	0.6453	0.8712

from DenseNet121 was when it had a dimension of 1024. This showed that dimension does not play a very significant role when combined with random forests.

A DenseNet121 performed better than the others again, with the results shown by the XGBoost algorithm in Table 7.4. Surprisingly these DenseNet121 results correspond to an NCA dimension of 70, which is different from the results seen earlier.

In the results given in Table 7.5, a SVM was used as a classifier, and the DenseNet121 again performed the best in accuracy and F1 measure, but the VGG16 had the highest ROC area. It was seen that the NCA dimensions of most of the architectures which gave the best results was 70, which might suggest that a reduction in dimensionality helped the model more.

The results from k-NN showed that the DenseNet121 performed the best in all the metrics. The dimension corresponding to this was 100, which was lower than the original dimension of 1024. In VGG19, the dimension was 70, which suggested that a lower dimension might lead to an improvement.

From the overall results in Table 7.6, it is seen that k-NN performed the best with most of the architectures. This shows that the NCA indeed transformed the features into another space where it was easier to separate them using k-NN and other algorithms.

TABLE 7.4
The Results of the Deep Features by Deploying XGBoost Classifier

Model	Accuracy	F1 Measure	Kappa	ROC Area
VGG16	0.787	0.8382	0.6702	0.8688
VGG19	0.7908	0.8521	0.6988	0.90641
ResNet152V2	0.77618	0.8186	0.6388	0.8808
DenseNet121	0.828	0.84	0.68	0.874
Inception_v3	0.756	0.7865	0.5722	0.8613
MobileNetV2	0.7775	0.8112	0.6252	0.866432
Xception	0.775	0.8187	0.637	0.8867

TABLE 7.5
The Results of the Deep Features by Deploying SVM Classifier

Model	Accuracy	F1 Measure	Kappa	ROC Area
VGG16	0.832	0.8784	0.7551	0.9524
VGG19	0.8257	0.8391	0.678	0.9265
ResNet151V2	0.823	0.8391	0.6767	0.901
DenseNet121	0.88	0.9	0.7999	0.945
Inception_v3	0.7908	0.7912	0.5764	0.879
MobileNetV2	0.8057	0.8722	0.7421	0.9311
Xception	0.8096	0.8527	0.6975	0.9366

TABLE 7.6

The Results of the Deep Features by Deploying *k*-NN Classifier

Model	Accuracy	F1 Measure	Kappa	ROC Area
VGG16	0.8578	0.8867	0.7721	0.88647
VGG19	0.8204	0.8724	0.742	0.8705
ResNet152V2	0.78	0.7807	0.5561	0.7805
DenseNet121	0.886	0.9127	0.8255	0.91288
Inception_v3	0.764	0.7844	0.5723	0.787
MobileNetV2	0.815	0.8522	0.7051	0.8531
Xception	0.82035	0.8523	0.7047	0.8523

7.4.6 DISCUSSION

The results showed that end to end training of the neural networks, although more costly, led to better performance metrics. The results also showed that when the features were directly fed into the network, it resulted in worse results, but when they were transformed into another hyperspace using an algorithm, like NCA, it improved the overall metric. The DenseNet121 consistently performed the best in all of the experiments, which seem to suggest that the weights were well optimized with a skip connection type architecture. Our final classifier thus had an accuracy of 0.886 when the features were taken from the images and transformed into a new plane, and *k*-NN was applied to them. However, it somewhat reduces the ROC area. The maximum ROC area was obtained at 0.9839 again by the DenseNet architecture when transfer learning was done directly. Most of the models showed that a reduction in the dimension did not cause a significant change in the accuracy of the model.

7.5 CONCLUSION

The world is facing one of the most significant challenges right now, and the world economy is also on the verge of collapse. It showcases the importance of more research in the field of medicine and why experts in multiple fields need to come together to fight the problems faced to come up with a solution. The most important thing that one can do is to avoid being in direct contact with the virus, and follow the social distancing guidelines given by the government. On signs of infection, people must be screened as quickly as possible to test whether they are positive or not. If they test positive for the virus, they should isolate themselves to prevent others from getting infected. Among the methods that are currently used, CT scans give one of the most detailed analyses of the problems in the specific areas of the lungs. We produced our results using a very small amount of data, which were not of a good quality as they were extracted from preprints and were not actual medical data. But the high scores obtained with these data highlight the importance of artificial intelligence in general. The high accuracy scores on less data might even stop a pandemic by early screening of the patients. The multi-disciplinary methodologies will help to stop the spread of the disease and move toward a healthier future.

REFERENCES

1. March 14 JB-LSE-C, 2020. 1st known case of coronavirus traced back to November in China [Internet]. livescience.com. [cited 2020 Sep 1]. Available from: https://www.livescience.com/first-case-coronavirus-found.html.
2. COVID-19 Map [Internet]. Johns Hopkins Coronavirus Resour. Cent. [cited 2020 Sep 1]. Available from: https://coronavirus.jhu.edu/map.html.
3. Timeline: WHO's COVID-19 response [Internet]. [cited 2020 Sep 1]. Available from: https://www.who.int/emergencies/diseases/novel-coronavirus-2019/interactive-timeline.
4. Wang M, Cao R, Zhang L, et al. Remdesivir and chloroquine effectively inhibit the recently emerged novel coronavirus (2019-nCoV) in vitro. *Cell Res.* 2020;30:269–271.
5. Yang X, Yu Y, Xu J, et al. Clinical course and outcomes of critically ill patients with SARS-CoV-2 pneumonia in Wuhan, China: a single-centered, retrospective, observational study. *Lancet Respir Med.* 2020;8:475–481.
6. Forecasting COVID-19 impact on hospital bed-days, ICU-days, ventilator days and deaths by US state in the next 4 months [Internet]. *Inst. Health Metr. Eval.* 2020 [cited 2020 Sep 2]. Available from: http://www.healthdata.org/research-article/forecasting-covid-19-impact-hospital-bed-days-icu-days-ventilator-days-and-deaths.
7. Culp WC. Coronavirus disease 2019: In-home isolation room construction. *Aa Pract.* 2020;14:e01218.
8. Ai T, Yang Z, Hou H, et al. Correlation of chest CT and RT-PCR testing for coronavirus disease 2019 (COVID-19) in China: A report of 1014 cases. *Radiology.* 2020;296:E32–E40.
9. Lauer SA, Grantz KH, Bi Q, et al. The incubation period of coronavirus disease 2019 (COVID-19) from publicly reported confirmed cases: Estimation and application. *Ann Intern Med.* 2020;172:577–582.
10. Muhammad LJ, Islam MdM, Usman SS, et al. Predictive data mining models for novel coronavirus (COVID-19) infected patients' recovery. *SN Comput Sci.* 2020;1:206.
11. Li Y, Xia L. Coronavirus disease 2019 (COVID-19): Role of chest CT in diagnosis and management. *AJR Am J Roentgenol.* 2020;214:1280–1286.
12. Islam MdZ, Islam MdM, Asraf A. A combined deep CNN-LSTM network for the detection of novel coronavirus (COVID-19) using X-ray images. *Inform Med Unlocked.* 2020;20:100412.
13. Fang Y, Zhang H, Xie J, et al. Sensitivity of chest CT for COVID-19: Comparison to RT-PCR. *Radiology.* 2020;296:E115–E117.
14. Van Gansbeke W, Vandenhende S, Georgoulis S, et al. SCAN: Learning to classify images without labels. ArXiv200512320 Cs [Internet]. 2020 [cited 2020 Sep 6]; Available from: http://arxiv.org/abs/2005.12320.
15. Yang X, He X, Zhao J, et al. COVID-CT-dataset: A CT scan dataset about COVID-19. ArXiv200313865 Cs Eess Stat [Internet]. 2020 [cited 2020 Sep 6]; Available from: http://arxiv.org/abs/2003.13865.
16. Irvin J, Rajpurkar P, Ko M, et al. CheXpert: A large chest radiograph dataset with uncertainty labels and expert comparison. ArXiv190107031 Cs Eess [Internet]. 2019 [cited 2020 Sep 6]; Available from: http://arxiv.org/abs/1901.07031.
17. Kanne JP, Little BP, Chung JH, et al. Essentials for radiologists on COVID-19: An Update—Radiology Scientific Expert Panel. *Radiology.* 2020;296:E113–E114.
18. Touvron H, Vedaldi A, Douze M, et al. Fixing the train-test resolution discrepancy: FixEfficientNet. ArXiv200308237 Cs [Internet]. 2020 [cited 2020 Sep 6]; Available from: http://arxiv.org/abs/2003.08237.

19. Sun W, Chen Z. Learned image downscaling for upscaling using content adaptive resampler. ArXiv190712904 Cs Eess [Internet]. 2019 [cited 2020 Sep 6]; Available from: http://arxiv.org/abs/1907.12904.

20. Tan M, Pang R, Le QV. EfficientDet: Scalable and efficient object detection. ArXiv191109070 Cs Eess [Internet]. 2020 [cited 2020 Sep 6]; Available from: http://arxiv.org/abs/1911.09070.

21. Edunov S, Ott M, Auli M, et al. Understanding back-translation at scale. ArXiv180809381 Cs [Internet]. 2018 [cited 2020 Sep 6]; Available from: http://arxiv.org/abs/1808.09381.

22. He K, Zhang X, Ren S, et al. Delving deep into rectifiers: Surpassing human-level performance on ImageNet classification. ArXiv150201852 Cs [Internet]. 2015 [cited 2020 Sep 6]; Available from: http://arxiv.org/abs/1502.01852.

23. Hu Q, Guan H, Sun Z, et al. Early CT features and temporal lung changes in COVID-19 pneumonia in Wuhan, China. *Eur J Radiol.* 2020;128:109017.

24. CDC. Communities, Schools, Workplaces, & Events [Internet]. *Cent. Dis. Control Prev.* 2020 [cited 2020 Sep 6]. Available from: https://www.cdc.gov/coronavirus/2019-ncov/community/general-business-faq.html.

25. Panwar H, Gupta PK, Siddiqui MK, et al. A deep learning and Grad-CAM based color visualization approach for fast detection of COVID-19 cases using chest X-ray and CT-scan images. *Chaos Solitons Fract.* 2020;110190.

26. Pirouz B, Shaffiee Haghshenas S, Shaffiee Haghshenas S, et al. Investigating a serious challenge in the sustainable development process: Analysis of confirmed cases of COVID-19 (new type of coronavirus) through a binary classification using artificial intelligence and regression analysis. *Sustainability.* 2020;12:2427.

27. Pirouz B, Shaffiee Haghshenas S, Pirouz B, et al. Development of an assessment method for investigating the impact of climate and urban parameters in confirmed cases of COVID-19: A new challenge in sustainable development. *Int J Environ Res Public Health.* 2020;17(8).

28. Chen Y-C, Lu P-E, Chang C-S, et al. A time-dependent SIR model for COVID-19 with undetectable infected persons. ArXiv200300122 Cs Q-Bio Stat [Internet]. 2020 [cited 2020 Sep 6]; Available from: http://arxiv.org/abs/2003.00122.

29. Guan W, Liang W, Zhao Y, et al. Comorbidity and its impact on 1590 patients with COVID-19 in China: a nationwide analysis. *Eur Respir J* [Internet]. 2020 [cited 2020 Sep 6];55. Available from: https://www.ncbi.nlm.nih.gov/pmc/articles/PMC7098485/.

30. Sun Q, Qiu H, Huang M, et al. Lower mortality of COVID-19 by early recognition and intervention: experience from Jiangsu Province. *Ann Intensive Care.* 2020;10:33.

31. Li L, Qin L, Xu Z, et al. Using artificial intelligence to detect COVID-19 and community-acquired pneumonia based on pulmonary CT: Evaluation of the diagnostic accuracy. *Radiology.* 2020;296:E65–E71.

32. Zhang K, Liu X, Shen J, et al. Clinically applicable AI system for accurate diagnosis, quantitative measurements, and prognosis of COVID-19 pneumonia using computed tomography. *Cell.* 2020;181:1423–1433.e11.

33. Kang H, Xia L, Yan F, et al. Diagnosis of coronavirus disease 2019 (COVID-19) with structured latent multi-view representation learning. *IEEE Trans Med Imag.* 2020;39:2606–2614.

34. Milletari F, Navab N, Ahmadi S-A. V-Net: Fully convolutional neural networks for volumetric medical image segmentation. ArXiv160604797 Cs [Internet]. 2016 [cited 2020 Sep 6]; Available from: http://arxiv.org/abs/1606.04797.

35. Bai HX, Wang R, Xiong Z, et al. Artificial intelligence augmentation of radiologist performance in distinguishing COVID-19 from pneumonia of other origin at chest CT. *Radiology.* 2020;296:E156–E165.

36. Tan M, Le QV. EfficientNet: Rethinking model scaling for convolutional neural networks. ArXiv190511946 Cs Stat [Internet]. 2019 [cited 2020 Sep 6]; Available from: http://arxiv.org/abs/1905.11946.

37. Fan D-P, Zhou T, Ji G-P, et al. Inf-Net: Automatic COVID-19 lung infection segmentation from CT images. ArXiv200414133 Cs Eess [Internet]. 2020 [cited 2020 Sep 6]; Available from: http://arxiv.org/abs/2004.14133.

38. Ardakani AA, Kanafi AR, Acharya UR, et al. Application of deep learning technique to manage COVID-19 in routine clinical practice using CT images: Results of 10 convolutional neural networks. *Comput Biol Med.* 2020;121:103795.

39. Voulodimos A, Protopapadakis E, Katsamenis I, et al. Deep learning models for COVID-19 infected area segmentation in CT images. *medRxiv.* 2020;2020.05.08.20094664.

40. Ronneberger O, Fischer P, Brox T. U-Net: Convolutional networks for biomedical image segmentation. ArXiv150504597 Cs [Internet]. 2015 [cited 2020 Sep 6]; Available from: http://arxiv.org/abs/1505.04597.

41. Long J, Shelhamer E, Darrell T. Fully convolutional networks for semantic segmentation. ArXiv14114038 Cs [Internet]. 2015 [cited 2020 Sep 6]; Available from: http://arxiv.org/abs/1411.4038.

42. Hu R, Ruan G, Xiang S, et al. Automated diagnosis of COVID-19 using deep learning and data augmentation on chest CT. *medRxiv.* 2020;2020.04.24.20078998.

43. Yousefzadeh M, Esfahanian P, Movahed SMS, et al. ai-corona: Radiologist-assistant deep learning framework for COVID-19 diagnosis in chest CT scans. *medRxiv.* 2020;2020.05.04.20082081.

44. Song Y, Zheng S, Li L, et al. Deep learning enables accurate diagnosis of novel coronavirus (COVID-19) with CT images. *medRxiv.* 2020;2020.02.23.20026930.

45. Wang S, Zha Y, Li W, et al. A fully automatic deep learning system for COVID-19 diagnostic and prognostic analysis. *medRxiv.* 2020;2020.03.24.20042317.

46. Huang G, Liu Z, van der Maaten L, et al. Densely connected convolutional networks. ArXiv160806993 Cs [Internet]. 2018 [cited 2020 Aug 30]; Available from: http://arxiv.org/abs/1608.06993.

47. Lin T-Y, Dollár P, Girshick R, et al. Feature pyramid networks for object detection. ArXiv161203144 Cs [Internet]. 2017 [cited 2020 Sep 6]; Available from: http://arxiv.org/abs/1612.03144.

48. Kumar D, Klefsjö B. Proportional hazards model: a review. *Reliab Eng Syst Saf.* 1994;44:177–188.

49. Mei X, Lee H-C, Diao K, et al. Artificial intelligence–enabled rapid diagnosis of patients with COVID-19. *Nat Med.* 2020;26:1224–1228.

50. Zhou L, Li Z, Zhou J, et al. A rapid, accurate and machine-agnostic segmentation and quantification method for CT-based COVID-19 diagnosis. *IEEE Trans Med Imag.* 2020;39:2638–2652.

51. Han Z, Wei B, Hong Y, et al. Accurate screening of COVID-19 using attention-based deep 3D multiple instance learning. *IEEE Trans Med Imag.* 2020;39:2584–2594.

52. Leshno M, Lin VYa, Pinkus A, et al. Multilayer feedforward networks with a nonpolynomial activation function can approximate any function. *Neural Netw.* 1993;6:861–867.

53. Agatonovic-Kustrin S, Beresford R. Basic concepts of artificial neural network (ANN) modeling and its application in pharmaceutical research. *J Pharm Biomed Anal.* 2000;22:717–727.

54. Rumelhart DE, Hinton GE, Williams RJ. Learning representations by back-propagating errors. *nature.* 1986;323:533.

55. Guo G, Wang H, Bell D, et al. KNN Model-based approach in classification. In: Meersman R, Tari Z, Schmidt DC (eds.), *Move Meaningful Internet Syst 2003 CoopIS DOA ODBASE.* Berlin, Heidelberg: Springer; 2003. pp. 986–996.

56. Alpaydin E. Voting over multiple condensed nearest neighbors. *Lazy Learn.* Springer; 1997. pp. 115–132.

57. Yang W, Wang K, Zuo W. Neighborhood component feature selection for high-dimensional data. *J Comput.* 2012;7(1):161–168.

58. Evgeniou T, Pontil M. Support vector machines: Theory and applications. In: Paliouras G, Karkaletsis V, Spyropoulos CD (eds.), *Mach Learn Its Appl Adv Lect* [Internet]. Berlin, Heidelberg: Springer; 2001 [cited 2020 Aug 30]. pp. 249–257.

59. Ali J, Khan R, Ahmad N, et al. Random forests and decision trees. *Int J Comput Sci Issues.* 2012;9(5):272.

60. Santhanam R, Uzir N, Raman S, et al. Experimenting XGBoost algorithm for prediction and classification of different datasets. *International Journal of Control Theory and Applications.* 2016; 9:651–62.

61. Gu J, Wang Z, Kuen J, et al. Recent advances in convolutional neural networks. ArXiv151207108 Cs [Internet]. 2017 [cited 2020 Sep 7]; Available from: http://arxiv.org/abs/1512.07108.

62. Khare SK, Bajaj V. Time-frequency representation and convolutional neural network-based emotion recognition. *IEEE Trans Neural Netw Learn Syst.* 2020.

63. Demir F, Şengür A, Bajaj V, et al. Towards the classification of heart sounds based on convolutional deep neural network. *Health Inf Sci Syst.* 2019;7:16.

64. Goldberger J, Hinton GE, Roweis ST, et al. Neighbourhood components analysis. In: Saul LK, Weiss Y, Bottou L (eds.), *Adv Neural Inf Process Syst 17* [Internet]. MIT Press; 2005 [cited 2020 Sep 7]. pp. 513–520. Available from: http://papers.nips.cc/paper/2566-neighbourhood-components-analysis.pdf.

65. Zeiler MD, Fergus R. Visualizing and understanding convolutional networks. ArXiv13112901 Cs [Internet]. 2013 [cited 2020 Sep 7]; Available from: http://arxiv.org/abs/1311.2901.

66. Demir F, Bajaj V, Ince MC, et al. Surface EMG signals and deep transfer learning-based physical action classification. *Neural Comput Appl.* 2019;31:8455–8462.

67. Simonyan, K., & Zisserman, A. (2014). Very deep convolutional networks for large-scale image recognition. *arXiv preprint arXiv:1409.1556.*

68. He K, Zhang X, Ren S, et al. Deep residual learning for image recognition. ArXiv151203385 Cs [Internet]. 2015 [cited 2020 Aug 30]; Available from: http://arxiv.org/abs/1512.03385.

69. Szegedy C, Vanhoucke V, Ioffe S, et al. Rethinking the inception architecture for computer vision. ArXiv151200567 Cs [Internet]. 2015 [cited 2020 Aug 30]; Available from: http://arxiv.org/abs/1512.00567.

70. Howard AG, Zhu M, Chen B, et al. MobileNets: Efficient convolutional neural networks for mobile vision applications. ArXiv170404861 Cs [Internet]. 2017 [cited 2020 Sep 7]; Available from: http://arxiv.org/abs/1704.04861.

71. Chollet F. Xception: Deep learning with depthwise separable convolutions. ArXiv161002357 Cs [Internet]. 2017 [cited 2020 Aug 30]; Available from: http://arxiv.org/abs/1610.02357.

72. Ioffe S, Szegedy C. Batch normalization: Accelerating deep network training by reducing internal covariate shift. ArXiv150203167 Cs [Internet]. 2015 [cited 2020 Sep 7]; Available from: http://arxiv.org/abs/1502.03167.

73. Srivastava N, Hinton G, Krizhevsky A, et al. Dropout: A simple way to prevent neural networks from overfitting. *J Mach Learn Res.* 2014;15:1929–1958.

74. Chen G, Chen P, Shi Y, et al. Rethinking the usage of batch normalization and drop-out in the training of deep neural networks. ArXiv190505928 Cs Stat [Internet]. 2019 [cited 2020 Sep 7]; Available from: http://arxiv.org/abs/1905.05928.

8 Suspicious Region Diagnosis in the Brain
A Guide to Using Brain MRI Sequences

Ayca Kirimtat, Ondrej Krejcar,
Ali Selamat, Enrique Herrera-Viedma,
Kamil Kuca, and Anis Yazidi

CONTENTS

8.1 INTRODUCTION

The brain is the furthermost multifaceted organ in the human body, and it is difficult to comprehend. Brain tumors are one of the most life-threatening diseases, and they can exist in different forms and within different types of tissues; therefore, there should be a classification method for understanding the behavior of these tissues. Medical image analysis is a highly complicated and challenging task because of the unhomogenous pixels and complicated backgrounds on the images. Additionally, manual diagnosis methods are time consuming and necessitate high domain expertise. Thus, with the dramatic increase in the number of patients suffering from tumor, researchers have dedicated their time to study these areas with the aim of finding reliable solutions and methods [1].

Magnetic resonance imaging (MRI) is a non-invasive imaging modality for diagnostic purposes. For surgery purposes, doctors might take the medical image as a guiding tool. Tumor segmentation is a fundamental task for tumor localization, and there have been a large number of works dealing with tumor segmentation in the brain area of the literature. Moreover, in order to generate tumor masks, there are image modality sequences, so-called T1W1, T1C, T2W2, and Fluid-attenuated inversion recovery (FLAIR), which uncovers enhanced tumor, edema, necrosis, and non-enhanced tumors within the whole brain area [1].

Since MRI scanners are highly complicated when used for abnormality detection in the brain area. Different lesion characteristics and features should be identified by the support of a differential diagnosis [2]. Therefore, this study aims to define practical procedures to correctly classify brain lesions and to enlighten scientific community on the important approaches found in the literature. When it comes to abnormality detection in the human brain, there exist numerous methods for the classification and segmentation of MRI images. We highlight these methods in detail in the following section.

This chapter is organized as follows: In Section 8.2, we compare and discuss related works on classification methods and sequence types for brain MRI images. In Section 8.3, brain MRI sequences are described and explained in detail with brain images from the hospital at the University of Hradec Kralove. Section 8.4 discusses the different sequence types and the figures given in the previous section. We give concluding remarks in the last section of this book chapter.

8.2 RELATED WORKS

Lately, a high number of scientific works have been dedicated to the segmentation and classification of tumors in the brain area. For this purpose, MR images have been widely used for making precise detections of any brain tumors [3]. Most of these studies were performed using various machine learning algorithms using MR images as modality. We show some of them in detail in Table 8.1. For instance, in the work by Zacharaki et al. [4], the main aim of the study was to study pattern classification techniques for various brain tumor types. The T1-weighted sequence was also used for the MRI images. A computer-based classification was used for differentiating the different MRI images. Moreover, the proposed method contained different phases such as a region of interest (ROI) description, image extraction, image selection and classification, and the image selection was using support vector machines.

Another study reported in [5] proposed a novel method to characterize tumors in the brain area. The originality of the proposed technique emanated from the usage of soft computing techniques, namely, fuzzy cognitive maps (FCMs). In addition to this, the grading model classification capacity of the FCMs was heightened by an activation Hebbian algorithm. In conclusion, the power of the proposed method was shown by the efficiency of the decision process that would be useful in clinical practices to characterize brain tumors.

According to the study in [6], an original convolutional neural network model was used for brain tumor detection through MR image classifications. BrainMRNet comprised an attention module based on hypercolumn technique, while a residual block was developed using a CNN. The foremost target of the work was the choice of the best and the most competent images in the collection. MRI images were selected with the aim of detecting brain tumors using a novel convolutional neural network model. The proposed BrainMRNet reached 96.05% arrangement achievement through the MR images of the brain area.

TABLE 8.1

Related Works on Existing Segmentation and Classification Methods and MRI Sequences for Brain Tumor

Reference	Year	Method	MRI Sequence	Journal/Conference
[5]	2008	Fuzzy C-Means	N/A	Applied Soft Computing
[4]	2009	Multiparametric framework	T1	Magnetic Resonance in Medicine
[9]	2012	Fuzzy clustering	FLAIR	Procedia Engineering
[10]	2017	Non-negative matrix factorization	T2	Computers in Biology and Medicine
[11]	2018	Deep Neural Networks [12], [13]	T1, T1C, T2 and FLAIR	Computer Methods and Programs in Biomedicine
[14]	2018	Machine Learning	T1, FLAIR	Computers in Biology and Medicine
[15]	2019	CNN and SVM	T1	Medical Hypotheses
[16]	2019	Fuzzy C-Means	T1	Journal of King Saud University-Computer and Information Sciences
[17]	2019	Convolutional Neural Network [18]	T1	Computerized Medical Imaging and Graphics
[19]	2019	Fuzzy C-Means	T1, T2, FLAIR	Biocybernetics and Biomedical Engineering
[20]	2019	Convolutional Neural Network	T1, T1C, T2 and FLAIR	Biocybernetics and Biomedical Engineering
[21]	2019	Convolutional Neural Network, Genetic Algorithm	T1	Biocybernetics and Biomedical Engineering
[7]	2020	Convolutional Neural Network	N/A	Medical Hypotheses
[6]	2020	BrainMRNet	N/A	Medical Hypotheses
[8]	2020	Convolutional Neural Network	T1C, FLAIR	Computers in Biology and Medicine
[1]	2020	K-means	T2, FLAIR	Materials Today: Proceedings
[3]	2020	Whale Harris Hawks optimization	T1, T1C, T2 and FLAIR	Journal of King Saud University-Computer and Information Sciences
[22]	2020	Deep Neural Networks	FLAIR	Journal of King Saud University-Engineering Sciences
[23]	2020	Convolutional Neural Network	T1, T2, FLAIR	Computers in Biology and Medicine
[24]	2020	Computer-aided Diagnosis	T1	Biomedical Signal Processing and Control
[25]	2020	Convolutional Neural Network	T1, T1C, T2 and FLAIR	Procedia Computer Science

(Continued)

TABLE 8.1 (CONTINUED)

Related Works on Existing Segmentation and Classification Methods and MRI Sequences for Brain Tumor

Reference	Year	Method	MRI Sequence	Journal/Conference
[26]	2020	Deep Learning	T1C	Biomedical Signal Processing and Control
[27]	2020	Convolutional Neural Network	T1, T1C, T2 and FLAIR	Biocybernetics and Biomedical Engineering
[28]	2020	Deep Neural Networks	T1, T1C, T2 and FLAIR	Pattern Recognition Letters
[29]	2020	Convolutional Neural Network	T1, T1C, T2 and FLAIR	Medical Image Analysis
[30]	2020	Discrete Wavelet Transform	T1, T1C, T2 and FLAIR	Biomedical Signal Processing and Control
[31]	2020	Bayesian Fuzzy Clustering	N/A	Biocybernetics and Biomedical Engineering
[32]	2020	Fuzzy C-Means, CNN	T1	Medical Hypotheses
[33]	2020	Discrete Wavelet Transform	T1, T1C, T2 and FLAIR	Pattern Recognition Letters
[34]	2020	Convolutional Neural Network	N/A	Expert Systems with Applications
[35]	2020	Convolutional Neural Network	N/A	Neural Networks
[36]	2020	K-Nearest Neighbor	N/A	Materials Today: Proceedings
[37]	2020	Convolutional Neural Network	T2	Pattern Recognition Letters
[38]	2020	Deep Neural Networks	T2	Medical Hypotheses

According to the study found in [7], the detection of brain tumors using computer-aided systems is a significant step for specialists. In comparison to traditional methods, in this article, a CNN model was used with the aim of analysis progression. In conclusion, the model had a high performance and classified the brain tumor images efficiently in comparison with the other models.

A study in [8] highlighted that MR technique provides non-invasive detailed brain tumor images. Also, automatic tumor segmentation was quite necessary for treatment planning. In order to reach these statements, a deep learning model was developed by the authors using a CNN-based model. In the segmentation models, T1-precontrast, T1-postcontrast, and FLAIR MRI images of 110 patients were used for training and evaluation. In conclusion, the authors proved with this work that deep learning models for MRI images provide a non-invasive method for tumor detection and segmentation.

Figure 8.1 shows a pie chart of classification and segmentation methods for brain MRIs. This pie chart belongs to studies between 2008 and 2020. Most of them belong to 2020. Among these methods, the most used was a CNN with 44% share.

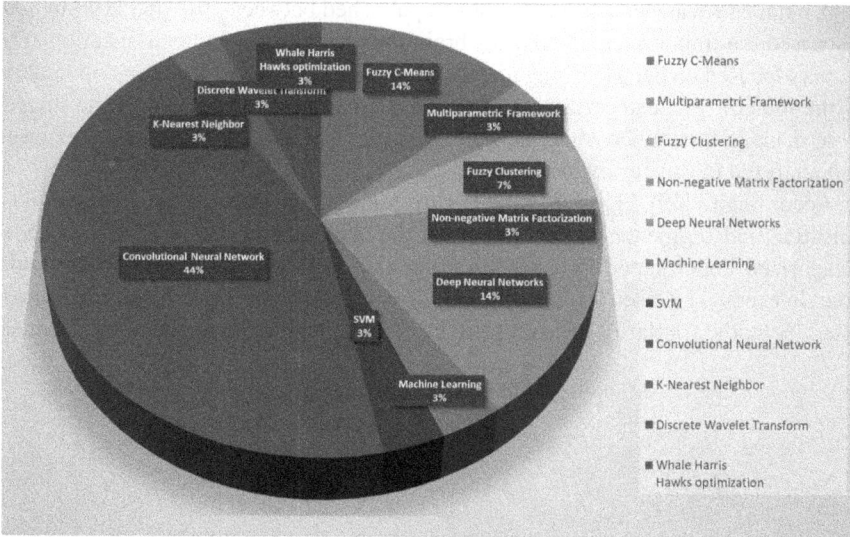

FIGURE 8.1 The pie chart for classification and segmentation methods for a brain tumor in MRI.

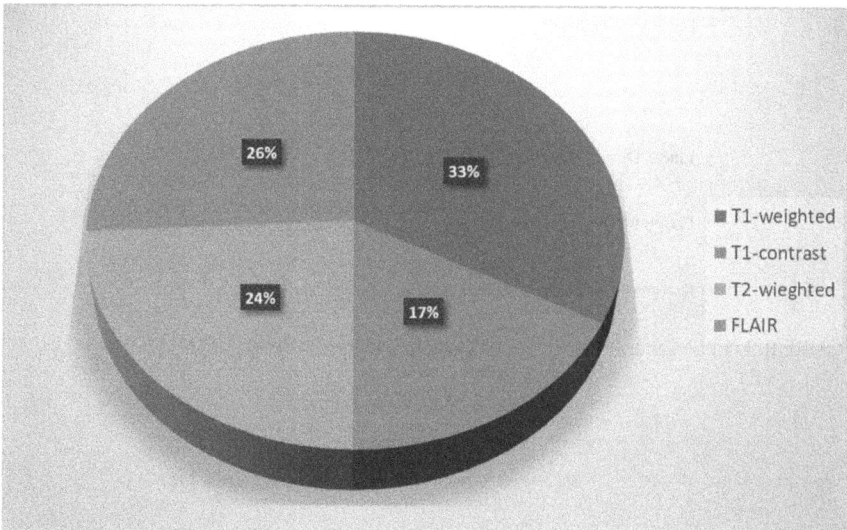

FIGURE 8.2 The pie chart for different brain MRI sequences.

The second was the FCMs and Deep Neural Networks (DNN) with 14% share. Some studies used specific methods for classification and segmentation, such as Whale Harris Hawks optimization and the non-negative matrix factorization.

On the other hand, Figure 8.2 introduces a pie chart for different brain MRI sequences used in earlier works. According to the pie chart, the most used sequence type was T1-weighted with 33%, the second most used was T2-weighted with 24%.

A detailed review article for the papers published between 2014 and 2019 by [39] discussed the importance of MRI for brain tumor diagnosis and classification. The authors focused on deep learning and metaheuristic approaches for the segmentation of the images. The categorization of segmentation and classification of the images were made in this review study as well. In total, 69 research articles were reviewed and analyzed in this review article.

According to [39], Figure 8.3 and Figure 8.4 show the general categories of classification and segmentation methods, and the classification methods are basically categorized under supervised and unsupervised methods. In supervised methods, domain expertise is used for labeling the correct class; on the other hand, in unsupervised methods, similarity measures are used for grouping the images. For instance,

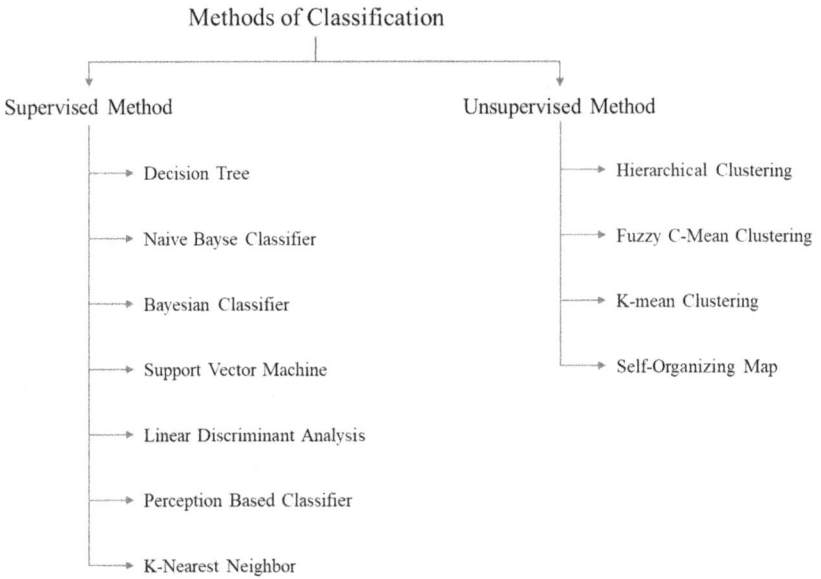

FIGURE 8.3 The general categories of classification procedures for MRI brain tumor [39].

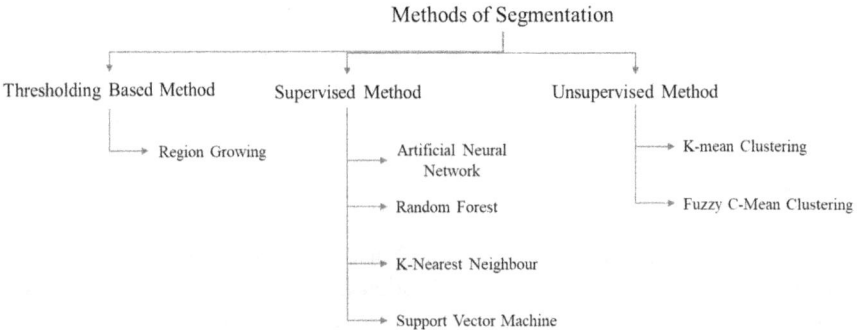

FIGURE 8.4 The general categories of segmentation procedures for MRI brain tumor [39].

in the study by [40], the authors suggested an automatic technique in order to easily recognize cancerous and non-cancerous regions in the brain MRI. A support vector machine classifier was implemented for comparison of the proposed framework of the features. In conclusion, the proposed approach could be easily applied for the identification of brain tumors in comparison with the existing methods in the literature.

Another study reported in [41] presented a semi-supervised learning method with the aim of training the DNN-based labeled images of so-called atlases. The authors implemented their technique on two datasets: open human brain and original marmoset brain images. In conclusion, the authors achieved superior and stable segmentation in comparison with the existing DNN-based methods. As a future study, the authors mentioned that their semi-supervised learning model could be improved and developed into a more sophisticated model.

According to [42], a deep learning-based framework was proposed with the aim of image segmentation of a brain tumor. The authors' proposed method involves the Stationary Wavelet Transform and a novel Growing Convolution Neural Network. Based on the experimental results obtained from this research, the presented methods achieved a great performance for brain tumor image segmentation in terms of some of the performance parameters, such as accuracy, PSNR, and MSE.

Recent studies have demonstrated a noteworthy growth for medical imaging in relation to brain tumors. Since the brain tumor is one of the most curable diseases in the human body, if recognized on time, more research should be carried out regarding this topic. MRI is one of the most extensively used methods for detecting abnormalities in the brain area.

8.3 BRAIN MRI SEQUENCES

Since MRI is extensively used in neurosurgery and neurology, various MRI sequences exist for brain imaging. T1W1 and T2W2 are two very common imaging modalities, which have two different relaxation times. Additionally, they are named T1W1 and T2W2 scanning because both are related to echo time and repetition time. For example, T1W1 images are produced from short repetitions and echo times. Moreover, the contrast and brightness of T1W1 images are formed by the specialities of the T1W1 tissue. On the contrary, T2W2 images are generated from longer repetition and echo times, and again the contrast and brightness of T2W2 images are formed by the specialities of the T2W2 tissue. Another very important sequence type is FLAIR, which resembles T2W2 images much more rather than T1W1 images. However, they are generated from much longer repetition and echo times [43, 44]. Based on a previous conference paper by the authors [43], the defined brain MRI modalities with images and sequence types are shown in Table 8.2, Table 8.3, Table 8.4, and Table 8.5. The images in these tables were attained from the Department of Radiology, University Hospital of Hradec Kralove, which included a MRI axial FLAIR sequence of a patient with a brain tumor.

The main structures of T1W1, T1C, T2W2, and FLAIR sequences are also compared in Table 8.6. According to this table, T2W2 and FLAIR sequences show a

TABLE 8.2
Examples of the Whole Lesion in the FLAIR Sequence [43]

Lesion Image	Sequence Type
	FLAIR
	FLAIR
	FLAIR

TABLE 8.3

The Examples of Enhanced Tumor in T1C and T1W1 Sequences [43]

Enhanced Tumor Image	Sequence Type
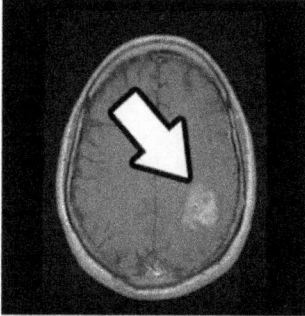	T1 contrast
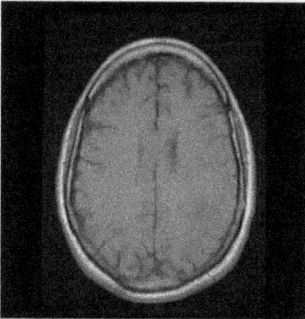	T1-weighted
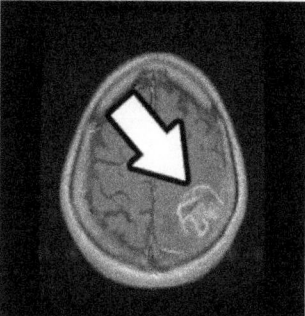	T1 contrast
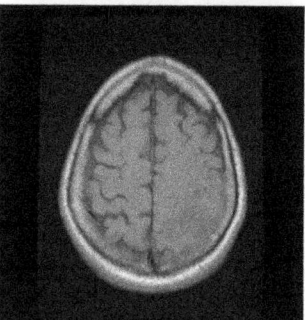	T1-weighted

TABLE 8.4

Examples of Necrosis in T2W2 and FLAIR Sequences [43]

Necrosis Image	Sequence Type
	T2-weighted
	FLAIR

non-enhanced tumor less brightly than other sequences, yet a necrosis is darker in T1W1 and FLAIR, while it is shown brighter in a T2W2 sequence. Actually, according to [43], a lesion is known as an unbalanced change in brain tissue. The whole lesion consists of a necrosis, edema, enhanced and non-enhanced tumor. The whole lesion area can be recognized with a FLAIR sequence. Moreover, a tumor is a mass in any brain tissue, which is generated by abnormal cells. In addition to these, a necrosis is known as the dying stage of the death cells, in other words it is an act of assassination. On the other hand, an edema can be found in a human brain as a swelling, and it has more effect than other lesions on the human body.

TABLE 8.5

Examples of Edema in T2W2 and FLAIR Sequences [43]

Edema Image	Sequence Type
	T2-weighted
	FLAIR

8.4 DISCUSSION

In the tables of the previous section and based on the figures obtained from the hospital of the University of Hradec Kralove, the real examples that belong to the patients in this hospital are shown with different MRI sequences, the so-called T1W1, T2W2, T1C, and FLAIR. In these figures, it is easy to see the contrast differences; for instance, we could see that the necrosis is bright or dark depending upon the sequence type.

Basically, there are four different MRI sequence types studied in this chapter, and they are separated from each other in terms of repetition time and echo time.

TABLE 8.6
The Key Features of the Presented Sequences [43]

	T1-weighted	T1 contrast	T2-weighted	FLAIR
Non-enhanced tumor	—	—	Less Bright	Less Bright
Enhanced tumor	—	Brighter than T1-weighted	—	—
Necrosis	Dark	—	Bright	Dark
Edema	—	—	—	Bright

TABLE 8.7
The Evaluation of Each Lesion Presence in Each Sequence [43]

	T1-weighted	T1 contrast	T2-weighted	FLAIR
Non-enhanced tumor	+	−	−	−
Enhanced tumor	+	+	+	+
Necrosis	−	−	+	+
Edema	−	−	−	+

These MRI sequences highlight the whole lesion that consists of an enhanced tumor, non-enhanced tumor, edema, and necrosis. For example, in Table 8.2, the samples of the whole lesion in a FLAIR sequence type are introduced, and the whole lesion could be detected in a bright color. In Table 8.3, the examples of enhanced tumors in T1W1 and T1C are shown in detail. The enhanced tumor can be identified in a bright color in T1-contrast; on the other hand, we could see the whole lesion in a bright color without presenting any contrast difference in a T1W1 sequence type. The other enhanced tumors presented in the same table can be seen. Furthermore, we can detect a necrosis in the T2W2 and FLAIR sequence types, and the necrosis is bright in the T2W2; however, it is dark in the FLAIR sequence. In Table 8.5, we can easily see the edema in the examples of the T2W2 and FLAIR sequences. In the T2W2 sequence, the edema around the necrosis was a bright color, whereas in the FLAIR sequence it was even brighter. The figures in these tables compare the lesions in terms of different colors, and the aim is to try to explore the sequence type for the recognition of the appropriate lesion. In this case, the best diagnosis of an enhanced tumor exists in T1C.

In Table 8.7, we compare each lesion presence with each sequence type; for instance, a non-enhanced tumor can be recognized well in the T1W1 sequence, yet it can not be recognized well in the T1C, T2W2 and FLAIR sequences. In addition to this, an enhanced tumor can be recognized well in all of the sequence types. On the other hand, a necrosis can be identified well in only the T2W2 and FLAIR sequences. Lastly, an edema can be recognized well only in the FLAIR sequence.

These tables and figures prove that each lesion in human brain MRIs can be recognized by the appropriate sequence type, and each of them is different from each other in terms of contrast. Moreover, various studies in the existing literature have taken the scientific lead in this topic; thus, we gathered them systematically into a literature matrix. Also, the MRI scanners are relatively complex when used to distinguish brain tumors appropriately; therefore, an accurate choice must be given on the correct sequence type. In this book chapter, we primarily defined and compared four different MRI sequence types as they were the most ideal sequence type in the previous studies that we gathered in the earlier section of this book chapter.

8.5 CONCLUSION

Tumor detection is already one of the most studied areas in medical imaging. This chapter presented a guideline for comprehending brain MRI sequences and classification methods. In the existing literature, there already exist powerful methods and approaches to perform classification and segmentation with brain MRI images. These methods could easily predict brain tumors and their variations. This book chapter also helps us understand the importance of tumor classification and segmentation approaches in the treatment period of any brain tumor. The classification and segmentation methods were gathered and presented in a detailed literature matrix, and most of these studies document the usage of convolutional neural network approach, especially in recent years. In addition to these, the number of studies in 2020 was relatively high, and this means that the significance of the topic is already at the top level. The presented methods and approaches also improved the correctness and toughness of brain MRI image classification and segmentation.

In this chapter we also presented and compared four MRI sequence types for identification of brain tumors in the human body, and they included T1W1, T1-contrast, T2W2, and FLAIR. We hope that the current survey will help the scientific community to identify the most promising directions towards achieving more accurate tumor detection in the brain. Different schemes remain yet to be developed using various MRI sequences for the sake of yielding superior classification and segmentation results.

Acknowledgments: The work and the contribution were supported by the project "Smart Solutions in Ubiquitous Computing Environments," Grant Agency of Excellence, University of Hradec Kralove, Faculty of Informatics and Management, Czech Republic (under ID: UHK-FIM-GE-2020).

REFERENCES

1. M. C. Trivedi, P. Tripathi and V. K. Singh, "Brain tumor segmentation in magnetic resonance imaging using OKM approach," *Materials Today: Proceedings*, p. S221478532035029X, Aug. 2020, doi:10.1016/j.matpr.2020.06.548.
2. M. Filippi et al., "Assessment of lesions on magnetic resonance imaging in multiple sclerosis: Practical guidelines," *Brain*, vol. 142, no. 7, pp. 1858–1875, Jul. 2019, doi:10.1093/brain/awz144.
3. D. Rammurthy and P. K. Mahesh, "Whale Harris Hawks optimization based deep learning classifier for brain tumor detection using MRI images," *Journal of King Saud*

University—Computer and Information Sciences, p. S1319157820304444, Aug. 2020, doi:10.1016/j.jksuci.2020.08.006.

4. E. I. Zacharaki et al., "Classification of brain tumor type and grade using MRI texture and shape in a machine learning scheme," *Magnetic Resonance Medicine*, vol. 62, no. 6, pp. 1609–1618, Dec. 2009, doi:10.1002/mrm.22147.

5. E. I. Papageorgiou et al., "Brain tumor characterization using the soft computing technique of fuzzy cognitive maps," *Applied Soft Computing*, p. 9, 2008.

6. M. Toğaçar, "BrainMRNet_ brain tumor detection using magnetic resonance images with a novel convolutional neural network model," *Medical Hypotheses*, p. 9, 2020.

7. A. Çinar, "Detection of tumors on brain MRI images using the hybrid convolutional neural network architecture," *Medical Hypotheses*, p. 8, 2020.

8. M. A. Naser and M. J. Deen, "Brain tumor segmentation and grading of lower-grade glioma using deep learning in MRI images," *Computers in Biology and Medicine*, vol. 121, p. 103758, Jun. 2020, doi:10.1016/j.compbiomed.2020.103758.

9. A.Rajendran and R. Dhanasekaran, "Fuzzy clustering and deformable model for tumor segmentation on MRI brain image: A combined approach," *Procedia Engineering*, vol. 30, pp. 327–333, 2012, doi:10.1016/j.proeng.2012.01.868.

10. Y. Li et al., "An advanced MRI and MRSI data fusion scheme for enhancing unsupervised brain tumor differentiation," *Computers in Biology and Medicine*, vol. 81, pp. 121–129, Feb. 2017, doi:10.1016/j.compbiomed.2016.12.017.

11. M. B. Naceur, R. Saouli, M. Akil and R. Kachouri, "Fully automatic brain tumor segmentation using end-to-end incremental deep neural networks in MRI images," *Computer Methods and Programs in Biomedicine*, vol. 166, pp. 39–49, Nov. 2018, doi:10.1016/j.cmpb.2018.09.007.

12. F. Demir, A. Şengür, V. Bajaj and K. Polat, "Towards the classification of heart sounds based on convolutional deep neural network," *Health Information Science and Systems*, vol. 7, no. 1, p. 16, Dec. 2019, doi:10.1007/s13755-019-0078-0.

13. F. Demir, V. Bajaj, M. C. Ince, S. Taran and A. Şengür, "Surface EMG signals and deep transfer learning-based physical action classification," *Neural Computing and Applications*, vol. 31, no. 12, pp. 8455–8462, Dec. 2019, doi:10.1007/s00521-019-04553-7.

14. S. Bonte, I. Goethals and R. Van Holen, "Machine learning based brain tumour segmentation on limited data using local texture and abnormality," *Computers in Biology and Medicine*, vol. 98, pp. 39–47, Jul. 2018, doi:10.1016/j.compbiomed.2018.05.005.

15. E. Sert, F. Özyurt and A. Doğantekin, "A new approach for brain tumor diagnosis system: Single image super resolution based maximum fuzzy entropy segmentation and convolutional neural network," *Medical Hypotheses*, vol. 133, p. 109413, Dec. 2019, doi:10.1016/j.mehy.2019.109413.

16. C. J. J. Sheela and G. Suganthi, "Automatic brain tumor segmentation from MRI using greedy snake model and fuzzy C-means optimization," *Journal of King Saud University—Computer and Information Sciences*, p. S1319157818313120, Apr. 2019, doi:10.1016/j.jksuci.2019.04.006.

17. Z. N. K. Swati et al., "Brain tumor classification for MR images using transfer learning and fine-tuning," *Computerized Medical Imaging and Graphics*, vol. 75, pp. 34–46, Jul. 2019, doi:10.1016/j.compmedimag.2019.05.001.

18. S. K. Khare and V. Bajaj, "Time-frequency representation and convolutional neural network-based emotion recognition," *IEEE Transactions on Neural Networks and Learning Systems*, pp. 1–9, 2020, doi:10.1109/TNNLS.2020.3008938.

19. S. Alagarsamy, K. Kamatchi, V. Govindaraj, Y.-D. Zhang and A. Thiyagarajan, "Multi-channeled MR brain image segmentation: A new automated approach combining BAT and clustering technique for better identification of heterogeneous tumors,"

Biocybernetics and Biomedical Engineering, vol. 39, no. 4, pp. 1005–1035, Oct. 2019, doi:10.1016/j.bbe.2019.05.007.

20. T. Yang, J. Song and L. Li, "A deep learning model integrating SK-TPCNN and random forests for brain tumor segmentation in MRI," *Biocybernetics and Biomedical Engineering*, vol. 39, no. 3, pp. 613–623, Jul. 2019, doi:10.1016/j.bbe.2019.06.003.

21. A. Kabir Anaraki, M. Ayati and F. Kazemi, "Magnetic resonance imaging-based brain tumor grades classification and grading via convolutional neural networks and genetic algorithms," *Biocybernetics and Biomedical Engineering*, vol. 39, no. 1, pp. 63–74, Jan. 2019, doi:10.1016/j.bbe.2018.10.004.

22. L. H. Shehab, O. M. Fahmy, S. M. Gasser and M. S. El-Mahallawy, "An efficient brain tumor image segmentation based on deep residual networks (ResNets)," *Journal of King Saud University - Engineering Sciences*, p. S1018363920302506, Jun. 2020, doi:10.1016/j.jksues.2020.06.001.

23. G. S. Tandel, A. Balestrieri, T. Jujaray, N. N. Khanna, L. Saba and J. S. Suri, "Multiclass magnetic resonance imaging brain tumor classification using artificial intelligence paradigm," *Computers in Biology and Medicine*, vol. 122, p. 103804, Jul. 2020, doi:10.1016/j.compbiomed.2020.103804.

24. S. S. Ghahfarrokhi and H. Khodadadi, "Human brain tumor diagnosis using the combination of the complexity measures and texture features through magnetic resonance image," *Biomedical Signal Processing and Control*, vol. 61, p. 102025, Aug. 2020, doi:10.1016/j.bspc.2020.102025.

25. D. Daimary, M. B. Bora, K. Amitab and D. Kandar, "Brain tumor segmentation from MRI images using hybrid convolutional neural networks," *Procedia Computer Science*, vol. 167, pp. 2419–2428, 2020, doi:10.1016/j.procs.2020.03.295.

26. N. Ghassemi, A. Shoeibi and M. Rouhani, "Deep neural network with generative adversarial networks pre-training for brain tumor classification based on MR images," *Biomedical Signal Processing and Control*, vol. 57, p. 101678, Mar. 2020, doi:10.1016/j.bspc.2019.101678.

27. S. Kumar and D. P. Mankame, "Optimization driven deep convolution neural network for brain tumor classification," *Biocybernetics and Biomedical Engineering*, vol. 40, no. 3, pp. 1190–1204, Jul. 2020, doi:10.1016/j.bbe.2020.05.009.

28. M. I. Sharif, J. P. Li, M. A. Khan and M. A. Saleem, "Active deep neural network features selection for segmentation and recognition of brain tumors using MRI images," *Pattern Recognition Letters*, vol. 129, pp. 181–189, Jan. 2020, doi:10.1016/j.patrec.2019.11.019.

29. M. Ben, M. A. Naceur, R. Saouli and R. Kachouri, "Fully automatic brain tumor segmentation with deep learning-based selective attention using overlapping patches and multi-class weighted cross-entropy," *Medical Image Analysis*, vol. 63, p. 101692, Jul. 2020, doi:10.1016/j.media.2020.101692.

30. R. Remya, K. P. Geetha and S. Murugan, "A series of exponential function, as a novel methodology in detecting brain tumor," *Biomedical Signal Processing and Control*, vol. 62, p. 102158, Sep. 2020, doi:10.1016/j.bspc.2020.102158.

31. P. M. Siva Raja and A. V. Rani, "Brain tumor classification using a hybrid deep autoencoder with Bayesian fuzzy clustering-based segmentation approach," *Biocybernetics and Biomedical Engineering*, vol. 40, no. 1, pp. 440–453, Jan. 2020, doi:10.1016/j.bbe.2020.01.006.

32. F. Özyurt, E. Sert and D. Avcı, "An expert system for brain tumor detection: Fuzzy C-means with super resolution and convolutional neural network with extreme learning machine," *Medical Hypotheses*, vol. 134, p. 109433, Jan. 2020, doi:10.1016/j.mehy.2019.109433.

33. J. Amin, M. Sharif, N. Gul, M. Yasmin and S. A. Shad, "Brain tumor classification based on DWT fusion of MRI sequences using convolutional neural network,"

Pattern Recognition Letters, vol. 129, pp. 115–122, Jan. 2020, doi:10.1016/j. patrec.2019.11.016.

34. M. Toğaçar, Z. Cömert and B. Ergen, "Classification of brain MRI using hyper column technique with convolutional neural network and feature selection method," *Expert Systems with Applications*, vol. 149, p. 113274, Jul. 2020, doi:10.1016/j. eswa.2020.113274.

35. H. Shahamat and M. Saniee Abadeh, "Brain MRI analysis using a deep learning based evolutionary approach," *Neural Networks*, vol. 126, pp. 218–234, Jun. 2020, doi:10.1016/j.neunet.2020.03.017.

36. K. S. Angel Viji and D. Hevin Rajesh, "An efficient technique to segment the tumor and abnormality detection in the brain MRI images using KNN classifier," *Materials Today: Proceedings*, vol. 24, pp. 1944–1954, 2020, doi:10.1016/j.matpr.2020.03.622.

37. D. R. Nayak, R. Dash and B. Majhi, "Automated diagnosis of multi-class brain abnormalities using MRI images: A deep convolutional neural network based method," *Pattern Recognition Letters*, vol. 138, pp. 385–391, Oct. 2020, doi:10.1016/j.patrec.2020.04.018.

38. Z. Ullah, M. U. Farooq, S.-H. Lee and D. An, "A hybrid image enhancement based brain MRI images classification technique," *Medical Hypotheses*, vol. 143, p. 109922, Oct. 2020, doi:10.1016/j.mehy.2020.109922.

39. A. Tiwari, S. Srivastava and M. Pant, "Brain tumor segmentation and classification from magnetic resonance images: Review of selected methods from 2014 to 2019," *Pattern Recognition Letters*, vol. 131, pp. 244–260, Mar. 2020, doi:10.1016/j.patrec.2019.11.020.

40. J. Amin, M. Sharif, M. Yasmin and S. L. Fernandes, "A distinctive approach in brain tumor detection and classification using MRI," *Pattern Recognition Letters*, p. S016786551730404X, Oct. 2017, doi:10.1016/j.patrec.2017.10.036.

41. R. Ito, K. Nakae, J. Hata, H. Okano and S. Ishii, "Semi-supervised deep learning of brain tissue segmentation," *Neural Networks*, vol. 116, pp. 25–34, Aug. 2019, doi:10.1016/j. neunet.2019.03.014.

42. M. Mittal, L. M. Goyal, S. Kaur, I. Kaur, A. Verma and D. Jude Hemanth, "Deep learning based enhanced tumor segmentation approach for MR brain images," *Applied Soft Computing*, vol. 78, pp. 346–354, May 2019, doi:10.1016/j.asoc.2019.02.036.

43. A. Kirimtat, O. Krejcar and A. Selamat, "Brain MRI modality understanding: A guide for image processing and segmentation," in *Lecture Notes in Computer Science*, 2020, vol. 12108, https://doi.org/10.1007/978-3-030-45385-5_63.

44. D. C Preston, "Magnetic Resonance Imaging (MRI) of the brain and spine: Basics," *Magnetic Resonance Imaging (MRI) of the Brain and Spine: Basics*, 2006. https ://casemed.case.edu/clerkships/neurology/Web%20Neurorad/MRI%20Basics.htm (accessed Jan. 04, 2020).

9 Medical Image Classification Algorithm Based on Weight Initialization-Sliding Window Fusion Convolutional Neural Network

Ankit Kumar, Pankaj Dadheech, S. R. Dogiwal,
Sandeep Kumar, and Rajani Kumari

CONTENTS

9.1 INTRODUCTION

As stated by the Global Healthiness Society, nearly 17.9 million individuals worldwide die each year, and the causes of death include many diseases associated with heart disease, which are caused by heart attacks and strokes [1]. There may be many genetic facets throughout which four generations last with a solitary form of cardiac ailment. Individually the leading ordinary causes of heart ailments are elevated blood pressure, fasting blood sugar, diabetes, cholesterol, BMI, and heart rate. People with diabetes have an increased probability of developing several different types of health issues, including heart disease. Over time diabetes increases the blood sugar, and elevated blood sugar harms the blood vessels of the heart and other organs, causing other health problems. This means that diabetes doubles the rate of heart disease. Unhealthy diet, lack of sleep, physical inactivity, depression, obesity, or family history could be factors causing a heart ailment. Monitor blood load, maintain cholesterol and triglyceride levels in control, maintain a balanced lifestyle, work out frequently, maintain cholesterol as well as triglyceride levels in management, and maintain fear is the protection of ticker disease. With the popularity of machine learning in the medical industry, we are able to help detect and prevent heart diseases. To enhance the quality of patients' lives, extracting useful information from medical data reports helps with better decision-making in data mining. An individual human being dies every 35 seconds owing to a heart ailment. They are affected by heart diseases in young people aged 20–30 years. Diagnosing heart ailments is an extremely important job in the field of healthcare.

Removal of evidence sets out the methodology and capacity to modify details on medical facts that are keen on the handy management method [2]. The classification consists of 13 medicinal specifications such as sexual characteristics, blood pressure, cholesterol, angina, heart rate, and obesity.

The heart is a fibrous body part that drives the blood around the body in sequence to deliver nutrients through the blood vessels, in addition to getting rid of the metabolic waste of the body [3]. Throughout a regular pulsation, blood circulates from tissues plus lungs keen on atria after that into ventricles. The wall within the heart vociferates the inter-ventricular septum to keep the blood on the left and right. Two spouts between the atria and ventricles keep blood from beginning on the atria towards the back. The tricuspid regulator opens hooked on the right ventricles in addition to the bicuspid regulator which opens hooked on the left ventricle. Muscular pain material in situ vociferate Chordae tedious grip valves in a strong ventricle contact. Blood going through the ventricles pass through the pulmonary valves in the right ventricles then the pulmonary trunk in addition to the aortic regulator linking the left ventricle as well as the aorta.

The heart power vociferates myocardium is differing in a distinctive outline in series drive blood well. The surface of the myocardium enfolds the region of the inferior section of the heart. They twist as well as pass in diverse directions to push blood throughout the heart. Leader cells produce the voltaic signal of the heart. Myocyte agreement vociferates the heart power like a cluster [4]. On the left as well as the right part, which functions jointly as a double pump, the heart is alienated keenly. The blood moves through the left ventricles, which contract and push blood out of the heart through the aorta to nourish the tissues. The primary branch of the aorta is the coronary artery which provides heart strength with oxygen and supplements. At the peak of the aorta, the artery penetrates to send the blood to head as well as arms and provide blood to the rest of the body. Heartbeats range from 60–100 BPM, i.e., 5 Quarts per minute.

9.1.1 HEART AILMENT

The heart needs strength to pump the blood; arteries provide blood to the heart muscle, in addition to the valves, which certify that the blood inside the heart is pumped in the correct route. Problems can occur in several of these areas.

9.1.1.1 Coronary Heart Ailment

There is an amount of work to manipulate the output of the heart, many of which are tapering or obstruction of the coronary arteries because of coronary artery disease, usually caused by atherosclerosis. Atherosclerosis (occasionally vociferate the "hardening" or else "pressing" of the arteries) produces cholesterol along with pinged deposits (called plaques) which lies along the internal ramparts of the arteries. This plaque may confine the bloodstream to the heart strength via closing the artery or with irregular arterial tone and role. With sufficient blood delivery, the heart is filled with oxygen along with the essential nutrients that are required for it to function correctly. This can be the reason for the chest pain called angina. If the blood delivers to a part of the heart that is completely blocked, or if the energy needed by the heart exceeds its blood supply, a heart attack can occur.

9.1.1.2 Peripheral Artery Diseases

Peripheral artery ailments are common circulatory problems in which tapering arteries lessen the stream of blood to the organs. When you have peripheral artery disease (PAD), your extremities—typically your feet—don't obtain enough blood to keep up with the demand.

9.1.1.3 Innate Heart Ailment

About 1% of surviving births show a deficiency in the heart structure that occurs during fetal development consisting of simple "holes" or narrowed heart valves in the heart. Faults may be in the heart ramparts, heart valves, and arteries, as well as the veins close to the heart [5]. These are capable of hindering the usual flow of blood through the heart. Blood flow may be slow, misdirected, or misplaced, or completely blocked.

9.1.1.4 Cardiomyopathy Diseases

Cardiomyopathy is known as a muscle ailment. With cardiomyopathy, the heart has difficulty pumping the blood. For this reason, blood can pool in the lungs and parts of the body. This manifests as swelling in the abdomen, feet, and legs. Cardiomyopathy can be acquired as a result of another disease or inherited [6].

9.1.1.5 Other Cardiovascular Diseases

- Heart malfunction: Heart malfunction, occasionally identified as heart failure, happens when the heart is unable to pump the blood. Certain conditions in your heart, such as narrow arteries (coronary artery ailment) or high blood pressure, make it extremely feeble [7].
- Arrhythmia: A cluster of cells in the heart vociferates a cardiac transmission classification, which uses electric pulses to manage the speediness as well as the pulse of each beat. A cardiac arrhythmia occurs when electrical impulses in the heart do not function properly [8]. For example, several individuals experience an abnormal heartbeat, which may possibly feel similar to a racing heart.
- Heart valve problem: In heart valve disease, one or more valves in the heart do not function properly. There are four hydrants in the heart that maintain the blood flow in. In some cases, one or more valves do not open or close appropriately. This can cause blood flow to your body to be obstructed through your heart [9].

9.1.2 Fact Removal

Fact removal is a procedure used for information discovery or mining of a large number of facts known as a KDD (knowledge discovery from data) [10]. Fact removal is used in healthcare and business, analyzing a huge set of facts to identify helpful and stable patterns. In the medical trade, in particular, fact removal can be used to reduce costs and increase efficiencies, improve the quality of a patient's life, and perhaps most importantly save the lives of new patients. Processes such as fact removal involve many functions, such as forecast, clustering, and time-series study.

The medical trade has many sources of data, such as electronic health check records, managerial information, and erstwhile benchmarking results. Nowadays fact removal is used to provide treatments at an effective cost, in order to help predictors use various diseases with clinical forecasts and recommendations from physicians in different decision-making [11, 12].

9.1.2.1 Classification Algorithm

This section presents an introduction to the classification algorithm used in our comparative study. As we stated in the introduction, the reason for this study is to scrutinize the stability and presentation of categorization algorithms for predicting patient status using previously known patient data along with cardiovascular statistics [13]. The classification algorithm employed was a decision tree, naïve Bayes, random forest, logistic regression, support vector machine (SVM), and k-NN.

Decision Tree: A hierarchy of decisions is seen as individual approaches intended for a creator of categorization. This is equivalent to a flowchart where the interior join describes the position of a feature, every branch acts for the resulting position with every leaf node representing a group label, and then all the attributes are decided after the calculation is performed. The passage as of source to leaf represents the categorization rule. In the medical field, the decision tree determines the sequence of the characteristics. First, it produces a set of solved cases; then the entire set is then divided into training sets and test sets. The decision tree is used to find the accuracy of the obtained solution.

Naïve Bayes: naïve Bayes is a prospect classifier based on Bayes' theorem, applying tough autonomy assumptions between the features. It is easy to construct a naïve Bayesian model without complex recurrence parameter estimates, which makes it useful in the medical field for the analysis of cardiac patients. In spite of its ease, the naïve Bayesian classifier frequently performs well, and it is extensively used as it frequently outperforms highly complicated categorization methods [14].

Random Forest: Heart ailments are the leading cause of early death worldwide. Numerous fact removal methods were used by researchers to assist physical condition professionals to forecast heart ailments. A random forest is a group learning algorithm for medical applications. It is the same as a decision tree, although the algorithm builds a forest of decision trees by way of the location of randomly selected features. Computing has the advantage of improving efficiency while improving forecast accuracy without a significant increase in computing costs. It can also predict thousands of other variables.

Support Vector Machine: The SVM projected by Vapnik as well as Cortes has been fruitfully useful to sex categorization exertion for a lot of researchers. A SVM classifier is a linear arrangement where different hyperplanes are selected to diminish the normal categorization mistakes of the hidden analysis pattern. A SVM consists of a robust arrangement so as to recognize two modules. In training the SVM classifies the picture analyzed into a group that has the greatest space at the closest point. The SVM training algorithm creates a replica that predicts whether the experiment figure falls into one group or a different group. A SVM requires an enormous quantity of preparation facts to choose a single range and computational price, while we still regulate ourselves in order to recognize a common currency. A SVM is a knowledge algorithm meant for categorization. It tries to discover the best stripping hyperplanes, so the normal categorization error for the hidden pattern is reduced. Inputs meant for linearly non-separable facts are mapped onto a high-dimensional attribute space where they are unconnected to hyperplanes. This outcrop in a high-dimensional attribute space can be improved by using a kernel. Further, a set of training samples in addition to an equivalent decision value −1, 1 SVM aim to discover the most distinct separation hyperplanes given by WTX + B so as to maximize the space between the two modules [15].

Logistic Regression: Logistic regression is essentially used to categorize low dimensional facts with non-dimensional limits. In addition to providing dissimilarity in the proportion of the dependent relative variable, and the position of a person varies according to its significance. So, the major purpose of logistic regression is to decide the outcome of every variable accurately. In addition, logistic regression is known as a logistic model that gives two categories of a definite component, such as beam or gloomy, slim or fit.

k-nearest neighbor: A k-NN is a basic calculation that stores all cases and orders new cases dependent on a likeness measure.

Classification methods are the most extensively used algorithms in the healthcare field because they help to predict the status of patients by classifying their records and locating the class that matches. Classification is referred to as a supervised learning technique that requires initial classifying of the data into classes or labels. These data are then entered into a classification algorithm in order for it to learn. In particular, relationships between attributes need to be explored by an algorithm to predict the outcome. In this progression, the arrangement calculation makes a classifier from a preparation set made up of dataset tuples and their comparable class marks. Each of the tuples that comprise a preparation set is alluded to as a classification or class. When a new case arrives the developed classification algorithm is used to classify it into one of the predefined classes. The term specifies "how to have a good algorithm" which is called the prediction accuracy. For example, trained medical databases will provide much more relevant information of a previously recorded patient, whether or not patients have heart disease [16].

In this work, this administered AI idea was used for making the predictions. The correlation of particular highlights utilizing diverse AI methods, like support vector machine, decision tree, random forest, naïve Bayes, and k-nearest-neighbor classifiers, were utilized for anticipating a coronary illness or not [17]. The examination was done with a few degrees of cross-approval, and a few rate split assessment strategies separately. The Cleveland and Statlog coronary illness dataset from the UCI AI archive was utilized for coronary illness predictions. The predictions were completed by an older model using arrangement calculations. Guileless mathematician used chance to predict coronary disease, SVM was used for order and relapse operations, random forest operates with a modified decision tree, k-NN was used to locate nearest neighbours. Such estimates suggest an alternative precision. In order to acquire greater accuracy, we sought to tune our methods, which would be useful for increasingly accurate forecasts.

This research addressed the challenge of improving predictive models for heart disease patients and providing timely feedback. A significant problem facing medical management is the provision of precious amenities at a sensible cost. Poor decisions concerning the patient can lead to unwanted as well as permanent consequences. Similarly, medical amenities should diminish health testing costs. A fact removal replica can achieve these results using real-world data furthermore facilitating correct decision creation [18].

The Cleveland and Statlog heart disease datasets from the UCI machine learning repository were exploited for the creation of heart ailment predictions in research work. The major purpose of this learning was to estimate whether a patient was affected by a heart ailment or was not using diverse machine learning algorithms but relying on eligible datasets. We discovered associations among dissimilar attributes, predictions, and drew performance comparisons as well. We forecast heart danger and possibilities for potential situations that assist in making informed choices after receiving an apparent thought about our intended evidence removal procedures and evaluating implications. Recently, the health industry released data on the predictions of health worldwide. When the medical data is huge, an machine learning algorithm is used for analysis.

9.2 REVIEW OF LITERATURE

Machine learning techniques are used for finding new rules and patterns, which are extracted from huge quantities of information. Fact removal helps to predict disease. Analysis of disease consists of the performance of a number of tests on patients. However, the data mining technique reduces the number of tests in healthcare. With a reduced number of tests, the wastage of time and money reduces. The data mining of heart disease is important as it provides support to doctors by informing them which attribute or feature is important, which helps in diagnosing heart ailment more efficiently. Many fact removal methods are used in the judgment of heart disease, including naïve Bayes, decision trees, neural networks, bagging algorithms, as well as support vector machines for viewing the dissimilar levels of accuracy.

Otoom et al. [2015] constructed an intelligent classifier for predicting heart problems by means of clinical data and machine learning algorithms, entered in by the user or physician. This analytic part is included in the mobile phone function with concurrent monitoring that continually monitors and raises the alarm when an emergency occurs. Our results suggest that the proposed diagnostic component proved to be successful with cross-validation tests, with a classification performance accuracy of over 85%. In addition, the monitoring algorithm provided a 100% detection rate [2].

Amin et al. [2013] presented a method for the forecast of heart problems by means of the main risk factors. The method included two pieces of fact removal equipment: neural networks and genetic algorithms. The implemented mixture scheme used a worldwide optimization for the benefit of a hereditary algorithm to initialize the neural system load. Knowledge is faster, further constant, and correct than back proliferation [3]. This scheme was implemented in Matlab in addition to predicting heart disease risk with an accuracy of 89%.

Desai et al. [12] utilized the Cleveland dataset to evaluate the precision of characterization models for the prediction of coronary illness. An investigation of parametric and non-customary methodologies for characterizing coronary illness was introduced. Two arrangement models, the back-dissemination neural system (BPNN) and calculated relapse (LR) were utilized in the examination. The created grouping model will help specialists to make more compelling clinical choices. A ten-overlap cross-approval strategy was utilized to gauge the impartial estimation of these arrangement models [12].

Dangare et al. [2012] broke down the prescient frameworks for coronary illness by utilizing a more prominent number of information attributes. In this framework, 13 qualities, for example, sex, circulatory strain, and cholesterol, were utilized to appraise the likelihood of the patient getting a coronary illness. Up until this point,

13 highlights have been utilized in the forecast. Two additional highlights were added to this examination paper, for example, corpulence and smoking. Information mining order systems on the coronary illness database, to be specific choice trees, naïve Bayes, and neural networks were investigated. The presentation of these strategies was looked at based on precision [16]. As indicated by our outcomes, the exactness of neural systems, choice trees, and naïve Bayes were 100%, 99.62%, and 90.74%, respectively. Our investigation proposed that of these three grouping models the neural system predicts coronary illness with the most elevated precision.

Parthiban et al. [2007] provided another methodology dependent on a coactive neuro-fluffy infusion framework and CANFIS that was displayed for the prediction of coronary illness. The CANFIS model, along with neural net versatile abilities and fluffy judgment subjective methodologies, was then included with hereditary calculations to break down occurrences of the disease [18]. The introduction of the CANFIS model prompted an assessment of direct execution and characterization exactness, and the outcomes indicated that the anticipated CANFIS model had a noteworthy potential for foreseeing coronary illness.

McAlister et al. [2019] states that potential to score for coronary disease in Andhra Pradesh. The investigation on 14 informational indexes from UCI information was performed utilizing ten cross-crease approvals. They produced class affiliation rules utilizing an element submittal choice measure, for example, SU, Ig, and a hereditary inquiry to decide which qualities contribute more toward coronary illness. Analytic tests must be taken by a patient. They utilized a cooperative approach to improve the exactness of the arrangement [19].

Habib et al. [2019] proposed an improved multilayer perceptron calculation that partitioned the dataset into a few subsets. At that point, the MLP calculation was performed exclusively for every subset and the outcomes were acquired from various subgroups using the greater likelihood rule of the democratic combiner. Lastly, the affectability, explicitness, exactness, and execution of these methods were estimated through ROC [20].

Prakash et al. [2019] created a best-fit arrangement calculation that had a noteworthy exactness over-characterization for coronary illness predictions [21]. The information was processed using an immense component choice strategy to find the datasets. The reduced information was then given for characterization. In the examination, the crossbreed property determination technique, joining together CFS and a channel subset assessment, gave more exact results. They proposed a component determination strategy calculation which was a half and half technique consolidating the CFS and Bayes hypothesis.

Mythili et al [2013] proposed a rule-based replica to evaluate the correctness of applying regulations to the entity results of logistic regression on a rule vector mechanism, a decision tree, as well as the Cleveland heart ailment record to predict heart problems. A rule-based loom is the generally used method that combines the results of several models. Rule-based models such as C4.5 have also been implemented, but never in conjunction with any predictive models. Thus, this paper presents a unique replica meant for the comparative study of rule-based algorithms for combining SVMs, decision trees, and logistic regressions [22].

The information set used by Vembandasamy et al. [2015] was a medical data set composed by the foremost diabetes research institutes in Chennai and had records of about 500 patients. The medical dataset requirement provided brief, unmistakable definitions meant for diabetes-related items. The naïve Bayes algorithm was the fact removal method to identify patients with heart disease. This paper analyzed attributes in addition to predicting heart diseases, thereby revealing a cardiovascular disease prediction scheme and HDPS based on the data mining approach [23].

Dalvi et al. [2016] forecast heart ailments in the medical field by means of data knowledge. There has been a lot of research completed linked to this difficulty, but the accurateness of the predictions still needed to be enhanced. As a consequence, this analysis centered on facility variety methods and algorithms where numerous heart ailment datasets were used to show rapid minor progress inaccuracy as an instrument; decision tree, logistic regression, logistic regression, SVM, naive Bayes, and random forest; algorithms are used as methods of attribute preference in addition to results are improved by accurate output [24].

Gandhi et al. [2015] developed a framework for classifying medical data where orthogonal limited safety projection (OLPP) was used to diminish attribute measurement. Once an attribute was lacking, it resolved a prediction based on the classification. The database consisted of 76 features; though, every dispersed test was consigned to using only 14 of them. The accurateness standards of the three datasets were 92.08% for the Cleveland dataset, 92.85% for the Hungarian dataset, and 94.93% for the Switzerland dataset. The sensitivity values for the three datasets were 0.902% for the Cleveland dataset, 0.994% for the Hungarian dataset, and 0.5% for the Switzerland dataset. Typical values for the three datasets were 0.920% for the Cleveland dataset, 0.811% for the Hungarian dataset, as well as 0.99% for the Switzerland dataset [25].

Repaka et al.[2019] used data on humanity as seen by the Registrar General of India that heart disease is a significant cause of death in India, and about 33 percent of deaths in rural areas in Andhra Pradesh are attributed to coronary heart disease. Therefore there is a need to expand the classifications used for predicting heart ailments. In this paper, we recommend resourceful associative categorization algorithms by means of hereditary approaches for the prediction of heart diseases. The most important motivation for using hereditary algorithms in the search for a high-level forecast system was that the revealed system was extremely predictive, by means of high predictive accurateness and elevated concepts of importance. Results suggest that classifier regimens primarily aid the prognosis of a heart ailment, which assists hospitals in their decisions [26].

Tuli et al. [2020] used classification models to identify which selective features played an important role in the prediction of heart ailments using the Cleveland and Statlog Project heart ailment datasets. The accuracy of the random forest algorithm in both classification and feature selection models was observed to be 90–95% based on three different percentage splits. The eight and six selected features incur minimal feature requirements to create an improved performance model. Therefore, omitting eight or six selected features may not lead to better performance for the prediction model [27].

Yilmaz et al. [2013] used a least-squares support vector machine (LS-SVM) with a binary judgment hierarchy for categorizing cardiotoco graphs to determine fetal position [28]. The attributes of a toll-SVM were optimized with a particle swarm optimization. The sturdiness of the process was checked by consecutively creased cross-validation. The presentation of the process was evaluated with the stipulations of an all-purpose classification accurateness. Receivers were also presented to analyze and envisage the presentation of the procedure trait analysis with the cobweb symbol process. Tentative results revealed that the planned process achieved an outstanding categorization accurateness pace of 99.62%.

Fact removal has played a significant role in intelligent medical systems. Relationships of real causes as well as special effects of diseases as seen in patients and can effectively test consumers by built-in apps. Huge databases can be implemented using the facts entered by the extendibility of the software. In addition to secret information connections, the belongings of those associations are preserved between the broad medical dataset so that not to test are satisfactorily investigated. This study was discovered by searching for frequent items using a candidate generation. Medical databases may contain sets of simultaneously observed diseases to be minimized through the non-candidate approach. Information on risk factors can be linked by healthcare professionals to recognize high-risk heart ailment patients. Arithmetical examinations in addition to fact removal techniques assists health care professionals in the judgment of heart ailments.

9.3   PROPOSED METHODOLOGY

9.3.1   DATASET STRUCTURE AND DESCRIPTION

The current research aims to improve the analysis of heart ailments by examining patient symptoms through fact removal categorization methods; to achieve this goal, a literature review to review data mining operations connected to the analysis of heart ailments was completed. After that, six classifiers (e.g., naïve Bayes, decision tree, support vector machine, k-nearest neighbor, logistic regression, and random forest) were selected to create the model with the maximum accuracy possible. We also explored precision scores, recall scores, F-scores, and false-negatives using a confusion matrix for every algorithm utilized [29]. Each technique had a different way of creating a classifier, which ensured that these techniques behaved differently and produced different results. To underline the practical feasibility of our approach, the selected classifier was implemented using machine learning, in all conducted experiments, to assess the performance of the algorithm where the cross-validation model was omitted, and where the original dataset differed—different training and testing data were divided into sets. In each validation cycle, an observation was conducted where the remaining observations served as training sets, and they were used to create the classifier model. This stride was repeated until all observations served as test observations. Accordingly, performance errors were calculated at each iteration, and the average performance was calculated at the end of the validation.

Two datasets were used in this study:

- The first was obtained from the Cleveland Clinic Foundation and contained 303 records, 297 complete and six with missing/unknown values. In this experiment, six missing values in the Cleveland dataset were cleaned and transformed by data preprocessing; only 297 instances were taken for this study. The dataset itself (297 instances) was divided as 80 % training data and 20% testing data [30].
- The second was the Statlog dataset that included 270 complete records. Originally, both of these datasets had 76 attributes, but they were preprocessed to produce 14 features in an attempt to reduce the number of variables; as a result, we compared these specific characteristics with other literature used. The dataset itself (270 instances) was divided as 80% training data and 20% testing data. The attribute "num" was the heart disease diagnosis attribute. It was classified as presence and absence. If it was present, the value of num would be low or medium, or high or very high. If it was absent, the value of num would be zero.

Figure 9.1 shows the flow chart of heart disease prediction. Our aim was a comparative study of two datasets of heart disease, which were the datasets of Cleveland and Statlog heart disease, to predict the presence or non-appearance of the disease or which, on the basis of fewer characteristics, affected the patient. The heart ailment datasets were obtained from the UCI machine learning library, which contained 14 attributes for prediction on the basis of a smaller number of attributes and with a faster performance for correct predictions.

9.3.2 Algorithm for Performance Evaluation

1) Load data and import libraries to calculate the number of records and missing values from the large infinite data.
2) Convert the large data into categorical valued data for feature selection and fix the missing values by applying data preprocessing.
3) Use the correlation matrix for depicting the relation between feature attribute targets and then find the strong and negative correlation between them.
4) Apply the visualization technique for data analysis of the feature selection of the relevant data using a standard scalar to fit and transform the data into 0 and 1 form, which is easy for predicting disease.
5) Split the dataset into an 80% training and 20% testing dataset to fit the parameters and assess the performance of the model, respectively.
6) Apply the machine learning algorithm on both datasets and compare the performance using the performance matrix for depicting which is a better algorithm.
7) After modeling and predicting with the machine learning classifier, check the accuracy score, precision, recall, and F-measure of both datasets.
8) Evaluate the results of both datasets and calculate the performance measure by plotting a graph of a different classifier.

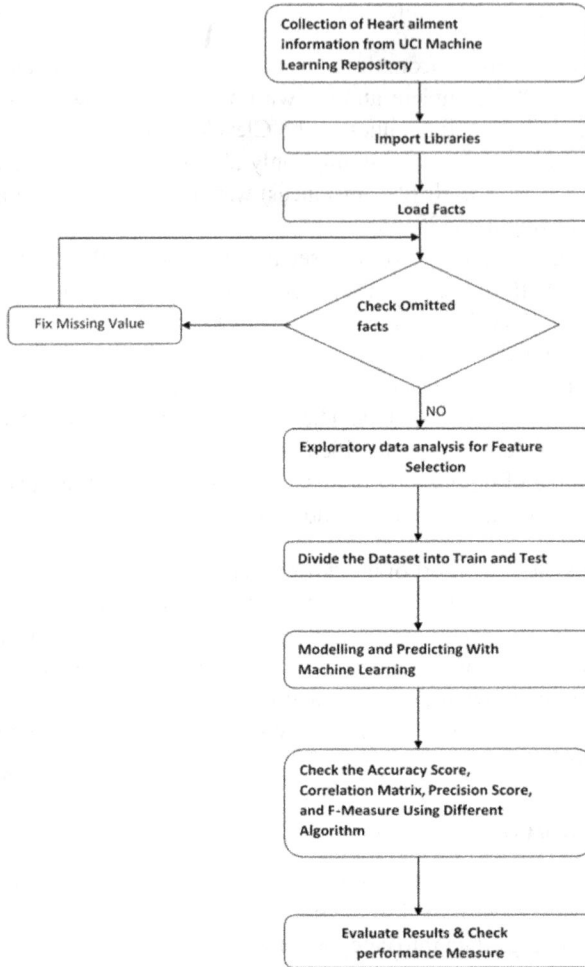

FIGURE 9.1 Flow chart of heart disease prediction.

9) In our result, a random forest gave a higher accuracy, which was 99.80% in the Cleveland dataset and 87.03% in the Statlog dataset over logistic regression, which was second-lowest.

9.3.3 DATA COLLECTION

A heart disease dataset is suitable to extract the features necessary to create an intelligent decision support system. Generally, medical data is not easily released by many health centers due to their confidentiality agreements. Therefore, this research work used standard heart ailment datasets from the UCI machine learning repository, including the Cleveland and Statlog heart ailment datasets from the UCI dataset [31]. The heart ailment dataset can be easily downloaded from the machine available at the UCI repository. We used 14 of them to obtain accurate results with a small number of attributes. The "goal" field refers to the occurrence of a heart ailment in the patient (target). It is numerically valued from 0 (no occurrence) to 4 (occurrence). The main advantage of using this dataset was that it could also be used by other manufacturers or health experts. Advice from other manufacturers may help improve the results of the proposed framework [32]. Table 9.1 describes the characteristics and their possible data types or values-selected in the heart disease dataset.

In this study, all the supposed attributes were transformed into numerical attributes.

9.3.4 DATA PREPROCESSING

The collected data are usually weakly organized and have missing values and out-of-range values. Datasets that do not encounter these types of problems may produce incorrect results. The selected data were examined for sound, discrepancies, and omitted values. In addition to extracting data from the dataset for training data, hard missing values were replaced by means of allocation occurrence, most close to those calculated by the nearest neighbor algorithm. However, these values must be treated

TABLE 9.1
Attribute Description

Number	Feature Name	Feature Information
1	Age	Age of patients in existence
2	Sex	1 = male, 0 = female
3	Cp	Value 1: Typical Angina, Value 2: Atypical Angina, Value 3: Non-Angina Pain, Value 4:Asymptotamatic
4	Trestbps	Resting Blood Pressure in mm Hg in admittance to the sanatorium
5	Chol	Serum Cholesterol of patients calculated in mg/dl
6	Fbs	Fasting blood sugar of the patient. If >120 mg/dl Value 1 = accurate, Value 0 = fake
7	Restecg	Value 0 = Usual, Value 1 = having ST-T Wave irregularity, Value 2 = Showing likely or exact Left Ventricular Hypertrophy by Estes' criterion
8	Thalach	Maximum Heart pace Achieved of Patient
9	Exang	Value 1= positively, Value 0 = negatively
10	Oldpeak	ST Depression induced by exercise relative to rest
11	Slope	Value 1 = Up Sloping, Value 2 = Flat, Value 3 = Down Sloping
12	Ca	quantity of foremost vessel{0-3}
13	Thal	Value 1 = Usual, Value 2 = Permanent Defected, Value 3 = Reversible Defect
14	Target	Numerical value 0 to 4

prior to use, as they may predict failure classification or inaccurate disease allocation [33]. For the preprocessing of data, there are several steps, such as cleaning, normalization, and transformation. The result of data preprocessing is an important dataset with minimal characteristics.

The presence of redundant records is the main drawback of healthcare providers. The existence of redundant instances in the dataset can lead to inappropriate learning, often biased toward certain data, which could raise ethical issues. To achieve better accuracy, redundant records must be removed from the data. Thousands of records in healthcare data are added to a relational database table, and records that fail to match the layout can be removed from the record set. By average value, if any instance or attribute falls outside this range then those instances are removed from the dataset [34]. Irrelevant and missing data are also removed from the record set. During preprocessing, the entire dataset is taken as one input, and various data preprocessing approaches are applied to reduce invalid instances from the dataset [22].

The preprocessing process is done as follows:

- Thousands of records from the healthcare data [35] are merged into a relational database table and records that fail to match the layout are removed from the record set.
- Records [36] that go beyond the range of the mean value are also removed.
- The record with the missing value is replaced with the average of the entire column.
- Removing instances [23] that have more than one value, that is, removing the multi-level attributes (Figure 9.2).

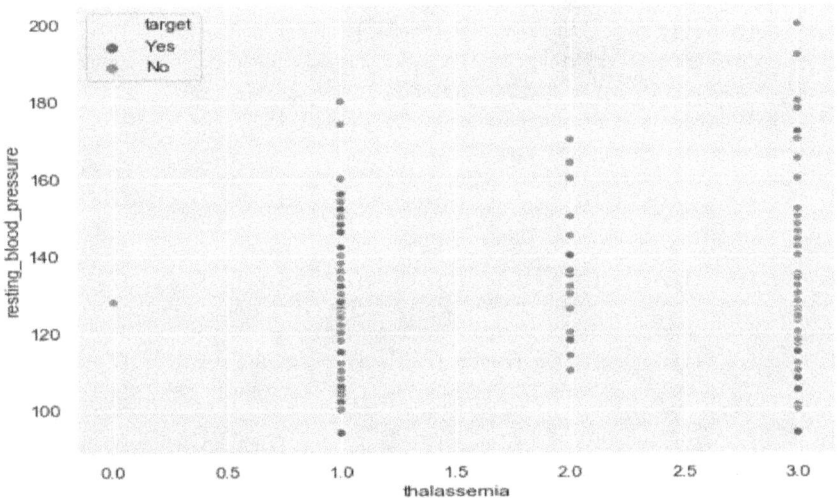

FIGURE 9.2   Thalessemia V/s resting blood pressure.

9.3.5 CORRELATION MATRIX

Correlation study is a process of numerical valuation used to learn the potency of an association among two, numerically calculated, continuous variables [37, 38]. The correlation matrix can be seen in Figure 9.3 and Figure 9.4.

Correlation scrutiny is a means of arithmetical assessment used to learning the potency of an association among two, numerically deliberate, incessant variables [24, 39]. The different values of Thalessemia V/s resting blood pressure can be seen in Figure 9.4, and correlation can be seen in Figure 9.5.

Matrix Correlation Analysis: There was no particular attribute that had an incredibly high association with our objective. Also several descriptions had a pessimistic association with the objective, and various were optimistic [40, 41].

9.4 EVALUATION AND RESULT

9.4.1 PERFORMANCE MEASURE

Five general performance measurements were used to evaluate the accuracy of the classification algorithms [25]. These measures were selected because they were widely used to assess the performance of categorization models. The first measure was memorized, as shown in Equation (9.1), which measures how good a binary

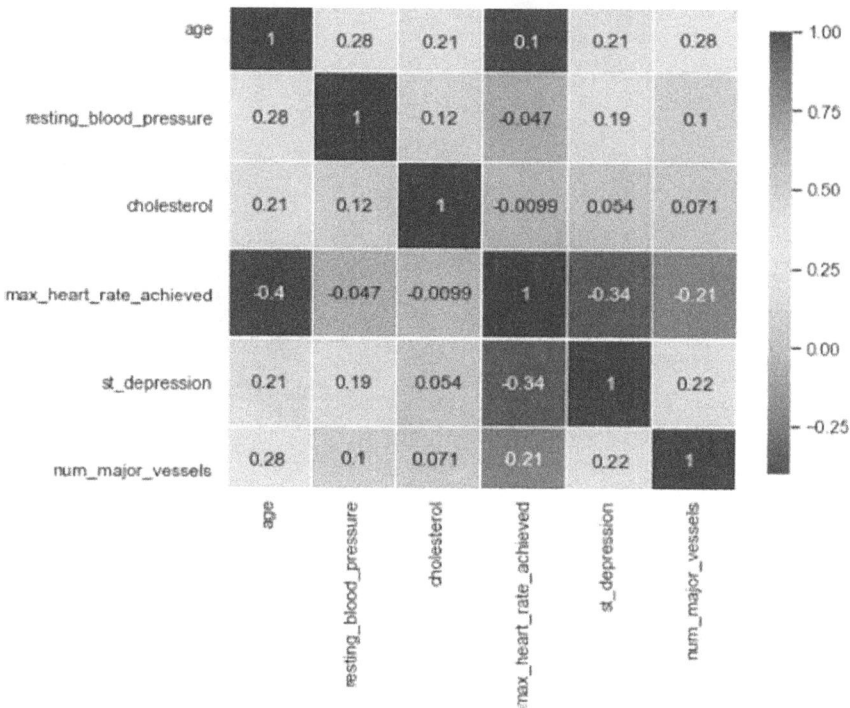

FIGURE 9.3 Correlation matrix.

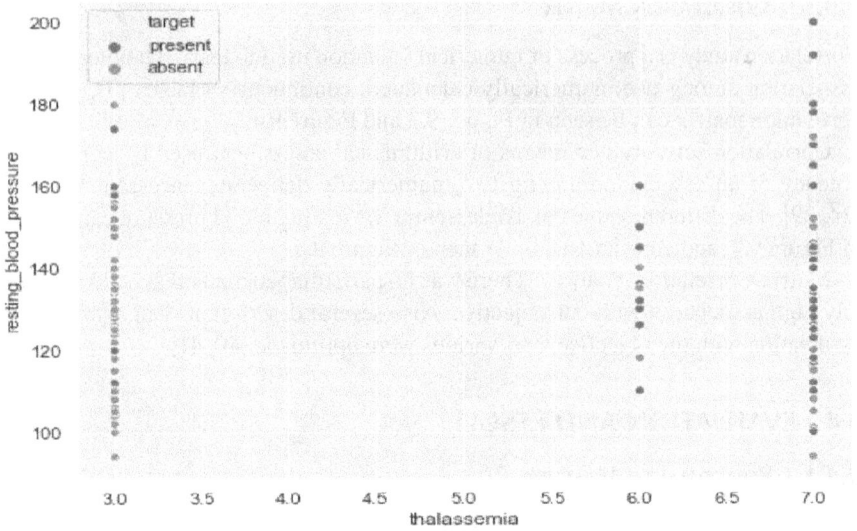

FIGURE 9.4   Thalessemia V/s resting blood pressure.

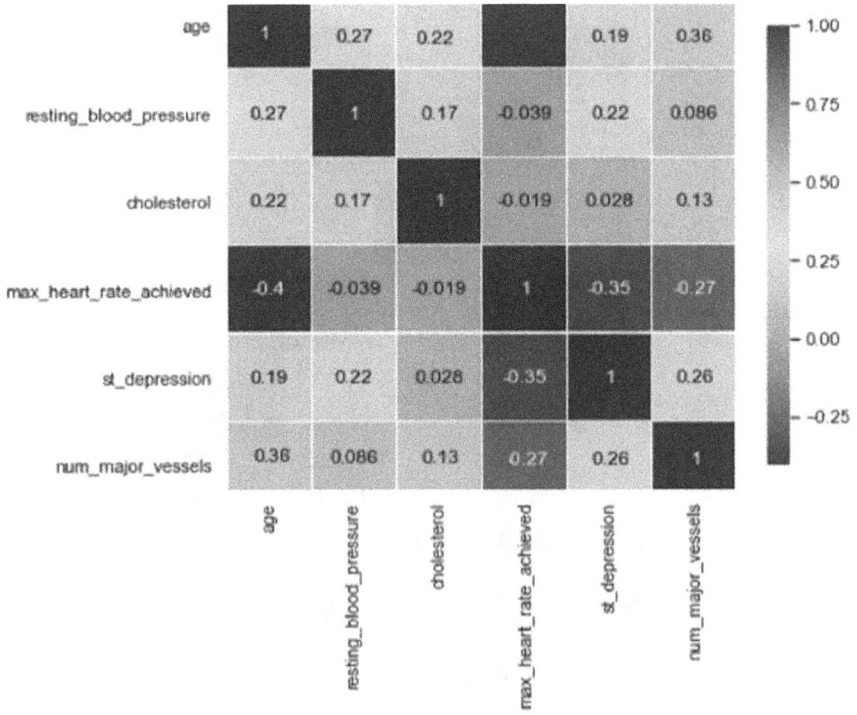

FIGURE 9.5   Correlation.

categorization test identifies the likelihood of correcting the classification members of the objective class [26]. The second measure is shown in Equation (9.2), which processes the likelihood that an optimistic prediction is accurate. The third assesses accuracy, as revealed in Equation (9.3), which measures the performance of classification. The fourth measure is uniqueness, as shown in Equation (9.4), which processes how well the binary categorization test identifies pessimistic cases [27, 42]. The last measure is the F-measure, as shown in Equation (9.5), which processes the likelihood that an optimistic prediction is accurate.

$$\text{Precision} = \frac{TP}{TP + FP} \tag{9.1}$$

$$\text{Recall} = \frac{TP}{TP + FN} \tag{9.2}$$

$$F1 = \frac{2*\text{Precision}*\text{Recall}}{\text{Precision} + \text{Recall}} \tag{9.3}$$

$$\text{Accuracy} = \frac{TP + TN}{TP + FN + TN + FP} \tag{9.4}$$

$$\text{Specificity} = \frac{TN}{TN + FP} \tag{9.5}$$

Where,
TP is figure of factual optimistic.
TN is figure of factual pessimistic.
FP is the figure of fake optimistic.
FN is figure of fake pessimistic.

9.4.2 EXPERIMENTAL RESULTS

In traditional healthcare data systems, doctors rely on patients' symptoms to diagnose heart disease. However, many signs and symptoms are not significant, which can lead to unwanted errors and thus affect the quality of health service provided. To overcome these issues, expert systems were developed to simulate health decision-making with better accuracy [28, 43]. This involves correlating different pieces of patient information after obtaining specific patterns. In this research, we implemented six classification algorithms for the prediction of heart ailments based on pre-selected characteristics.

For each dataset, we ran six classification algorithms, which omitted one cross-validation. Table 9.2 shows that all the categorization algorithms were predictive in addition to giving accurate responses regarding the patient's condition. In particular,

TABLE 9.2

Performance Measure Using Statlog Dataset

Performance Measure	Decision Tree	Naïve Bayes	k-NN	SVM	Logistic Regression	Random Forest
Accuracy	0.851851	0.740740	0.6851851	0.759259	0.777777	0.8703703
Support	54	54	54	54	54	54
Precision	0.85	0.74	0.69	0.76	0.78	0.87
Recall	0.84	0.74	0.68	0.75	0.78	0.86
F-Measure	0.85	0.74	0.69	0.76	0.78	0.87

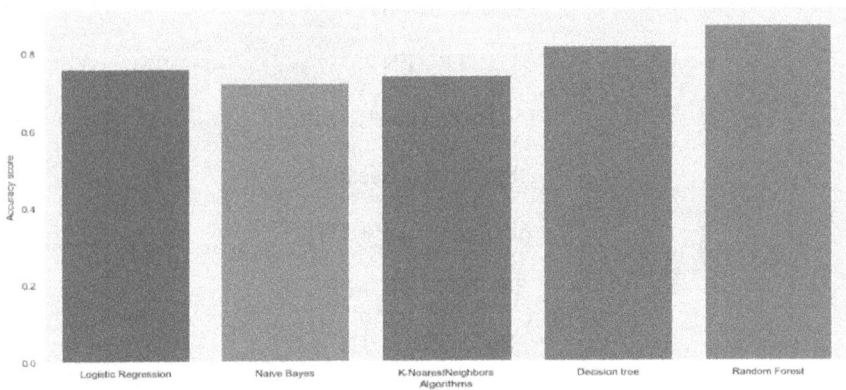

FIGURE 9.6   Statlog dataset accuracy score graph.

the recall measurements suggested that all the classification models [44] could work well and the Cleveland dataset, but unfortunately it wasn't established which classification algorithm produced more accurate results. Similarly, the uniqueness measure could not tell us which classification algorithms gave better results. However, the remaining performance measurements (F-measure, accuracy, specificity) had the same trend that confirmed that the decision tree was the most correct classifier among all the classification algorithms.

The Statlog Dataset Accuracy Score Graph is shown in Figure 9.6. Similarly we ran six algorithms on the Cleveland dataset. The results obtained for this dataset were quite similar to those obtained for the Statlog dataset. The random forest was most accurate algorithm confirmed by the results of accuracy, precision, and F-measure. Both recall and support measure cannot tell us any useful feedback because they cannot distinguish among algorithm [45, 46].

Another important issue that should be emphasized in this study was the ranking stability of the classification algorithms over both datasets as well as the crossways manifold performance measure. Table 9.3 shows the performance measure using the Cleveland dataset. Position constancy to facilitate this algorithm should create the correct results and be used with all datasets and crossways assessment measures.

TABLE 9.3

Performance Measure Using Cleveland Dataset

Performance Measure	Decision Tree	Naïve Bayes	k-NN	SVM	Logistic Regression	Random Forest
Accuracy	0.786885	0.901639	0.868852	0.859259	0.852459	0.918032
Support	61	61	61	61	61	61
Precision	0.79	0.90	0.87	0.85	0.85	0.88
Recall	0.79	0.90	0.87	0.85	0.85	0.87
F-Measure	0.79	0.79	0.87	0.85	0.85	0.87

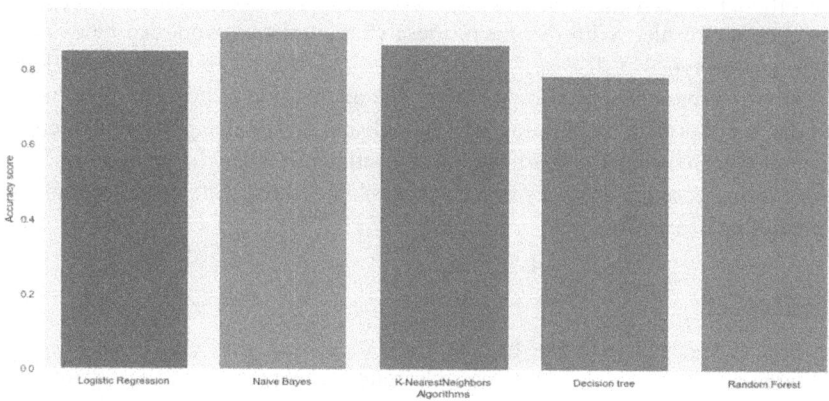

FIGURE 9.7   Cleveland dataset accuracy score graph.

Figure 9.7 shows the Cleveland and dataset accuracy score graph. From our result, we could observe that the random forest was ranked first across all evaluation measures in both datasets, followed by logistic regression in the second position. This gave us an important conclusion on the role of the random forest in classifying data.

9.5   CONCLUSION AND FUTURE SCOPE

9.5.1   CONCLUSION

Motivated by the increasing rates of many different heart ailments, scientists are using fact removal strategies for the scrutiny of heart diseases. The strategy of fact removal for healthcare specialists in the scrutiny of heart ailments has been a definite achievement, the use of a data mining system recognizes the appropriate treatments for heart disease patients. The major reason of our report was to compare accurateness and analyze the reasons following the disparity of dissimilar algorithms. Accurate predictive safety measures provide correct healthcare; similarly, using comparative studies of heart disease on various datasets has demonstrated favorable results in detecting heart disease. We used the Cleveland and Statlog datasets for heart diseases which consisted of 303 and 270 instances, respectively, and split the facts into two sections, which are the training and testing datasets. We used 14 attributes and implemented six dissimilar algorithms to analyze accurateness. For this, in the present work, we successfully achieved a high accuracy using the random forest classifier for heart disease analysis.

9.5.2   FUTURE SCOPE

In the future, automated heart disease prediction systems may be implemented in remote areas such as rural areas to replicate human specialists for diagnosis. The prediction system is appropriate to support healthcare experts in the judgment of heart ailments. The characteristics of the disease should be enhanced to achieve more accurate results. With the use of these characteristics, work can be extended to detect other types of disease, such as cancer and arthritis, but early detection of any chronic disease should also be noted. The automation of heart ailment predictions can be improved by building a GUI, website, or mobile application. Genuine information from hospitals, healthcare organizations, and healthcare agencies could be used to increase the accuracy of data technologies along with other technologies for predicting heart ailments.

REFERENCES

1. N. S. C. Reddy, S. Shue Nee, L. Z. Min, and C. X. Ying. 2019. Classification and feature selection approaches by machine learning techniques: Heart disease prediction, *International Journal of Innovative Computing* Vol. 9, No. 1, pp. 39–46, doi:10.11113/ijic.v9n9.210.
2. A. F. Otoom, E. E. Abdallah, Y. Kilani, A. Kefaye, and M. Ashour. 2015. Effective diagnosis and monitoring of heart disease, *International Journal of Software Engineering and its Applications*, Vol. 9, No. 1, pp. 143–156, doi:10.14257/ijseia.2015.9.9.12.
3. S. U. Amin, K. Agarwal, and R. Beg. 2013. Genetic neural network based data mining in prediction of heart disease using risk factors, *IEEE Conference on Information and Communication Technologies 2013*, pp. 1227–1231, doi:10.1109/CICT.2013.6558288.
4. J. Hippisley-Cox, C. Coupland, Y. Vinogradova, J. Robson, M. May, and P. Brindle. 2007. Derivation and validation of QRISK, a new cardiovascular disease risk score for the United Kingdom: Prospective open cohort study, *British Medical Journal*, Vol. 335, No. 7611, pp. 136–141, doi:10.1136/bmj.39269.471806.55.
5. A. Patel, S. Gandhi, S. Shetty, and B. Tekwani. 2017. Heart disease prediction using data mining, *International Research Journal of Engineering and Technology (IRJET)*, Vol. 4, No. 1, pp. 1705–1707.
6. H. E. Kuper, M. Marmot, and H. Hemingway. 2002. Systematic review of prospective cohort studies of psychosocial factors in the etiology and prognosis of coronary heart disease, *Seminars in Vascular Medicine*, Vol. 2, No. 3, pp. 267–314, doi:10.1055/s-2002-35401.
7. T. Marikani, and K. Shyamala. 2017. Prediction of heart disease using supervised learning algorithms, *International Journal of Computer Applications*, Vol. 165, No. 5, pp. 41–44, doi:10.5120/ijca2017913868.

8. G. Shanmugasundaram, V. Malar Selvam, R. Saravanan, and S. Balaji. 2018. An investigation of heart disease prediction techniques, *IEEE International Conference on System, Computation, Automation, and Networking, ICSCA 2018*, pp. 1–6, doi:10.1109/ICSCAN.2018.8541165.

9. A. Golande, and T. P. Kumar. 2019. Heart disease prediction using effective machine learning techniques, *International Journal of Recent Technology and Engineering (IJRTE)*, Vol. 8, No. 1, Special Issue 4, pp. 944–950.

10. M. A. Jabbar, B. L. Deekshatulu, and P. Chandra. 2012. Prediction of risk score for heart disease using associative classification and hybrid feature subset selection, *International Conference on Intelligent Systems Design and Applications ISDA*, pp. 628–634, doi:10.1109/ISDA.2012.6416610.

11. N. Kumar, and S. Khatri. 2017. Implementing WEKA for medical data classification and early disease prediction. *3rd International Conference on Computational Intelligence & Communication Technology (CICT)*, pp. 1–6, doi:10.1109/CIACT.2017.7977277.

12. S. D. Desai, S. Giraddi, P. Narayankar, N. R. Pudakalakatti, and S. Sulegaon. 2019. Back-propagation neural network versus logistic regression in heart disease classification. *In Advances in Intelligent Systems and Computing*, Vol. 702, pp. 133–144, doi:10.1007/978-981-13-0680-8_13.

13. K. S. Purushottam, and R. Sharma. 2015. Efficient heart disease prediction system using decision tree, *International Conference on Computing, Communication & Automation ICCCA 2015*, pp. 72–77, doi:10.1109/CCAA.2015.7148346.

14. K. Dsouza, and Z. Ansari. 2018. Big data science in building medical data classifier using naïve bayes model, *IEEE International Conference on Cloud Computing in Emerging Markets (CCEM)*, pp. 76–80. doi:10.1109/CCEM.2018.00020.

15. A. Daftaribesheli, M. Ataei, and F. Sereshki. 2011. Assessment of rock slope stability using the Fuzzy Slope Mass Rating (FSMR) system, *Applied Soft Computing*, Vol. 11, No. 8, pp. 4465–4473, https://doi.org/10.1016/j.asoc.2011.08.032.

16. C. S. Dangare, and S. S. Apte. 2012. Improved study of heart disease prediction system using data mining classification techniques, *International Journal of Computer Applications*, Vol. 47, No. 10, pp. 44–48, doi:10.5120/72280076.

17. A. Cooper, G. Lloyd, J. Weinman, and G. Jackson. 1999. Why patients do not attend cardiac rehabilitation: Role of intentions and illness beliefs, *Heart*, Vol. 82, No. 2, pp. 234–236, doi:10.1136/hrt.82.2.234.

18. L. Parthiban, and R. Subramanian. 2007. Intelligent heart disease prediction system using CANFIS and genetic algorithm, *World Academy of Science, Engineering and Technology, International Journal of Medical, Health, Biomedical, Bioengineering and Pharmaceutical Engineering*, Vol. 1, pp. 278–281.

19. F. A. McAlister et al. 2019. A comparison of four risk models for the prediction of cardiovascular complications in patients with a history of atrial fibrillation undergoing non-cardiac surgery, *Anaesthesia*, Vol. 2019, pp. 27–36, doi:10.1111/anae.14777.

20. S. Habib, M. B. Moin, S. Aziz, K. Banik, and H. Arif. 2019. Heart failure risk prediction and medicine recommendation using exploratory data analysis, *1st International Conference on Advances in Science, Engineering and Robotics Technology (ICASERT)*, pp. 1–6.

21. S. Prakash, K. Sangeetha, and N. Ramkumar. 2019. An optimal criterion feature selection method for prediction and effective analysis of heart disease, *Cluster Computer*, Vol. 22, pp. 11957–11963, https://doi.org/10.1007/s10586-017-1530-z.

22. T. Mythili, D. Mukherji, N. Padalia, and A. Naidu. 2013. A heart disease prediction model using SVM-decision trees-logistic regression (SDL), *International Journal of Computer Applications*, Vol. 68, No. 16, pp. 11–15, doi:10.5120/11662-7250.

23. K. Vembandasamy, R. Sasipriya, and E. Deepa. 2015. Heart diseases detection using naive bayes algorithm, *International Journal of Innovative Science Engineering & Technology*, Vol. 2, No. 9, pp. 441–444.

24. P. K. Dalvi, S. K. Khandge, A. Deomore, A. Bankar, and V. A. Kanade. 2016. Analysis of customer churn prediction in telecom industry using decision trees and logistic regression, *Symposium on Colossal Data Analysis and Networking (CDAN)*, Indore, pp. 1–4, doi:10.1109/CDAN.2016.7570883.

25. M. Gandhi, and S. N. Singh. 2015. Predictions in heart disease using techniques of data mining, *1st International Conference on Futuristic Trends on Computational Analysis and Knowledge Management ABLAZE 2015*, pp. 520–525, doi:10.1109/ABLAZE.2015.7154917.

26. A. N. Repaka, S. D. Ravikanti, and R. G. Franklin. 2019. Design and implementing heart disease prediction using naives bayesian, *International Conference on Trends in Electronics and Informatics (ICOEI)*, pp. 292–297, doi:10.1109/icoei.2019.8862604.

27. S. Tuli et al. 2020. HealthFog: An ensemble deep learning based smart healthcare system for automatic diagnosis of heart diseases in integrated IoT and fog computing environments, *Future Generation Computer Systems*, Vol. 104, pp. 187–200, doi:10.1016/j.future.2019.10.043.

28. E. Yilmaz, and Ç. Kilikçier. 2013. Determination of fetal state from cardiotocogram using LS-SVM with particle swarm optimization and binary decision tree, *Computational and Mathematical Methods in Medicine*, Vol. 2013, pp. 1–8, doi:10.1155/2013/487179.

29. B. M. Al-maqaleh. 2017. Intelligent predictive system using classification techniques for heart disease diagnosis, *International Journal on Computer Science & Engineering (IJCSE)*, Vol. 6, No. 6, pp. 145–151.

30. S. Palaniappan, and R. Awang. 2008. Intelligent heart disease prediction system using data mining techniques, *AICCSA 08–6th IEEE/ACS International Conference on Computer Systems and Applications*, pp. 108–115, doi:10.1109/AICCSA.2008.4493524.

31. A. U. Haq, J. P. Li, M. H. Memon, S. Nazir, R. Sun, and I. Garciá-Magarinō. 2018. A hybrid intelligent system framework for the prediction of heart disease using machine learning algorithms, *Mobile Information Systems*, Vol. 2018, doi:10.1155/2018/3860146.

32. J. S. Sonawane, and D. R. Patil. 2014. Prediction of heart disease using multilayer perceptron neural network, *International Conference on Information Communication and Embedded Systems (ICICES2014)*, pp. 1–6.

33. P. P. Sunila, and N. Godara. 2012. Decision support system for cardiovascular heart disease diagnosis using improved multilayer perceptron, *International Journal of Computers & Applications*, Vol. 45, No. 8, pp. 12–20.

34. M. M. Bhajibhakare, N. Shaikh, and D. Patil. 2019. Heart disease prediction using machine learning, *International Journal for Research in Applied Science & Engineering Technology (IJRASET)*, Vol. 7, No. XII, pp. 455–460.

35. R. Roy, and K. T. George. 2017. Detecting insurance claims fraud using machine learning techniques, *Proceedings of IEEE International Conference on Circuit, Power and Computing Technologies, ICCPCT 2017*, pp. 531–536.

36. J. A. A. G. Damen, L. Hooft, E. Schuit et al. 2016. Prediction models for cardiovascular disease risk in the general population: Systematic review, *BMJ*, Vol. 353, pp. 1–11, doi:10.1136/bmj.i2416.

37. S. C. Virgeniya, and E. Ramaraj. 2019. Predictive analytics using rule based classification and hybrid logistic regression (HLR) algorithm for decision making, *International Journal of Scientific & Technology Research*, Vol. 8, pp. 1509–1513.

38. O. Ye, J. Deng, Z. Yu, T. Liu, and L. Dong, 2020. Abnormal event detection via feature expectation subgraph calibrating classification in video surveillance scenes, *In IEEE Access*, Vol. 8, pp. 97564–97575, doi:10.1109/ACCESS.2020.2997357.

39. B. Tarle, and S. Jena. 2016. Improved artificial neural network for dimension reduction in medical data classification, *International Conference on Computing Communication Control and Automation (ICCUBEA)*, Pune, pp. 1–6, doi:10.1109/ ICCUBEA.2016.7860033.

40. P. S. Kohli, and S. Arora. 2018. Application of machine learning in disease prediction, *4th International Conference on Computing Communication and Automation (ICCCA)*, Greater Noida, India, pp. 1–4, doi:10.1109/CCAA.2018.8777449.

41. S. S. Bano, and S. Mary. 2019. Heart disease prediction system using genetic algorithm, *International Journal for Research in Applied Science & Engineering Technology (IJRASET)*, Vol. 7, No. 6, pp. 2178–2182, doi:10.22214/ijraset.2019.6366.

42. C. B. C. Latha, and S. C. Jeeva. 2019. Improving the accuracy of prediction of heart disease risk based on ensemble classification techniques, *Informatics in Medicine Unlocked*, Vol. 16, doi:10.1016/j.imu.2019.100203.

43. C. Ordonez et al. 2001. Mining constrained association rules to predict heart disease, *Proc. - IEEE International Conference on Data Mining, ICDM*, pp. 433–440, doi:10.1109/icdm.2009.989549.

44. S. K. Khare, and V. Bajaj. 2020. Time-frequency representation and convolutional neural network-based emotion recognition [published online ahead of print, 2020 Jul 31]. *IEEE Transactions on Neural Networks and Learning Systems 2020*, doi:10.1109/ TNNLS.2020.3008938.

45. F. Demir, A. Şengür, V. Bajaj et al. 2019. Towards the classification of heart sounds based on convolutional deep neural network. *Health Information Science and Systems*, Vol. 7, p. 16, https://doi.org/10.1007/s13755-019-0078-0.

46. F. Demir, V. Bajaj, M. C. Ince et al. 2019. Surface EMG signals and deep transfer learning-based physical action classification. *Neural Computing and Applications*, Vol. 31, pp. 8455–8462, https://doi.org/10.1007/s00521-019-04553-7.

10 Positioning the Healthcare Client in Diagnostics and the Validation of Care Intensity

Marjo Rissanen

CONTENTS

10.1 INTRODUCTION

Care intensity can be regarded as one of the most fundamental concepts in health-care and a vital research area in clinical practice within the information system schema. Proper diagnostics are a key part of professional care intensity levels. The digital revolution and the general knowledge of patients are changing traditional roles in healthcare [1]. Boosted knowhow also gives the patient more competence to perform a care intensity assessment of their personal case. Translational design refers to the intense cooperation across the translational spectrum with the purpose of increasing stakeholder (e.g., patients, providers, and experts) engagement throughout the process and ensuring their perspectives and views are taken into consideration when designing care schemes [2]. The role of healthcare clients as feedback givers makes them co-designers in health system improvement only if their contribution is taken into account. Their active role in the midst of care processes is likewise seen as important. Care intensity and adequate diagnostics as a vivid development area requires a deeper exploration of the methods and tools that could advance practice-oriented research. Feedback systems that focus on care intensity issues and connected monitoring systems form an interesting but also a huge management challenge. Information system design and its context contain three key cycles: relevance, which refers to the system environment, its requirements, and testing schema; rigor, which integrates knowledge bases and expertise needed in design; and actual activities in the design process [3]. Intensity evaluation carried out by patients connects the environmental requirements of supporting informative systems and healthcare delivery as well as the identified problems. Such evaluation requires more serviceable feedback systems in healthcare. Thus, it is also essential to understand those knowledge bases that guide these primary development attempts. Ethical and care intensity-related frames play an important role in this context. The question is about the functionality of system dynamics and the aims of connected translational design.

Health informatics and clinical decision support (CDS) tools have the potential to support diagnostic processes [4]. There are various AI-related technologies to improve medical diagnostics and treatment [5]. The purpose of artificial intelligence (AI), in diagnostics for example, is to enhance the user's performance, not to replace human work entirely [6]. Final treatment decisions and communication are also done by humans [7]. However, the advanced forms of AI at present already have the ability to accurately diagnose certain diseases and conditions [8]. AI can save money by decreasing the use of resources while improving client outcomes [9]. However, getting the full potential of AI-related healthcare technology requires a multifaceted understanding of the complex healthcare system [5]. CDS tools and AI with human interpretation form only one part of the strategy toward more accurate diagnostics. Beside emergent and more efficient technologies, there is a need for strategies and quality policies that enhance the current situation and take care of identified problems. Better informed and engaged healthcare clients should be in focus, and their role in cooperation should get more attention.

10.2 DIAGNOSTIC PROCESS IN CHANGE

10.2.1 PATIENT AS AN ACTIVE PARTNER IN DIAGNOSTICS

Current trends promote profound communication between patients and healthcare providers. The purpose of patient engagement is to improve health outcomes by enabling patients and their families to contribute valuable input into diagnostics and decision-making [4]. Kraetcshmer et al. [10] noticed that more than 60% of patients preferred shared decision-making. Nowadays, patient participation is a basic condition for good care and a key factor in the redesign of healthcare processes [11, 12]. Moreover, patient loyalty appears highest when patients and care providers cooperate in diagnosis and treatment decision-making [13]. Nevertheless, the role of patients in safer diagnostic processes requires more consideration and debate.

Patient involvement improves diagnostic safety [14], and more cooperative consumers may reduce the risk of diagnostic errors [15]. Two-way communication also improves clinical reasoning [16], and the patients themselves can play a critical role in bringing to light diagnostic errors [17]. Feedback on diagnostic errors aids physicians in their work [1]. In their article "The Cost of Satisfaction," Fenton et al. [18] noticed that higher patient satisfaction is connected with higher inpatient use and drug expenditures and, surprisingly, with lower emergency department use and increased mortality. The result is interesting. Are the most satisfied patients perhaps those who never question doctors' decisions? In uncertain situations, patients typically try to seek a second opinion or obtain a second opinion on their own, sometimes even with the encouragement of their own physicians [4]. These processes are more straightforward because of the enhanced knowledge level of healthcare clients [19]. Healthcare organizations should provide more intensive opportunities for patients to take a role in the diagnostic process.

10.2.2 DIAGNOSTICS AS A KEY ROLE IN PROFESSIONAL CARE INTENSITY

Amenable mortality means deaths that are avoidable in the presence of timely and effective care [20]. The care intensity perceived by patients refer to such signals that should be investigated properly. Information related to care intensity relates to process intensifiers. Care intensity level is evaluated by examining aspects such as the prevalence or type of specific service units, like testing intensity (e.g., laboratory, radiology services, diagnostic imaging), interventions, medical treatments or treatment styles, treatment styles in intensive care, hospital days, health systems, models in physiotherapy, physical activity and exercise intensity, or medicine recipes. Therefore, the degree or type to which healthcare clients receive different kinds of interventions and services during their care episodes or consultancy appointments gives information about case-connected intensity level [21].

Problems in coordination may complicate access to an adequate care level [19] and, consequently, the success in care coordination also reflects aspects of care intensity. Communicative interaction with a patient and between health professionals or the dynamics of supporting information channels play a major role in professional

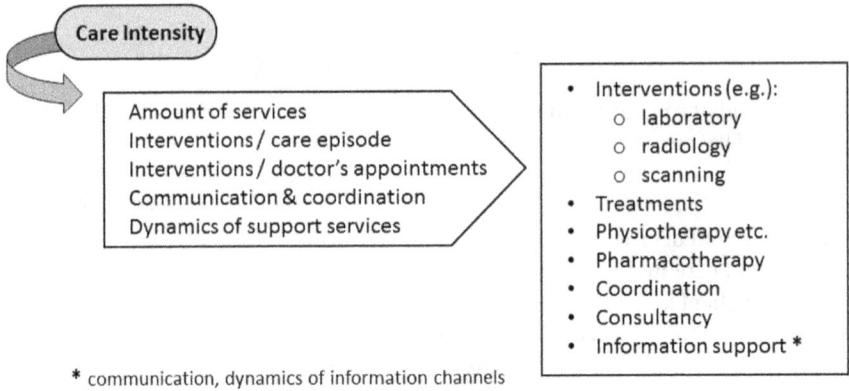

FIGURE 10.1   Elements integrated into the care intensity level.

care, and these can likewise be regarded as issues connected to the intensity of care (Figure 10.1). The purpose here is for the patient to get the right diagnosis, optimal care, and relevant information to support the care alongside the process and with all connected sub-processes like diagnostics, treatments, curative options, recovery phases, and prevention [21].

Higher care intensity is often connected to better health outcomes [22, 23, 24] and lower mortality [25, 26]. This is also true in intensive care units [27]. However, this connection is not always straightforward [18]. Care intensity is not only connected with process- and customer quality. Justice means the duty to treat individuals fairly, free from bias, and based on the medical need [28]. Thus, care intensity has a decisively deep connection with the ethical principles of care [29].

An adequate care intensity level requires first and foremost accuracy in diagnostics. Care intensity level in diagnostics is easily perceived as a task that is primarily connected with the prevalence of clinical investigations and tests. However, the success of communicative procedures between patient and doctor and the intensity of professional negotiation and consultations between health professionals play remarkable roles in producing the right diagnostics. A diagnostic management team represents, for example, a model where experts primarily based in the clinical laboratory advise physicians on the selection and interpretation of questions in diagnostics [30]. Consumer-targeted informatics, which is targeted for preventive healthcare, is a tool that supports clients in controlling possible symptoms of diseases and aids in inspiring clients to seek professional help in uncertain situations. Feedback systems integrated into diagnostics also give information for evaluative purposes in quality management and affect the intensity of diagnostics in corrective purposes. Health informatics have a supporting role throughout the care process for professionals (EHR with enriched functions, CDS, AI, feedback data), and thus the dynamics and utilization rates of these supporting services also have a role in the care intensity scheme in diagnostics. Therefore, care intensity in diagnostics is not only an aspect integrated predominantly in process and customer quality, which is actualized through communicative procedures (doctor–patient, doctor–doctor) and

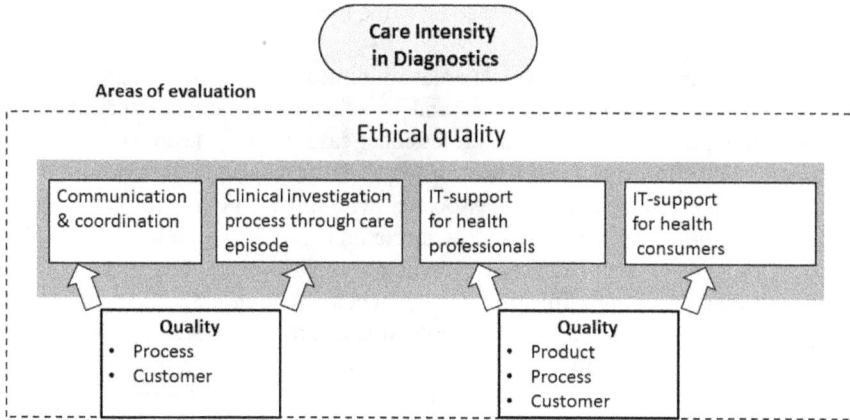

FIGURE 10.2 Care intensity evaluation in the area of diagnostics.

clinical investigation intensities (diagnostic investigations, clinical examinations, decision-making processes). Thus, the intensity in clinical practice is also connected to product quality emphasis (information technology support) and finally appears as an ethical aspect of care (justice in clinical practice) [29] (Figure 10.2). All these areas need their evaluation protocols when enhancing service quality in the area of diagnostics.

10.2.3 UNCERTAINTY AND TOO LOW CARE INTENSITY IN DIAGNOSTICS

Uncertainty in diagnostics refers to the inability to provide an accurate explanation of a patient's health problem [31]. Medical errors are one of the five most common causes of death [32]. Diagnostic errors are a leading type of malpractice claim, which is more likely to be associated with deaths than other types of medical errors [33, 34]. According to estimates, at least 1 in every 20 outpatients experience a diagnostic error, such as missed, delayed, or incorrect diagnoses every year and sometimes with upsetting consequences [4, 31, 35]. However, detecting diagnostic errors is demanding in clinical practice [4], partially because of the lack of an effective measurement praxis [35]. Disease complications and delayed diagnoses are extremely expensive and cause patients additional stress and dissatisfaction [4, 30].

There are various reasons for diagnostic errors, including the uncertainty of clinicians in ordering tests or interpreting test results [36]. Furthermore, testing intensity can be too low (not enough specific testing set or no tests at all) to find the right diagnosis. Treatment intensity has traditionally been the highest for pharmacotherapy and correspondingly significantly lower for laboratory examinations and imaging, which are, however, important diagnostic tools—doctors in primary care order laboratory tests for slightly less than one-third of patient visits [36]. Medical imaging has a solid role in diagnostics. For many conditions, imaging is the only non-invasive diagnostic method available [4]. Its low utilization in diagnostics takes place for many reasons. Uncertainty in selecting the appropriate test and improper radiology education at medical school represent some of the mentioned professionalism-related reasons [4].

Savings are easily made in critical areas of clinical practice which play a minor role in the overall cost policy [30]. The random reduction of the laboratory budget is one such example [30] that can risk professional care intensity levels. Efficiency in healthcare is not attainable alone with cost-focused thinking and reform [37]. Missed or delayed diagnoses are ultimately associated with losses in other categories of the budget [4]. Laboratory testing costs form approximately 3% of all clinical costs [30]. However, diagnostic errors contribute as much as 70% of medical errors, which means that the process of clinical reasoning needs attention [38]. Random savings in key investigation resources can risk professional care with costly and devastating consequences [30]. "An efficient healthcare system will be able to produce more care, or be less expensive, all else being equal, than an inefficient system"; however, inability to measure quality and efficiency often poses a problem in implementation science [39]. Health professionals themselves form the most substantial cost factor. Personnel costs form the major portion of total costs in hospitals. Emergent technologies that aid health professionals to reorganize their time management and work protocols are future aids for controlling the cost policy in healthcare [40]. Research in health economics should focus more intensively on these areas of major costs and potential areas of major savings. New, less invasive, investigation and treatment technologies likewise offer possibilities for better diagnostics, shorter care periods with minimized complication risk, and safer care. Therefore, modern technology may have an advantageous effect even on cost policy.

Health IT is a potential tool for improving diagnostics [4]. Even if health informatics has a remarkable role in advancing outcome quality in healthcare [41], diagnostic delays can also be associated with problems in health informatics [42]. Many of the reasons for health-information-related diagnostic delays are similar to problems in managing communication or notifications of test results, poor interoperability or data visualization, information overload, technical problems, and data entry problems [42].

10.2.4 NOTICING GROUPS WITH A RISK OF UNDER-DIAGNOSIS

The healthcare of senior citizens is insufficient and disorganized in many countries [43]. Therefore, under-diagnosis and under-treatment increase the amenable mortality rate among elderly patients. The occurrence of the disease is less detectable because classical and general symptoms among older people are often missed, making the diagnostic task more complex [44]. Care intensity of the elderly population is particularly challenging due to the potential multi-problem nature of elderly patients [45] and the built-in bureaucracy of care systems (e.g., time constraints on admission). Some reasons for under-diagnosis are related to attitudes, for instance, in the case of Alzheimer's disease [46]. Respiratory dysfunctions [47, 48, 49], skeletal disorders [50, 51], endocrinological diseases [52], and heart diseases [53] are examples of under-diagnosis among older patient groups [54]. Inappropriate medication and medication errors also cause problems among the elderly [55].

Problems in coordination may complicate access to specialty care among chronically ill elderly patients [19] even if proper coordination and access to community health services can reduce hospital demand [56]. Less prominent treatment intensity is noted among the elderly [57] despite the availability of less invasive and safer treatment and anesthetic procedures. Physical condition or advanced age can also be seen as an obstacle to efficient care procedures because of inadequate preoperative assessment [58]. Elderly patients should not be excluded from intensive care because clinical outcomes are also favorable among them [59], and a potential link between increased treatment and improved survival has been observed [25]. In addition, adequate palliative care services may sometimes be unavailable [60]. Among the elderly, the right diagnosis is therefore only the first step. The evaluation of treatment policy and its adequacy analysis requires attention among the elderly population and in general palliative care policies. Adequate diagnostics and access to medically justified care among the elderly population are measures that tell if values are constructed around equity in clinical practice in healthcare.

The development of participatory communication between the medical patient and the relatives (family) is important in improving the care intensity for the elderly population [61]. There is even a greater need for applications that empower this collaboration as the proportion of the elderly population increases. Supporting ITs for older clients have various known challenges in interface design. Most of the applications focus on the independent elderly [62]. An analysis of these factors also indicates how seriously ethical dilemmas and intensity aspects are integrated into health policy and daily clinical praxis.

10.3 ACCESS TO CARE AND DOCTOR APPOINTMENT

10.3.1 Awareness of the Problem and Referral for Treatment

In diagnostics, time is essential and some disorders must be diagnosed immediately [4]. As is known, getting the client to the right place of treatment by the right experts as quickly as possible is often one of the most critical stages in professional diagnostics and care. Improving access to specialty care is connected with the proper coordination between primary and specialty care levels, and this issue forms a critical task in the delivery of health services to ensure that patients are diagnosed properly and receive timely and effective treatment [19]. Poor coordination between hospitals is associated with greater costs as well [63]. Problems in accessibility can be based on clinical decision-making, in information management, management in patient flows, or in the monitoring of the quality of care and system performance [19]. Improvement in the access to specialty care, for instance, among under-diagnosed, non-emergent, or patients with chronic conditions, requires more attention [19].

Nowadays, people often monitor their own health intensively and usually independently seek a great deal of information about their health problems. Even self-diagnostic applications are used to some extent. When a person concludes that it is necessary to contact a doctor, in public healthcare he or she usually contacts office staff to schedule an appointment with the doctor. In some service models the process then continues to the nurse's appointment, where the need for further medical attention is assessed. The primary purpose is to avoid any unnecessary doctor consultations. In such a case where the patient is eventually referred to specialist care, the situation will involve additional bureaucracy and a waste of time for everybody.

The private sector has a policy that includes a direct channel for first-time contact with a physician without intermediaries whenever a problem arises. This policy allows the patient to immediately discuss their problem with a specialist. If the situation so requires, the doctor may endorse the patient directly to a specialist in a specific field or start the process otherwise. This type of access decreases bureaucracy when the so-called intermediate medical contact (non-specialist, nurse, or general practitioner) is omitted, or the process can start immediately.

Evaluation mechanisms are needed to monitor the meaningfulness of different approaches in access policy. Parallel models could also be available in public healthcare. Such models could offer two first-contact possibilities: a possibility of a doctor consultation and a possibility of a nurse consultation. Not all problems require medical expertise. Most of the time, the client can assess by themselves if their first contact requires a doctor consultation or if in fact the nurse can handle the situation. Finally, in demanding cases, nurses would also guide the patient to further care. Parallel ways in access policy may decrease unnecessary bureaucracy and prevent delays in critical situations (Figure 10.3).

The doctor, who is specially trained to handle the initial stage, could accomplish the screening work and treatment guidance. In public health, guidance for access and treatment in the first-contact stage is normally provided by health professionals who are less trained in both primary healthcare and specialist care. Especially in specialist care and emergency care, the care guidance decision may be critical to ensure rapid diagnosis and selection of the right treatment site. To illustrate, if and when an older patient with a chronic disease is brought to a hospital, the placement decision between different units (conservative vs emergency supervision) may also be made by a less trained employee within the current practice, which may jeopardize patient safety. For the same reason, the queuing systems of emergency patients should be controlled and supported by high-level specialists. This approach also requires evaluative mechanisms, which follow that adequate quality could be actualized in clinical practices via the access policy. Although the patient may finally provide feedback on excessive bureaucracy or mistakes in his or her situation, it is rare that such feedback will no longer benefit the patient.

There are various examples of improved access practices. Referral management via centralized intake has been implemented to improve access to specialty care in Canada. Specialty-based clinics offer improved urgency-based access to specialists. A group of academic gastroenterologists adopted a centralized referral intake system known as a central triage, which means an entry model (SEM) for referrals rather than the traditional system of individual practitioners managing their own referrals and queues [64]. In a broader scope, this idea would work throughout healthcare. A model, where so-called first-time appointment physicians with a wide range of expertise in different specialties and the coordination of care would make an initial assessment on the patient's situation, could be beneficial. While the procedure

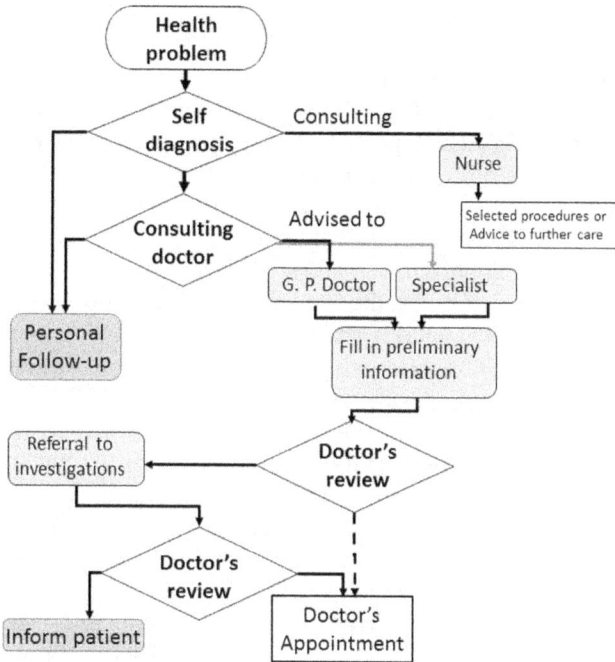

FIGURE 10.3 Parallel and alternative access to care.

where a doctor makes an initial assessment may sound expensive, the practice can ultimately result in a significant reduction in bureaucracy and enhanced diagnostic accuracy, which would mean savings for the same reason.

Beside more flexible access to doctor appointments, there is a need to create opportunities to consult general practitioners and specialists regarding questions that might bother laymen but do not require immediate access to a physical appointment. In such non-emergent situations, virtual services, possibly strengthened by AI applications, could answer the client's consultancy question. Such professional response to the presented problem may be clarifying enough and make a separate physical appointment unnecessary. However, such services can also advise clients if a specific investigation might be preferable in the client's situation. Such quick advisory comment may prevent possible delays in diagnostics. Deploying these kinds of services to healthcare may be cost-effective as well. Such services may also complement traditional phone consultancy services provided by nurses or physicians to serve patients.

10.3.2 Cooperative Preparing for a Doctor Appointment

Despite values that underline patients' essential position, the patients seem to be easily overlooked in medical consultations [65]. Additionally, within the consultation room, the relationship between doctor and patient remains unequal, not only because of the doctor's behavior but also because of the patients' own perspective on what makes the consultation effective [66]. The doctors' responsibility in opening the patient's agenda is emphasized [66]. Today, the patient is given many different instructions on how to prepare for a doctor's appointment so that the solution to the problem is as straightforward as possible. However, such guidance does not always enhance the situation if time limits complicate the appointment.

The first contact between patient and doctor can be a "too ad hoc situation" for both parties. Today, the normal situation builds up so that most often the doctor receives only superficial information about the symptoms of a patient before the first appointment. This means that the first appointment is when the doctor becomes acquainted with the patient and his problem properly for the first time. Given that consultancy appointments are often scheduled and have limited time, problems may arise if a patient has more than one problem. Time constraints may mean that in the situation where there are several problems, only a few of them can be handled in a given time [31]. As a result, the overall picture of the patient's actual status may be blurred. Owing to time limits, the overall situation may be interpreted as a set of separate problems, and the potential syndrome and synergistic effect of symptoms may remain unnoticed. This may lead to a situation where symptoms are treated without seeing and treating their causes. The patient may also attempt to focus on the most difficult or disturbing problem and forget to mention things that could shed more light on the actual and most serious cause of the problem. In such a case, the actual diagnosis may be disrupted and the patient is referred to the so-called palliative care processes, which could in turn mean prolonged diagnosis and treatment, additional costs, and stress to the patient.

Technology provides opportunities for both the patient and the doctor to prepare in advance for the doctor consultation. However, how to develop such approaches and methods that are safe and attractive for customers is another question. The described streamlined model represents in cases of straightforward health problems well-known telemedicine applications, where the patient and the consulting specialist do not even meet each other in physical context. In a streamlined model, a prospective snapshot of the patient's situation, which the doctor will review in advance before the actual appointment, could diminish such problems that may disturb the efficiency in consultations. A form with structured and open-ended questions to be completed by the patient well in advance should be made available to the patient. This may be done from the patient's own mobile phone before the actual appointment with the physician. More detailed and comprehensive prospective information allows the physician to examine the patient's situation carefully prior to the actual appointment. Prospective information may make it possible that selected clarifying tests (laboratory and imaging) can be done in advance. These prospective data produced by the customer may be a temporary file, which will be replaced by appropriate documentation as the situation becomes clearer. The prospective file should include key points to help the physician understand the patient's situation.

However, not all patients are fully able or even interested in encoding their prospective data to digital interfaces, and there is no point in being pressured to work only digitally. For these patients, they should be offered the opportunity to write their prospective data with paper and pen before their consultation time or to deliver their information in advance. In addition, nurses may help input the data into digital form or generally aid in creating the prospective data in cooperation with the patient. When the prospective information is finally in digital form, computerized decision support tools can be utilized even with AI before the actual consultation time.

Good doctor–patient communication is a challenge and a reason for the majority of client complaints [67]. Many ethical dilemmas come from inadequate doctor–patient communication [28]. If the prospective phase is done properly, then the time during the actual appointment can be used for the patient's clinical examination, communication, analysis, and a more detailed investigation. In case necessary routines (problem clarification, brief history, tests) are already under consideration, the collaborative patient will be the focus in the examination room. Disorganized patient information may cause cognitive burden in an appointment situation [4]. A streamlined model may also reduce the cognitive burden of clinicians because they can prepare for the appointment thoughtfully in advance. In addition, many problems may be resolved on the basis of the delivered pre-informative data and completed laboratory and imaging tests such that a separate doctor's visit may not even be necessary in each case.

If preparatory assignments are done on both sides before the doctor appointment, this may in part decrease problems in diagnostic accuracy and give both sides the needed analysis time. These changes in general routines may sound laborous and costly. However, such a model may increase the accuracy in diagnostics, which typically means increased cost efficiency. Some problems may be resolved even without physical appointment. However, such changes in protocols require that preparatory assignments must be taken seriously by both sides in practice (Figure 10.4).

However, note that such procedures do not fit all patients' styles. Similar to the development of healthcare training applications, "one size does not fit all" [68], and so the design of health protocols should also have space for choices. Some patients may prefer the traditional model because of security reasons. As is known, no healthcare data security system is foolproof in spite of the new technologies [69, 70]. The

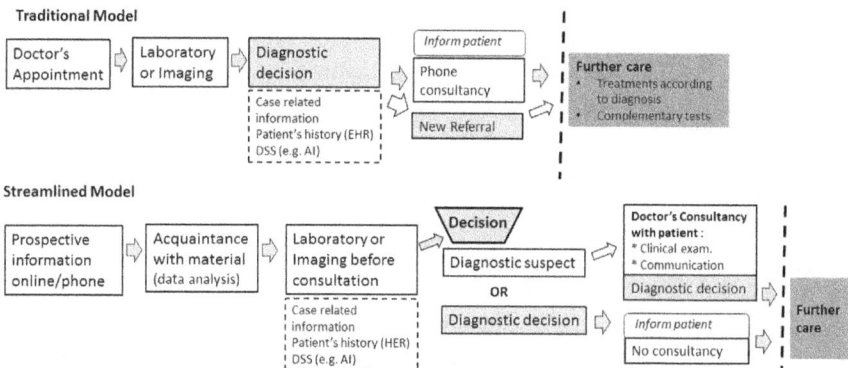

FIGURE 10.4 Parallel models for a doctor appointment.

question should be about presenting an option that the patient can consider when trying to streamline one's case and appointment policy.

10.3.3 Supporting Patients' Self-Care and Preventive Care

Preventive healthcare is seen as the most cost-effective treatment policy. In the digital age, prophylactic applications play an important role in the correct treatment of patients. Some diseases may progress extensively before presenting noticeable symptoms. This means that the diagnosis can be delayed because reasons are missed that may require a test appointment. Certainly, key laboratory findings do not always indicate or give clear and direct indications of latent illnesses. However, in some cases, monitoring of key blood characters could lead the patient to a doctor's appointment and to receive a more accurate workout diagnosis before the overall situation worsens. Mass screening techniques with various blood sample tests are used to identify patients with abnormalities, such as those with cardiac, celiac, or colorectal disorders. Screening-type laboratory controls would defend their position, especially with older patient groups or certain risk groups, particularly due to their cost-efficiency and easy accomplishment. If early symptoms can be detected, then the process to the right diagnosis and optimal care can be straightforward and fast enough.

10.4 FEEDBACK POLICY IN DIAGNOSTICS AND INTENSITY IMPROVEMENT

10.4.1 Hearing Customers' Voice and Experience

Patient feedback forms a critical part of quality management in care. Patient experience is defined as "the sum of all interactions, shaped by an organization's culture, that influence patient perceptions, across the continuum of care" [71]. Patient experience is associated with the ethical aspects of care [72]. Feedback-related information supports and complements organizations' own intensity monitoring data. Health consumers today can give more feedback than ever before about different topics connected to care by using various feedback channels. Systems that facilitate feedback assist with the diagnostic process [73] and, therefore, feedback from patients and their families should inform practice more efficiently [74]. Care intensity is a task in which evaluation should happen throughout the healthcare process. For synergy reasons, the coordination of intensity issues likewise requires attention to health policies. All types of multifaceted patient feedback could be harnessed more efficiently to utilize practice in hospital settings [75].

Perceived non-satisfactory level of care intensity and connected quality problems as well as concerns about the accuracy of diagnostics form the major topics of complaints in healthcare. Such issues deal with disagreements over treatment decisions, insufficient information, lack of confidence, unavailability of physicians [76, 77] or nursing care [78], and lack of communication [77, 79]. Moreover, the opinions of elderly patients and their connection to design seem problematic [80]. Patient-reported outcomes do not always provide information describing the cause of poor care; therefore, there is a need to integrate and interpret outcome measures with other outcome data [81].

10.4.2 FEEDBACK CHANNELS AND FORMATS

It is proposed that health providers should offer their patients multiple feedback channels [82, 83]. A chance for anonymous feedback [84] has its place as well. Safety feedback systems at all levels in healthcare and their effective use form a dynamic part of quality policy [85, 86]. In general, digital health technologies include multiple possibilities for communication, such as apps, text messages, emails, internet, chats, and voice agents, all of which can serve also as channels for feedback. Aside from the various instruments for feedback, systems should also be able to respond to the given feedback [82].

Given that standard questions can fail to capture all the needed information, additional narrative (open questions) information is considered to be equally useful [87]. Customized hospital-specific feedback systems are also needed [84]. Lengthy and time-consuming feedback forms may discourage patients from giving feedback [84]. Patients have mixed views about online feedback systems [82]. They are concerned with aspects of accessibility, privacy, and security as well as the real impact of complaints on the practice [88]. Specific feedback given just-in-time aids in identifying problems and can lead to improvements in a timely manner with greater validity [84, 87].

10.4.3 DIFFICULTIES IN CAPTURING THE CARE INTENSITY FEEDBACK OF PATIENTS

It is important that patient feedback that connects intensity issues in the data is captured by certain phases of care episodes in an accurate way. There are many reasons why the feedback on perceived care intensity level does not actualize in a satisfactory way in current health practices. When feedback is given through formal channels, there are less structured formats to answer, especially for care intensity purposes. This means that such feedback information that would be needed for improvement actions does not exist. If feedback forms are multifaceted, then gathering all kinds of feedback on perceived care and specific issues of care intensity may not get enough emphasis.

The purpose of open questions is to complement the given information, but the patient is not necessarily able to specify details in a way that could lead to improvements. When improvements take place, specific information is the most valuable basis for identifying and justifying changes in health protocols. Timely feedback with a clear indication of the source of the problem aids the provider in making improvements [81]. Therefore, a more in-depth phase-related categorization can aid consumers when they give their feedback about the perceived care intensity level and related problems. Structured and detailed information offers such knowledge which can ease the work of quality auditors when they operate with feedback information and perceived problems of care intensity.

Diagnostics forms the most critical part of the care process, which means that its accuracy should be evaluated as a common procedure in all clinical practices. Diagnostics, for example, is a phase that contains several items. Feedback on diagnostics should be seen as a standard procedure, and so clients should be encouraged to give feedback on the diagnostic process after diagnostic decisions. If the patient complains about non-satisfactory diagnostics, then the question is as follows: Which part or item in the diagnostics phase represents the non-satisfactory level? Where and how a diagnostic error occurred must be identified in clinical reasoning [38]. Checklists for health professionals are an approach to increase diagnostic safety [4]. The patient is not always aware that all the details are necessary for improvement actions, or he/she is not able to mention the needed attributes specifically enough if these options are not presented clearly in question form or in an application. The quick response of patients in the diagnostic phase may be especially valuable in university hospitals, which also function in an educational capacity. Clients may have questions and may need added information especially in cases when the diagnosis remains open or is uncertain.

Not all patients have enough spirit to give feedback in written form through formal feedback channels, such as apps or tablets. Speech-enabled healthcare technology is time-saving and can increase accuracy [83]. In addition, simple systems for quick feedback (press button for agreement/disagreement) have their place. Intensity feedback can and should actually be spontaneous in daily engagement in clinical practice and given directly to health professionals in the midst of the care process. Although this is actually the most preferable way for discussion, it is not always possible because patients may also need time for their own deeper analysis.

10.4.4 Feedback as a Normal Procedure

Intensity feedback and its integration into health records should be regarded as a normal procedure. Patients may feel that giving negative feedback connected to care intensity issues means they are criticizing the given care and, consequently, they may get less quality care because of their complaints or notices [89]. Such attitudes may diminish later if care intensity evaluation is connected as a normal process to health protocols and formal patient feedback. In this respect, information system development in an intensity feedback area and its connection to patients' data in real time is one of the most urgent development issues. The opinion of patients about professional care intensity level may sometimes be totally different from that predicted in the patient's case by professionals. This phenomenon is not surprising because health professionals may also differ in their perception of patients' care intensity [90]. Patients feel that they have limited possibilities to influence their care process [91]. Therefore, health professionals must be prepared to adapt to a new intensity cooperation. This issue can motivate health consumers to engage more efficiently in their health matters and take a more active and serious role in their care as well as in care intensity issues and its evaluation.

10.4.5 Various Areas of Patient Feedback Connecting Care Intensity

The healthcare process contains different phases. Typically, the diagnostic phase is followed by the care or treatment phase. Afterward, the recovery and follow-up phases are planned for new preventive steps. In general, intensity evaluation focuses on the diagnosis phase in terms of its comprehensiveness and accuracy, on the care and follow-up phases in terms of the overall degree of professionalism, and on the preventive phase in terms of coordination and efficiency (Figure 10.5). The patient can also evaluate the success of the coordinative actions during the whole process.

The actual care process is often managed between multiple units. This makes its coordination challenging even for professionals. For this purpose, enough structured evaluation forms of perceived care intensity make it possible to capture the opinions of consumers during the whole process. The intensity evaluation must gather all professionals needed in care, including medical care given by physicians, care given by other health professionals, and general support services that integrate intensity issues.

Consumer-targeted informative systems have a supporting role in care intensity issues in different phases of a health episode; available relevant medical information about health issues forms one critical area in care intensity support. Connected information has a role in patient activation and connects health outcomes [92]. Poor system availability and usability [93] of supportive information as well as a lack of support for elderly users [94] may, however, decrease the potential of connected information as a process intensifier [21]. These systems aid consumers in care intensity issues by delivering connected information in each phase, and thus their added utility value as well as aspects of clinical practice and interventions should be evaluated by patients. It is underlined that design of HIT solutions should focus on the following core areas: clinical effectiveness, patient safety, and patient experience [72]. Digital aid support should be integrated with ease in the same phases recognizable in a care episode (prevention and general information, information connected to care and treatment and post-operative care, and issues of coordination) (Figure 10.6). The patient has to evaluate how successfully information offered at certain phases supports the desired knowledge level, and how easily phase-connected feedback can be

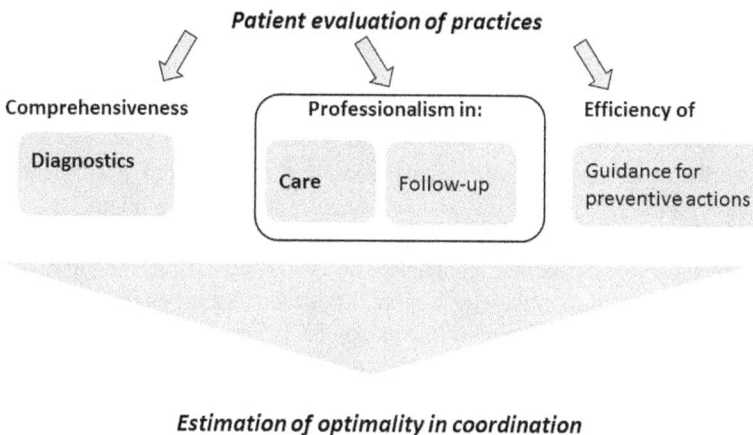

Patient evaluation of practices

Comprehensiveness

Diagnostics

Professionalism in:

Care   Follow-up

Efficiency of

Guidance for preventive actions

Estimation of optimality in coordination

FIGURE 10.5   Evaluation content in phases in clinical practice.

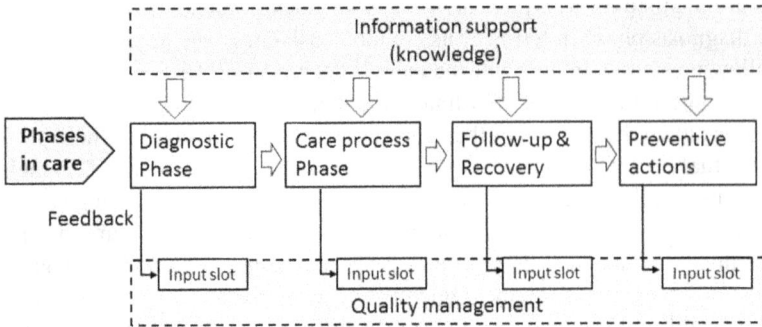

FIGURE 10.6 Information system support in knowledge sharing and focus of patient evaluation.

TABLE 10.1

Areas of Consumer Evaluation of Care-Related Health Information Systems

Character	Content for evaluation
Clarity	Are needed issues and topics easy to find?
Usability	Flexibility of the information and feedback channel
Understandability	Clarity of content and presentation
Comprehensiveness	How complete guidance, information is; are treatment and curative options, and complication risks clearly presented?
Accuracy	How specific is the content?
Coordination	Logic in presentation

posted. Table 10.1 contains attributes [95] that are seen as meaningful when patients evaluate the functionality of these assistance systems in an intensity scheme.

10.4.6 JUST-IN-TIME FEEDBACK

If the feedback on aspects of care intensity is given after the care period, then much valuable information may be missed, and such information would not be beneficial for the patient in question, or the feedback would not reach the people or units that were targeted by the patient's feedback. The question is about the delay and may decrease the motivation of patients to give proper feedback. Feedback given just-in-time can utilize the patient in an efficient way [87]. Negative messages that connect care intensity may be urgent and require timely management; therefore, their connection to the patient's health information is necessary. A patient may give all kinds of feedback connected to other necessary services during the healthcare process. However, not all aspects may be connected to care intensity issues but to other service aspects instead. This type of feedback, as well as all kinds of positive feedback, would be handled on a daily or weekly basis or after the patient leaves the hospital. Such feedback does not always need a direct or immediate connection to the patient's health information or even an immediate response.

10.4.7   HOW TO ORGANIZE AND MANAGE THE INTENSITY-RELATED EVALUATION OF PATIENTS?

Organizational culture and systems in healthcare often move too slowly toward an efficient response and action taking [96], and the lack of organizational structures for handling complaints lowers the efficient responses to patient feedback. Less than one-third of the complainants were satisfied with the statements from health-care providers in a Swedish county about their complaints [77]. There is a need to more properly understand the "big picture" when handling feedback data [96]. Tools and methods to synthesize information better for practical purposes are likewise essential [97]. The evaluative feedback of patients on perceived care intensity levels should be managed and analyzed in a way that could lead to actions in situational cases. Timely, sufficient detailed data and flexible channels can aid in this challenge.

However, wider problems in care protocols or informative systems often require extensive corrective actions. By recognizing such internal problems, healthcare providers demonstrate that they are prepared to change routines [77]. It is important that the intensity evaluations given by patients are not only on paper or in an information system context. Finally, feedback information should not be regarded as a subjective opinion given by non-professionals because a patient is ultimately the expert when it comes to experience in his/her personal case [40].

10.4.8   HOW TO MATURATE SYSTEMS FOR THE INTENSITY EVALUATION OF PATIENTS?

The development process for more sensitive feedback forms and channels specifically targeted to research more precise aspects of perceived care intensity may differ between institutions. Organizations have their own plans on how to organize feedback systems better to serve patients and health professionals. The following table presents a step-by-step representation describing such a process at the macro level. The team with multidisciplinary knowledge may be the most innovative in terms of defining and testing novel ideas. For this purpose, professionals are needed from the required medical specialty areas as well as specialists from administration and information system management.

The team formulates first a mission for the evolution plans of patients' intensity evaluation, which covers clinical practice, connected IT systems, and their management. An analysis of the current feedback system should at least cover the aspects of usability, utility, flexibility, up to date, and security policy. In system improvement, comparisons between the different practices in patient feedback systems may bring new insights for achieving further development. New routines may require tutorial programs for health professionals. The question might also be related to a change of values, which requires a multifaceted approach from coaching programs. Even if

TABLE 10.2

Steps for Intensity Feedback System Maturation

1	Gathering a group of professionals from various specialties
2	Mission statement and tools (values, methods, evidence, decisions)
3	Analysis of current intensity feedback systems (weaknesses)
4	Comparisons: examples of praxis in intensity evaluation
5	Development of good practices, customizations
6	Development of connected models
7	Model validations
8	Model integration to system development
9	Model integration to clinical practices
10	Model integration to management routines
11	Development of tutoring, instructional programs, and processes
12	Stabilizing new protocols for intensity evaluation
13	Regular process evaluation and improvement

protocols are changed on a minor level, enhanced models and protocols may require validation rounds and regular evaluation. Care intensity evaluation is its own section and should be differentiated from general feedback systems because of its "just-in-time" nature. It is important for team members to understand the critical role of such evaluation in quality improvement (Table 10.2).

10.4.9 ENHANCING THE MEDICAL EXPERTISE OF CLIENTS

In modern society, the healthcare client's own competence enables the patient to take a more active role in his or her own care process or give feedback when necessary. It would be desirable for the patient to have the broadest possible knowledge of his or her own health problem and not only to be better equipped with regard to one's own diagnosis or suspected condition. It is good that the patient has the competence to evaluate the decision of a physician, to a certain degree, in an insecure situation. Evaluation guidelines may sometimes be designed according to the evaluation criteria of other subject areas and may thus, e.g., ignore the assessment of specific features of health websites [98]. However, there are also customized organization-specific information channels available. In the past, homes usually had comprehensive sets of medical books from which laymen used to obtain very exact information about their problems. Nowadays, patient-developed medical knowledge databases should be constructed to provide the most exact and comprehensive knowledge, which does not differ from the knowledge level of doctors. When specific medical vocabularies are combined with databases, not all information may need to be translated into the so-called vernacular. However, narrower formats are also needed. A compact presentation form gives an overview, general guidelines, and core information for those who are mainly interested in diffused and compact information sets [54].

The traditional book format with a detailed table of contents provides one model to present medical information with an easy-to-adopt structure. Explanatory video clips for subjects such as surgical procedures, investigations, physiotherapy aids, and views of care environments can then be added to enrich the traditional presentation format. Similar specificity is also required for patient-specific drug databases, which should be provided with glossaries that illuminate pharmacological terms. Doctors and nursing staff may still use professional language and idioms when discussing with the patient. Hence, it is only a positive thing that the patient also becomes somewhat familiar with professional terminology through the advisory material.

10.4.10 CONSUMER-CENTERED FEEDBACK SYSTEMS FOR INTENSITY ENHANCEMENT

Care intensity and diagnostics represent an issue in which the responsibility for evaluation is a primary task of health professionals. However, care intensity and its essence, namely, proper diagnostics, is primarily connected with the patient in question. Therefore, the patient has an important role in this process. When the adequacy problems of care intensity are taken into consideration just-in-time, communicative processes can be enhanced [99]. Misunderstandings can also be better avoided if unclear issues are handled just-in-time. The purpose is for feedback systems to support all areas of care (prevention, access to care, diagnostics, care process, follow-up, and coordination). Achieving this requires the development of more specific feedback systems that integrate each part of the episode. The development of more sophisticated feedback systems requires new technical properties and more intense content design.

Care intensity evaluation is a task for all parties, and therefore its deeper consideration in all levels of management is of primary importance. Patient feedback about care intensity issues often happens randomly in the moment. It is often given in situations when serious mistakes in care are noticed. Therefore, only a few patients are ready to give any feedback at all. Minor problems that connect intensity issues could easily remain in the patient's mind. In addition to instant feedback possibilities, detailed structuring is likewise needed in the feedback data interface. Such feedback options may be necessary in demanding cases and will ease patients' concerns about giving information that is specific enough for deeper analysis and improvements in feedback management. Care intensity evaluation, however, is not limited to the complaints or feedback given by consumers. It should include normal quality evaluation and management routines, even if everything appears fine. This is a responsibility also of professionals other than those involved in clinical practice. A deeper role for patients in the evaluation of care intensity is a development challenge for the more sophisticated information systems in healthcare.

As previously stated, the perceived care intensity level is the most common complaint area in customer feedback. This item should be extracted from the overall patient feedback in its own phase, and its management needs special attention. Equally important is how efficiently and fast this feedback information can be connected to patient data and observed in care procedures. Intensity evaluation of care is a shared challenge in healthcare because it does not only connect clinical practice.

At present, information support systems are a strong part of care intensity in the knowledge sharing of different phases of care. The real utility value of consumer-targeted information support systems is thus a meaningful evaluation target. All information connected to the feedback on intensity issues can be monitored and saved as "big data," which directs health professionals make quality improvements in intensity issues. More detailed evidence of intensity-related information with supporting analysis tools aid in recognizing the problems of care and can decrease rates of amenable mortality. This all results in improved synergy policy.

10.5   CONCLUSIONS

Adequate diagnostics in healthcare is a multifaceted challenge that is an essential part of the professional care intensity level. It is connected deeply with service protocols, IT, feedback, and quality systems. The healthcare client has a remarkable role as well. Cooperation with patients should be valued more dynamically in this change process. Therefore, this multidimensional challenge cannot be approached with narrow and restricted strategies. The question concerns a shared development challenge that connects many items and aspects in health policy. Synergy thinking is required when aiming for remarkable improvements in the area of diagnostic safety. Diagnostic accuracy should have a central position in the ideas around care intensity in healthcare.

REFERENCES

1. Richards, Tessa, Angela Coulter, and Paul Wicks. "Time to deliver patient centred care." *BMJ* 350 (2015): h530.
2. Czajkowski, Susan M., Minda R. Lynch, Kara L. Hall, Brooke A. Stipelman, Lynne Haverkos, Harold Perl, Marcia S. Scott, and Mariela C. Shirley. "Transdisciplinary translational behavioral (TDTB) research: opportunities, barriers, and innovations." *Translational Behavioral Medicine* 6, no. 1 (2016): 32–43.
3. Hevner, Alan R. "A three cycle view of design science research." *Scandinavian Journal of Information Systems* 19, no. 2 (2007): 4.
4. Committee on Diagnostic Error in Health Care, Board on Health Care Services, Institute of Medicine, and The National Academies of Sciences, Engineering, and Medicine. *Improving Diagnosis in Health Care*. Edited by Erin P. Balogh, Bryan T. Miller, and John R. Ball. Washington, DC: National Academies Press (US), 2015. http://www.ncbi.nlm.nih.gov/books/NBK338596/.
5. Elkefi, Safa, Hongwei Wang, and Onur Asan. "Organizational considerations from HFE to speed up the adoption of AI-related technology in medical diagnostics." In *Proceedings of the International Symposium on Human Factors and Ergonomics in Health Care*, vol. 9, no. 1, pp. 230–234. Los Angeles, CA: SAGE Publications, 2020.
6. Park, Yoonyoung, Gretchen Purcell Jackson, Morgan A. Foreman, Daniel Gruen, Jianying Hu, and Amar K. Das. "Evaluating artificial intelligence in medicine: phases of clinical research." *JAMIA Open* 3, no.3 (2020): 326–331.
7. McCowan, Ashley L. "Technology in healthcare: how artificial intelligence will revolutionize the profession." *Kentucky Journal of Undergraduate Scholarship* 4, no. 1 (2020): 6.
8. Nsoesie, Elaine O. "Evaluating artificial intelligence applications in clinical settings." *JAMA Network Open* 1, no. 5 (2018): e182658–e182658.

9. Houlton, Sarah. "How artificial intelligence is transforming healthcare." *Prescriber* 29, no. 10 (2018): 13–17.

10. Kraetschmer, Nancy, Natasha Sharpe, Sara Urowitz, and Raisa B. Deber. "How does trust affect patient preferences for participation in decision-making?" *Health Expectations* 7, no. 4 (2004): 317–326.

11. Longtin, Yves, Hugo Sax, Lucian L. Leape, Susan E. Sheridan, Liam Donaldson, and Didier Pittet. "Patient participation: current knowledge and applicability to patient safety." In *Mayo Clinic Proceedings*, vol. 85, no. 1, pp. 53–62. Elsevier, 2010.

12. Wallerstein, Nina. "What is the evidence on effectiveness of empowerment to improve health." Copenhagen: WHO Regional Office for Europe 37 (2006).

13. Chang, Chia-Wen, Ting-Hsiang Tseng, and Arch G. Woodside. "Configural algorithms of patient satisfaction, participation in diagnostics, and treatment decisions' influences on hospital loyalty." *Journal of Services Marketing* 27, no. 2 (2013): 91–103.

14. Carman, Kristin L., Pam Dardess, Maureen Maurer, Shoshanna Sofaer, Karen Adams, Christine Bechtel, and Jennifer Sweeney. "Patient and family engagement: a framework for understanding the elements and developing interventions and policies." *Health Affairs* 32, no. 2 (2013): 223–231.

15. Lindsay, S., and H. J. M. Vrijhoef. "A sociological focus on 'expert patients.'" *Health Sociology Review: The Journal of the Health Section of the Australian Sociological Association* 18, no. 2 (2009): 139–144.

16. Schiff, Gordon D. "Minimizing diagnostic error: the importance of follow-up and feedback." *The American Journal of Medicine* 121, no. 5 Supplement (2008): S38–S42.

17. Ward, Jane K., and Gerry Armitage. "Can patients report patient safety incidents in a hospital setting? A systematic review." *BMJ Quality & Safety* 21, no. 8 (2012): 685–699.

18. Fenton, Joshua J., Anthony F. Jerant, Klea D. Bertakis, and Peter Franks. "The cost of satisfaction: a national study of patient satisfaction, health care utilization, expenditures, and mortality." *Archives of Internal Medicine* 172, no. 5 (2012): 405–411.

19. Greenwood-Lee, James, Lauren Jewett, Linda Woodhouse, and Deborah A. Marshall. "A categorisation of problems and solutions to improve patient referrals from primary to specialty care." *BMC Health Services Research* 18, no. 1 (2018): 986.

20. Cylus, Jonathan, Irene Papanicolas, and Peter C. Smith. "Identifying the causes of inefficiencies in health systems." *Eurohealth* 23, no. 2 (2017): 3–7.

21. Rissanen, Marjo. "Intensity thinking as a shared challenge in consumer-targeted eHealth." In *International Conference on Health Information Science*, pp. 183–192. Cham: Springer, 2018.

22. Burke, Leah A., and Andrew M. Ryan. "The complex relationship between cost and quality in US health care." *AMA Journal of Ethics* 16, no. 2 (2014): 124–130.

23. Taylor, N. "High-intensity acute hospital physiotherapy for patients with hip fracture may improve functional independence and can reduce hospital length of stay [synopsis]." *Journal of Physiotherapy* 63, no. 1 (2017): 50.

24. Hsu, B., D. Merom, F. Blyth, V. Naganathan, D. Handelsman, and R. Cumming. "Temporal relationship between physical activity, exercise intensity, and mortality in older men." *Innovation in Aging* 1, no. suppl_1 (2017): 1052–1052.

25. Lerolle, Nicolas, Ludovic Trinquart, Caroline Bornstain, Jean-Marc Tadié, Audrey Imbert, Jean-Luc Diehl, Jean-Yves Fagon, and Emmanuel Guérot. "Increased intensity of treatment and decreased mortality in elderly patients in an intensive care unit over a decade." *Critical Care Medicine* 38, no. 1 (2010): 59–64.

26. Pronovost, Peter J., Derek C. Angus, Todd Dorman, Karen A. Robinson, Tony T. Dremsizov, and Tammy L. Young. "Physician staffing patterns and clinical outcomes in critically ill patients: a systematic review." *JAMA* 288, no. 17 (2002): 2151–2162.

27. Yang, Q., J. L. Du, and F. Shao. "Mortality rate and other clinical features observed in Open vs closed format intensive care units: a systematic review and meta-analysis." *Medicine* 98, no. 27 (2019): e16261.

28. Mueller, Paul S., C. Christopher Hook, and Kevin C. Fleming. "Ethical issues in geriatrics: a guide for clinicians." In *Mayo Clinic Proceedings*, vol. 79, no. 4, pp. 554–562. Elsevier, 2004.

29. Rissanen, Marjo. "Ethical quality in eHealth: a challenge with many facets." In *International Conference on Health Information Science*, pp. 146–153. Cham: Springer, 2015.

30. Verna, Roberto, Adriana Berumen Velazquez, and Michael Laposata. "Reducing diagnostic errors worldwide through diagnostic management teams." *Annals of Laboratory Medicine* 39 (2019): 121–124.

31. Bhise, Viraj, Suja S. Rajan, Dean F. Sittig, Robert O. Morgan, Pooja Chaudhary, and Hardeep Singh. "Defining and measuring diagnostic uncertainty in medicine: a systematic review." *Journal of General Internal Medicine* 33, no. 1 (2018): 103–115.

32. Rahimi, E., S. H. Alizadeh, A. R. Safaeian, and N. Abbasgholizadeh. "An investigation of patient safety culture: the beginning for quality and safety improvement plans in patient care services." *Journal of Health* 11, no. 2 (2020): 235–247.

33. Tehrani, Ali S. Saber, HeeWon Lee, Simon C. Mathews, Andrew Shore, Martin A. Makary, Peter J. Pronovost, and David E. Newman-Toker. "25-year summary of US malpractice claims for diagnostic errors 1986–2010: an analysis from the National Practitioner Data Bank." *BMJ Quality & Safety* 22, no. 8 (2013): 672–680.

34. Donaldson, Molla S., Janet M. Corrigan, and Linda T. Kohn, eds. *To Err is Human: Building a Safer Health System.* Vol. 6. Washington, DC: National Academies Press, 2000.

35. Singh, Hardeep, Ashley N. D. Meyer, and Eric J. Thomas. "The frequency of diagnostic errors in outpatient care: estimations from three large observational studies involving US adult populations." *BMJ Quality & Safety* 23, no. 9 (2014): 727–731.

36. Hickner, John, Pamela J. Thompson, Tom Wilkinson, Paul Epner, Megan Shaheen, Anne M. Pollock, Jim Lee, Christopher C. Duke, Brian R. Jackson, and Julie R. Taylor. "Primary care physicians' challenges in ordering clinical laboratory tests and interpreting results." *Journal of the American Board of Family Medicine* 27, no. 2 (2014): 268–274.

37. Abaluck, Jason, Leila Agha, Chris Kabrhel, Ali Raja, and Arjun Venkatesh. "The determinants of productivity in medical testing: intensity and allocation of care." *American Economic Review* 106, no. 12 (2016): 3730–3764.

38. Royce, Celeste S., Margaret M. Hayes, and Richard M. Schwartzstein. "Teaching critical thinking: a case for instruction in cognitive biases to reduce diagnostic errors and improve patient safety." *Academic Medicine* 94, no. 2 (2019): 187–194.

39. Wagner, Todd H., Alex R. Dopp, and Heather T. Gold. "Estimating downstream budget impacts in implementation research." *Medical Decision Making* 40, no. 8 (2020) : 959–67. 0272989X20954387.

40. Rissanen, M. "Translational health technology and system schemes: enhancing the dynamics of health informatics." *Health Information Science & Systems*, 8, no. 1 (2020): 1–10.

41. Kolodner, Robert M., Simon P. Cohn, and Charles P. Friedman. "Health information technology: strategic initiatives, real progress: there is nothing 'magical' about the strategic thinking behind health IT adoption in the United States." *Health Affairs* 27, no. Suppl 1 (2008): w391–w395.

42. Powell, Lauren, Dean F. Sittig, Kristin Chrouser, and Hardeep Singh. "Assessment of health information technology-related outpatient diagnostic delays in the US veterans

affairs health care system: a qualitative study of aggregated root cause analysis data." *JAMA Network Open* 3, no. 6 (2020): e206752–e206752.

43. Marcusson, Jan, Magnus Nord, Huan-Ji Dong, and Johan Lyth. "Clinically useful prediction of hospital admissions in an older population." *BMC Geriatrics* 20, no. 1 (2020): 1–9.

44. Jarrett, Pamela G., Kenneth Rockwood, Daniel Carver, Paul Stolee, and Sylvia Cosway. "Illness presentation in elderly patients." *Archives of Internal Medicine* 155, no. 10 (1995): 1060–1064.

45. Adams, Wendy L., Helen E. McIlvain, Naomi L. Lacy, Homa Magsi, Benjamin F. Crabtree, Sharon K. Yenny, and Michael A. Sitorius. "Primary care for elderly people: why do doctors find it so hard?" *The Gerontologist* 42, no. 6 (2002): 835–842.

46. Tsolaki, Magda, Sakka Paraskevi, Nicolaos Degleris, and Sofia Karamavrou. "Attitudes and perceptions regarding Alzheimer's disease in Greece." *American Journal of Alzheimer's Disease & Other Dementias* 24, no. 1 (2009): 21–26.

47. Hwang, Eui-Kyung, Hyun Jung Jin, Young-Hee Nam, Yoo Seob Shin, Young-Min Ye, Dong-Ho Nahm, and Hae-Sim Park. "The predictors of poorly controlled asthma in elderly." *Allergy, Asthma & Immunology Research* 4, no. 5 (2012): 270–276.

48. Dunn, R. M., P. J. Busse, and M. E. Wechsler. "Asthma in the elderly and late-onset adult asthma." *Allergy* 73, no. 2 (2018): 284–294.

49. Stellefson, Michael L., Jonathan J. Shuster, Beth H. Chaney, Samantha R. Paige, Julia M. Alber, J. Don Chaney, and P. S. Sriram. "Web-based health information seeking and eHealth literacy among patients living with chronic obstructive pulmonary disease (COPD)." *Health Communication* 33, no. 12 (2018): 1410–1424.

50. Ostergaard, Peter J., Matthew J. Hall, and Tamara D. Rozental. "Considerations in the treatment of osteoporotic distal radius fractures in elderly patients." *Current Reviews in Musculoskeletal Medicine* 12, no. 1 (2019): 50–56.

51. García, Francisco Javier Sánchez, Jorge Alberto de Haro Estrada, and Herman Michael Dittmar Johnson. "Pelvic insufficiency: underdiagnosed condition, therapeutic diagnostic review." *Coluna/Columna* 17, no. 2 (2018): 151–154.

52. Dombrowsky, Alex, Benjamin Borg, Rongbing Xie, James K. Kirklin, Herbert Chen, and Courtney J. Balentine. "Why is hyperparathyroidism underdiagnosed and undertreated in older adults?" *Clinical Medicine Insights: Endocrinology and Diabetes* 11 (2018): 1179551418815916.

53. Jung, B., Cachier, A., Baron, G., et al. "Decision-making in elderly patients with severe aortic stenosis: why are so many denied surgery?" *European Heart Journal* 26 (2005): 2714–2720.

54. Rissanen, Marjo. "Ways for enhancing the substance in consumer-targeted eHealth." In *International Conference on Health Information Science*, pp. 306–317. Cham: Springer, 2019.

55. Alkan, Ali, Arzu Yaşar, Ebru Karcı, et al. "Severe drug interactions and potentially inappropriate medication usage in elderly cancer patients." *Supportive Care in Cancer* 25, no. 1 (2017): 229–236.

56. Bird, Stephen R., William Kurowski, Gillian K. Dickman, and Ian Kronborg. "Integrated care facilitation for older patients with complex health care needs reduces hospital demand." *Australian Health Review* 31, no. 3 (2007): 451–461.

57. Boumendil, Ariane, Philippe Aegerter, Bertrand Guidet, and CUB-Rea Network. "Treatment intensity and outcome of patients aged 80 and older in intensive care units: a multicenter matched-cohort study." *Journal of the American Geriatrics Society* 53, no. 1 (2005): 88–93.

58. Audisio, Riccardo A. "Shall we operate? Preoperative assessment in elderly cancer patients (PACE) can help A SIOG surgical task force prospective study." *Critical Reviews in Oncology/Hematology* 65, no. 2 (2008): 156–163.

59. Oh, Dong Kyu, Wonjun Na, Yu Rang Park, Sang-Bum Hong, Chae-Man Lim, Younsuck Koh, and Jin-Won Huh. "Medical resource utilization patterns and mortality rates according to age among critically ill patients admitted to a medical intensive care unit." *Medicine* 98, no. 22 (2019): e15835.

60. Gerlich, Miriam G., Katharina Klindtworth, Peter Oster, Mathias Pfisterer, Klaus Hager, and Nils Schneider. "'Who is going to explain it to me so that I understand?' Health care needs and experiences of older patients with advanced heart failure." *European Journal of Ageing* 9, no. 4 (2012): 297–303.

61. Giovanna Vicarelli, Maria, and Micol Bronzini. "From the 'expert patient' to 'expert family': a feasibility study on family learning for people with long-term conditions in Italy." *Health Sociology Review* 18, no. 2 (2009): 182–193.

62. Paiva, Joseane O. V., Rossana M. C. Andrade, Pedro Almir M. de Oliveira, Paulo Duarte, Ismayle S. Santos, Aline L. de P. Evangelista, Rebecca L. Theophilo, Luiz Odorico M. de Andrade, and Ivana Cristina de H. C. Barreto. "Mobile applications for elderly healthcare: a systematic mapping." *PloS one* 15, no. 7 (2020): e0236091.

63. Barnett, Michael L., Nicholas A. Christakis, et al. "Physician patient-sharing networks and the cost and intensity of care in US hospitals." *Medical Care* 50, no. 2 (2012): 152.

64. Novak, Kerri L., Sander Veldhuyzen Van Zanten, and Sachin R. Pendharkar. "Improving access in gastroenterology: the single point of entry model for referrals." *Canadian Journal of Gastroenterology* 27 (2013): 633–635.

65. Bensing, Jozien M., Myriam Deveugele, Francesca Moretti, Ian Fletcher, Liesbeth van Vliet, Marjolein Van Bogaert, and Michela Rimondini. "How to make the medical consultation more successful from a patient's perspective? Tips for doctors and patients from lay people in the United Kingdom, Italy, Belgium and the Netherlands." *Patient Education and Counseling* 84, no. 3 (2011): 287–293.

66. Mazzi, Maria Angela, Michela Rimondini, Wienke GW Boerma, Christa Zimmermann, and Jozien M. Bensing. "How patients would like to improve medical consultations: insights from a multicentre European study." *Patient Education and Counseling* 99, no. 1 (2016): 51–60.

67. Kee, Janine W. Y., Hwee Sing Khoo, Issac Lim, and Mervyn Y. H. Koh. "Communication skills in patient-doctor interactions: learning from patient complaints." *Health Professions Education* 4, no. 2 (2018): 97–106.

68. Rissanen, M. "Customer's voice in eHealth evaluation." In *4th International Future-Learning Conference on Innovations in Learning for the Future 202: e-Learning*, Istanbul, Nov. 14–16, 2012, 230–244. Istanbul: Istanbul University Rectorate publication No.: 5115.

69. Chiou, Shin-Yan, and Ching-Hsuan Lin. "An efficient three-party authentication scheme for data exchange in medical environment." *Security and Communication Networks* 2018 (2018) [Online].

70. Cios, Krzysztof J., Bartosz Krawczyk, Jacquelyne Cios, and Kevin J. Staley. "Uniqueness of medical data mining: how the new technologies and data they generate are transforming medicine." *arXiv* (2019): arXiv-1905.

71. The Beryl Institute. "Defining patient experience." (2020). https://www.theberylinstitute.org/page/DefiningPatientExp.

72. McCarthy, Stephen, Paidi O'Raghallaigh, Simon Woodworth, Yoke Yin Lim, Louise C. Kenny, and Frédéric Adam. "Embedding the pillars of quality in health information technology solutions using 'integrated patient journey mapping' (IPJM): case study." *JMIR Human Factors* 7, no. 3 (2020): e17416.

73. El-Kareh, Robert, Omar Hasan, and Gordon D. Schiff. "Use of health information technology to reduce diagnostic errors." *BMJ Quality & Safety* 22 Suppl 2, no. Suppl 2 (2013): ii40–ii51.

74. Coulter, Angela, Louise Locock, Sue Ziebland, and Joe Calabrese. "Collecting data on patient experience is not enough: they must be used to improve care." *BMJ* 348 (2014): 2225.

75. Jones, Jennifer, Julian Bion, Celia Brown, Janet Willars, Olivia Brookes, Carolyn Tarrant, and PEARL Collaboration. "Reflection in practice: how can patient experience feedback trigger staff reflection in hospital acute care settings?" *Health Expectations* 23, no. 2 (2020): 396–404.

76. Wofford, Marcia M., James L. Wofford, Jashoda Bothra, S. Bryant Kendrick, Amanda Smith, and Peter R. Lichstein. "Patient complaints about physician behaviors: a qualitative study." *Academic Medicine* 79, no. 2 (2004): 134–138.

77. Skålén, Charlotta, Lena Nordgren, and Eva-Maria Annerbäck. "Patient complaints about health care in a Swedish County: characteristics and satisfaction after handling." *Nursing Open* 3, no. 4 (2016): 203–211.

78. Aiken, Linda H., Douglas M. Sloane, Jane Ball, Luk Bruyneel, Anne Marie Rafferty, and Peter Griffiths. "Patient satisfaction with hospital care and nurses in England: an observational study." *BMJ Open* 8, no. 1 (2018): e019189.

79. Reader, Tom W., Alex Gillespie, and Jane Roberts. "Patient complaints in healthcare systems: a systematic review and coding taxonomy." *BMJ Quality & Safety* 23, no. 8 (2014): 678–689.

80. Brocklehurst, Paul R., Gerald McKenna, Martin Schimmel, Anastassia Kossioni, Katarina Jerković-Ćosić, Martina Hayes, Cristiane da Mata, and Frauke Müller. "How do we incorporate patient views into the design of healthcare services for older people: a discussion paper." *BMC Oral Health* 18, no. 1 (2018): 61.

81. Greenhalgh, Joanne, Sonia Dalkin, Kate Gooding, Elizabeth Gibbons, Judy Wright, David Meads, Nick Black, Jose Maria Valderas, and Ray Pawson. "Functionality and feedback: a realist synthesis of the collation, interpretation and utilisation of patient-reported outcome measures data to improve patient care." *Health Services and Delivery Research* 5, no. 2 (2017): 1–280.

82. Boylan, Anne-Marie, Amadea Turk, Michelle Helena van Velthoven, and John Powell. "Online patient feedback as a measure of quality in primary care: a multimethod study using correlation and qualitative analysis." *BMJ Open* 10, no. 2 (2020): e031820.

83. Debnath, Saswati, and Pinki Roy. "Study of speech enabled healthcare technology." *International Journal of Medical Engineering and Informatics* 11, no. 1 (2019): 71–85.

84. Gowda, Naveen, Abhinav Wankar, Sanjay Kumar Arya, et al. "Feedback system in healthcare: the why, what and how." *International Journal of Marketing Studies* 12, no. 1 (2020).

85. Benn, Jonathan, Maria Koutantji, Laura Wallace, Peter Spurgeon, Mike Rejman, Andrew Healey, and Charles Vincent. "Feedback from incident reporting: information and action to improve patient safety." *BMJ Quality & Safety* 18, no. 1 (2009): 11–21.

86. Asprey, Anthea, John L. Campbell, Jenny Newbould, Simon Cohn, Mary Carter, Antoinette Davey, and Martin Roland. "Challenges to the credibility of patient feedback in primary healthcare settings: a qualitative study." *British Journal of General Practice* 63, no. 608 (2013): e200–e208.

87. DeRosis, Sabina, Domenico Cerasuolo, and Sabina Nuti. "Using patient-reported measures to drive change in healthcare: the experience of the digital, continuous and systematic PREMs observatory in Italy." *BMC Health Services Research* 20 (2020): 1–17.

88. Patel, Salma, Rebecca Cain, Kevin Neailey, and Lucy Hooberman. "Exploring patients' views toward giving web-based feedback and ratings to general practitioners in England: a qualitative descriptive study." *Journal of Medical Internet Research* 18, no. 8 (2016): e217.

89. Söderberg, Siv, Malin Olsson, and Lisa Skär. "A hidden kind of suffering: female patient's complaints to Patient's Advisory Committee." *Scandinavian Journal of Caring Sciences* 26, no. 1 (2012): 144–150.

90. van Oostveen, Catharina J., Hester Vermeulen, Els J. M. Nieveen van Dijkum, Dirk J. Gouma, and Dirk T. Ubbink. "Factors determining the patients' care intensity for surgeons and surgical nurses: a conjoint analysis." *BMC Health Services Research* 15, no. 1 (2015): 1–8.

91. Krasniqi, Hanife, Rose-Mharie Åhlfeldt, and Anne Persson. "Patients' experiences of communicating with healthcare-an information exchange perspective." In *15th International Symposium on Health Information Management Research (ISHIMR 2011)*, pp. 241–251. University of Zurich, 2011.

92. Barello, Serena, Stefano Triberti, Guendalina Graffigna, Chiara Libreri, Silvia Serino, Judith Hibbard, and Giuseppe Riva. "eHealth for patient engagement: a systematic review." *Frontiers in Psychology* 6 (2016): 2013.

93. Øvretveit, J. "Digital technologies supporting person-centered integrated care-A perspective." *International Journal of Integrated Care* 17, no. 4 (2017): 6.

94. Makai, Peter, Marieke Perry, Sarah H. M. Robben, Henk J. Schers, Maud M. Heinen, Marcel G. M. Olde Rikkert, and René F. Melis. "Evaluation of an eHealth intervention in chronic care for frail older people: why adherence is the first target." *Journal of Medical Internet Research* 16, no. 6 (2014): e156.

95. Sommerville, Ian. *Software Engineering* (5th ed.). Harlow: Addison-Wesley, 1995.

96. Sheard, Laura, Rosemary Peacock, Claire Marsh, and Rebecca Lawton. "What's the problem with patient experience feedback? A macro and micro understanding, based on findings from a three-site UK qualitative study." *Health Expectations* 22, no. 1 (2019): 46–53.

97. Leslie, Hannah H., Lisa R. Hirschhorn, Tanya Marchant, Svetlana V. Doubova, Oye Gureje, and Margaret E. Kruk. "Health systems thinking: a new generation of research to improve healthcare quality." *PLOS Medicine* 15, no. 10 (2018): 1–4.

98. Ayani, Shirin, Frahnaz Sadoughi, Reza Jabari, Khadijeh Moulaei, and Hassan Ashrafi-Rizi. "Evaluation criteria for health websites: critical review." *Frontiers in Health Informatics* 9, no. 1 (2020): 44.

99. Brown, Hilary, Deborah Davidson, and Jo Ellins. "NHS west midlands investing for health real-time patient feedback project." Birmingham Health Services Management Centre, University of Birmingham, 2009.

11 Computer-aided Diagnosis (CAD) System for Determining Histological Grading of Astrocytoma Based on Ki67 Counting

Fahmi Akmal Dzulkifli,
Maryam Ahmad Sharifuddin,
Mohd Yusoff Mashor, and Hasnan Jaafar

CONTENTS

11.1 BACKGROUND STUDY

The brain is one of the most important and complex organs in the human body. It is located inside the skull, which is close to our sensory organs for vision, hearing, smell, taste, and balance. The brain controls most of our body and mind functions. Besides that, the brain controls basic functions such as walking and talking, and superior functions such as remembering and thinking. The brain also controls vital functions like breathing and heart rate. Normally, the body controls the process of cell growth and division. As normal cells grow old, these cells will die, and the new cells will generate and take their place. However, this process sometimes goes contrarily, where the "old" cells do not die as they should, and the "new" cells remain to proliferate. This expansion of the neoplastic cell proliferation will form a tumor mass.

A brain tumor refers to a collection of abnormal neoplastic cells within the brain that may be benign or malignant. Neoplasia can be characterized as the abnormal growth and proliferation of abnormal tissues that may develop into a tumor [1]. A benign tumor typically grows slowly, and these abnormal cells often remain at the original site [2]. In most cases, this is not serious, and the tumor can be completely removed by surgery. Malignant tumors, however, can invade new tissues or metastasize into other distant organs [2]. In 2020, the American Cancer Society (ACS) estimated 23,890 new cases of malignant brain and nerve tumors were diagnosed among males and females in the United States [3]. The database also revealed that, in that year, there were an estimated 18,020 mortalities that occurred due to malignant brain and nerve tumors [3]. This value showed an increment from the previous database, which recorded 16,700 estimated mortalities in 2017 [4]. The ACS also reported that malignant brain and nerve tumors would become one of the ten leading new types of cancer in adolescents and young adults. In Malaysia, the occurrence of brain tumors has shown an increasing trend from year to year. The Malaysian National Cancer Registry, from 2012 to 2016, showed that malignant brain and nerve tumors were the second most common cancer among children under 14 years old, with 14.6% among males and 15.8% among females [5]. According to the GLOBOCAN 2018 database, 695 mortalities from 754 cases in Malaysia were reported due to malignant brain and nerve tumors [6].

Generally, brain tumors can be divided into two categories. These categories are known as primary brain tumors and metastatic brain tumors [7]. The primary brain tumor is defined as a tumor originally derived from the neoplastic cells of the brain. A metastatic or also known as a secondary brain tumor is a tumor that begins to develop elsewhere in the body and then spreads to the brain to form a new tumor. A primary brain tumor can be divided into two types: glioma and non-glioma. A glioma tumor is a tumor that grows from a glial cell. Glial cells act as supportive tissues in the brain, and they are responsible for providing support and protection for the neurons. Astrocytes, oligodendrocytes, ependymal cells, Schwann cells, satellite cells, and microglia are examples of supporting tissues in the brain [1]. Examples of glioma tumors include astrocytoma, oligodendroglioma, ependymoma, and brain stem glioma. Non-glioma tumors are tumors that form and arise from cells within the brain that are not glial cells. Examples of types of non-glioma tumors are meningioma, medulloblastoma, craniopharyngioma, and pineal gland and pituitary gland tumors.

11.1.1 Overview of Astrocytoma

Astrocytoma is a type of glioma tumor that can develop in the brain or spinal cord. For astrocytoma brain tumors, the tumor arises in the star-shaped cell (astrocytes) that functions as supporting tissue for the nerve cells. The signs and symptoms of astrocytoma are varying and depend on the size and location of the tumor. The symptoms may occur when the tumor puts pressure on the nerve of the brain and interferes with the normal functions of the brain. Headaches, seizures, memory loss, and changes in behavioral or cognitive functions are the most common general and early symptoms of astrocytoma [1]. In terms of treatment, it depends on the location, type, and size of the tumor. Each type has a different procedure for treatment. The standard treatment is to perform surgery for removing the tumor. Another treatment is the use of radiation therapy or chemotherapy. Usually, these two treatments will be performed if the tumor cannot be fully removed by surgery [1].

11.1.2 Overview of Ki67 and Its Characteristics

Ki67 is a nuclear antigen that responds to a monoclonal antibody, MIB-1. The Ki67 is a well-known independent prognostic and predictive indicator for evaluating the proliferation of cancer cells [2]. The Ki67 labeling index (LI) is defined as the percentage of immunoreactive tumor cell nuclei. The Ki67 LI functions as a marker for measuring normal and abnormal cell proliferation in various human tumors. Generally, Ki67 is associated with tumor cell proliferation and growth. A higher proliferation rate is one of the characteristics of cancer cells. Therefore, the Ki67 becomes an excellent marker for identifying the cells that proliferate actively in normal and tumor cell populations. Ki67 protein only exists in growing and dividing phases of the cell cycle (GI, S, G2, and M), but is absent during the resting phase (G0) [2]. This fact makes Ki67 a useful proliferation marker as cancer cells grow and divide aggressively. The Ki67 gene expression starts at the G1 phase, and the expression increases during the S phase and reaches the highest expression during metaphase, which is in the M phase [3]. Ki67 expression will begin to decrease during the anaphase and telophase stages.

11.1.3 Histological Criteria and Types of Astrocytoma

Tumor grading is typically used to characterize the aggressiveness of a tumor based on cell morphology and how the cell looks under a microscope. Tumor grading is one of the methods to classify cancer cells. Grading is essential for determining treatment decisions and the evaluation of a patient's prognosis. The grading refers to the morphological characteristics of the tumor cells, especially to their degree of anaplasia [4]. The two most common characteristics used to assess a tumor are known as cytological features and architectural features. The cytological features are often associated with the shape and size of the nucleus, the nucleus-to-cytoplasm ratio, and the relative number of mitotic index [5]. The architectural features include the histological structure of the tumor and the tumor boundaries [5].

According to the latest WHO classification, astrocytoma is classified into four categories: grade I (pilocytic astrocytoma), grade II (diffuse astrocytoma), grade III (anaplastic astrocytoma), and grade IV (glioblastoma) [6]. The WHO's grading criteria are based on the presence or absence of four histological parameters. These parameters consist of nuclear atypia, mitosis, microvascular proliferation, and necrosis [7]. Grade I (pilocytic astrocytoma) is known to be a slow-growing type and is unlikely to spread to other parts of the brain. The "pilocytic" specifies cells with hair-like and bipolar processes [8]. Based on the WHO classification, this grade is associated with long-term survival, with a ten-year survival rate of over 90% [8]. In Grade II (diffuse astrocytoma), the WHO classifies this according to the presence or absence of an isocitrate dehydrogenase (IDH) mutation [6]. Referring to the WHO's new criteria, diffuse astrocytoma is classified into two major subtypes, which are IDH-mutant and IDH-wild. From the histology features, this grade has a low mitotic activity with less than 3 per 10 high power field (HPF) [9]. Other features are slight cellularity, no endothelial proliferation, and uniform cells (closely resembling mature resting or reactive nonanaplastic astrocytes [9]).

Grade III is known as anaplastic astrocytoma. The histology features include increased mitotic activity with more than 3 per 10 HPF, moderate hypercellularity, anaplasia, and nuclear pleomorphism [9]. Grade IV astrocytoma is also known as glioblastoma, where this grade is the most common and aggressive among the glial tumors. These tumors are growing rapidly and commonly spread into nearby brain tissues. Histological features of this type comprise of marked hypercellularity, nuclear atypia, microvascular proliferation, and necrosis [10].

11.2 PREVIOUS STUDIES RELATED TO THE PATHOLOGICAL DIAGNOSIS

The application of CAD to pathology practice has demonstrated a lot of usefulness, especially in diagnosing different types of diseases with different tissue biopsies. Besides assisting pathologists in evaluating the histologic and cytological features of a cell, the role of the CAD system has also been broadened into the detection and quantification of cells for tumor grading purposes. The recent trend shows an increasing number of new tumors and cancer cases per year. Thus, it is essential to have a proper system that can diagnose the disease effectively.

11.2.1 MANUAL OR CLINICAL DIAGNOSIS

The manual diagnosis technique remains the gold standard for assessing prognosis and patient outcome. The histopathologic analysis is performed by manually observing the sample slides using conventional microscopy. The identification and interpretation of a cell are based on the pathologist's experience. This technique is also practiced for the ki67 quantification process, which is called an "eye-balling" estimation. This technique is fast and straightforward, where the pathologists only need to scan the slides and estimate the number of positive tumor cells instead of counting them individually. However, this technique has low accuracy and reliability. Furthermore, manual detection can lead to misidentification, misjudgment, and diagnostic errors.

Another quantification technique is by counting the cells individually on a digital or printed image. This technique produces high reliability and accuracy. However, it is a tedious and time-consuming technique. Additionally, this technique also requires an additional cost for printing the captured image.

11.2.2 ALTERNATIVE DIAGNOSIS

Today, with the rapid development of new devices and technological advances for pathological usage, the aforementioned issues appear to have been tackled. Recent studies have shown an increment of CAD development in diagnosing various pathological images.

Ştefănescu et al. [11] designed a CAD framework based on fractal analysis and neural network modeling of confocal laser endomicroscopy-generated colon mucosa images. The CAD comprised of three modules. The first module was to calculate the fractal dimension and lacunarity of each image. The second module was the application of a gray-level co-occurrence matrix for texture analysis purposes. In the third module, a marching square and linear interpolation algorithm were used to extract the relevant anatomical features from the normal colon images. Seven imaging features were used to classify between normal and cancerous colonic mucosa. The cross-entropy was calculated to measure the effectiveness of the proposed system in classifying between normal and cancer samples. As a result, from these seven features, only three features (contrast, homogeneity, and feature number) showed significant differences between the normal and cancer samples. The proposed system was able to produce a good result with a low cross-entropy value of 1.17 and 15.48% for diagnostic accuracy errors.

Misawa et al. [12] proposed a CAD system to analyze colorectal lesions based on endocytoscopic images. The CAD system comprised of three phases: image processing and quantification, machine learning, and diagnosis output. In the first phase, the system converted the color space of the input images into a grayscale color space. Then, noise reduction was applied to eliminate the existing noises in the images. Next, the proposed system segmented the image using a two-step thresholding technique to obtain the vessels. From the extracted vessels, the system automatically calculated the maximum vessel diameter, the ratio of the maximum and minimum vessel diameter, and the area of the vessel. For the machine learning phase, the CAD system used a support vector machine (SVM) to classify if the vessels were either nonneoplastic or neoplastic. For the third phase, the CAD system displayed the results of classification and the percentage confidence of the diagnosis. Based on 100 sample images, the CAD system was able to achieve 65 images with a "high-confidence" diagnosis. The diagnostic accuracy obtained from the "high-confidence" images was high with 96.9%.

Supriyanti et al. [13] developed a CAD system that automatically screened the shape and size of the leukocyte cells for identifying the abnormal cells in leukemia. The system consisted of a few stages. First, the system automatically cropped the input image by applying a template matching technique to obtain the leukocyte cell nucleus. Second, the system segmented the cropped image using a thresholding technique. Several morphological image operations (includes of erosion, dilation, opening, and closing) were employed to get the shape of the cell nucleus. Then, the system measured the diameter length of the cell. This value later was compared with the manual diameter calculation, which was performed using the Pythagorean formula. The result showed the proposed system was able to achieve low percentage error values with an average of less than 5%.

Win et al. [14] presented a CAD system for identifying cancer cells in cytological pleural effusion (CPE) images. This CAD system consisted of seven steps: pre-processing, cell nuclei segmentation, post-processing, isolation overlapped cell nuclei, feature extraction, feature selection, and classification. In the pre-processing step, the input image was resized first into 1024 x 1024 pixels. Then, the resized image was enhanced using image intensity adjustment and a median filter for noise reduction. Next, the system segmented the cell nuclei regions using a simple linear iterative clustering (SLIC) superpixels and k-means clustering technique. Later the system eliminated unwanted objects by applying a morphological opening and closing operation. This step was continued with the identification and isolation of overlapped cell nuclei. During this step, the system found the overlapped nuclei first by using a SVM classifier. Afterward, a concavity analysis was implemented for separating these cell nuclei. There were 201 features related to the morphometric, colorimetric, and textural features which were identified and extracted. A hybrid simulated annealing coupling artificial neural network was then used to select suitable features. For the last step, the system classified the cells into benign or malignant by implementing a bootstrap aggregating decision tree. According to the study, the proposed system was able to deliver a promising result, with an average accuracy of 98.7%, a sensitivity of 87.97%, and a specificity of 99.4%.

On top of detecting cancer cells, the application of CAD in the pathology field is also used in mycobacterial detection. Lo et al. [15] designed a CAD system that automatically detected bacillus in the whole-slide image of acid-fast stained mycobacteria. For this study, two different color spaces of sample images were used, namely color and grayscale images. The purpose of having different color spaces of the sample images was to evaluate the function of color features in identifying bacilli. In the beginning, the system cropped the input images into an image block with an approximate size of 20×20 pixels. The image block datasets were then randomly partitioned into an 80% training set and 20% testing set. A transfer learning based on a deep convolutional neural network (DCNN) was implemented to generate the probability of each image block containing a bacillus. If the probability was higher than 0.5, it indicated the image contained a bacillus. As a result, the proposed system was able to achieve a high accuracy for the colored sample images with 95.3%. For the greyscale sample images, the accuracy was 73.8%.

11.3 COMPUTER-AIDED DIAGNOSIS (CAD) SOFTWARE

The main objective of designing this CAD software was to assist pathologists by reducing their time and workload when counting the Ki67 cells before determining the tumor grade. The software was built using the MATLAB app designer. This app designer was an upgraded version of the existing graphical user interface (GUI) that was also provided in MATLAB, which is known as GUIDE. In terms of their function, both software have the same purpose, which allows the developer to create professional apps or software based on the features and tools available for easy use by the end-user. However, this app designer has more benefits when compared to GUIDE. This new app designer can be built as a standalone software, which does not require anything or any extra software to run. Contrasted to GUIDE, the user needs to install the MATLAB first before running the application. For this reason, the app designer is more popular nowadays among developers since it only requires minimum storage and works as a standalone software.

Additionally, the features and tools in this new GUI interface have also been expanded. Thus, it makes the apps more practical, interactive, and professional. Another improvement in this app is software availability. Previously in GUIDE, developers were not able to share their work or apps with the user. With the app designer, once the developer has finished designing the app, they can share their work in three ways. First, they can design their apps to operate on websites by creating a web app. Second, they can create the app as a standalone application that can run on a personal desktop. Third, they can also create an app installation file to share their apps among the MATLAB community.

11.3.1 LAYOUT OF THE CAD SOFTWARE

In general, CAD software needs to perform three main tasks. The software will first count the positive and negative Ki67 cells. Several image processing techniques will be carried out to count the Ki67 cells. Then, using the results from the count, the system will calculate the percentage of the Ki67 index. The third task is to determine the tumor grade based on the percentage value of the Ki67 index. Figure 11.1 displays the layout of the CAD software.

The MATLAB app designer is a high-level GUI development tool provided by MATLAB that allows the developer to create a GUI design. As shown in Figure 11.1, several component tools were implemented to design the CAD software. The tools involved in this GUI design were buttons, panel boxes, editable boxes, and axes.

Axes tools were used to create visualization and analysis of data plots. In Figure 11.1, the green arrows present the axes' components. For this CAD software, the axes function was to display three different types of images, which were input image, the resultant image of counting positive Ki67 cells, and the resultant image from counting negative Ki67 cells. The input image was displayed after the user

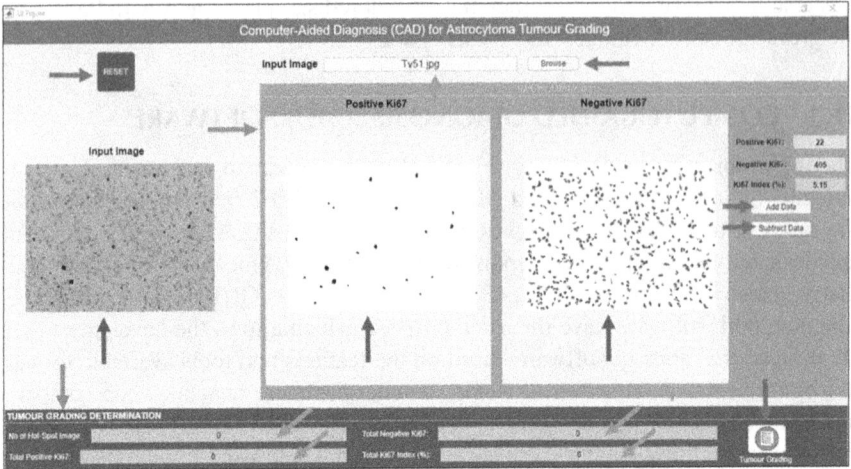

FIGURE 11.1   Layout of an automated CAD software.

selected the image that needed to be analyzed in a folder. For the images of positive and negative Ki67 cells, these images appeared after the system finished the process of analyzing and counting the cells with several image processing techniques.

Next is the button component. The button is a user interface (UI) component that responds when the user presses and releases it. From Figure 11.1, five button components (showed with purple arrows) were used in designing the GUI interface. The first button was the "RESET" button. This button needed to be pressed first before moving to the analyzing process. The function of this button was to make sure the system cleared all the memory to avoid issues such as miscalculation and miscounting of the Ki67 cells. Once the user pressed this button, the system displayed its default appearance, as shown in Figure 11.1. At this moment, the system displayed zero value to all result boxes to indicate all the previous data had been deleted. At the top of the CAD software, there was a button named "Browse". After the user clicked the button, the system opened a file directory window to allow the user to find the location of the file that stored the image to select it for the analysis. The next two buttons were in the "IMAGE ANALYSIS" panel box, and consisted of the "Add Data" and "Subtract Data" buttons. Usually, pathologists use and analyze three or four "hot-spot" histopathological images before deciding the tumor grade. Therefore, it was necessary to have these two buttons to facilitate the user in analyzing and determining tumor grading. The "Add Data" button was used to add all the current results obtained in the "IMAGE ANALYSIS" panel box with the results stored in the "Tumour Grading Determination" panel box. The "Subtract Data" button was used to subtract all the results in the "TUMOUR GRADING DETERMINATION" panel box with the current results obtained in the "IMAGE ANALYSIS" panel box. The "Subtract Data" button was pressed if the user was not satisfied with the current results obtained or if they inadvertently pressed the "Add Data" button without analyzing an image. The last button was located at the bottom of the GUI interface. This button was defined as the "Tumour Grading" button. After the user was satisfied with all the results, he or she would click this button for determining the tumor grading. Then, a pop-up window would appear and show the grade of the astrocytoma tumor to the user.

Another tool used in designing the CAD software was the panel box. The panel box was a container that grouped all the UI components for performing a specific process. As shown in Figure 11.1, two-panel boxes (orange arrow) were created, known as the "IMAGE ANALYSIS" and "TUMOUR GRADING DETERMINATION" panel boxes. The "IMAGE ANALYSIS" panel box was the place where all the analysis processes starting from enhancing image quality to the process of counting Ki67 cells and the calculation of the Ki67 index were involved. Several image processing techniques like image enhancement, color deconvolution, and feature extraction were involved in this process. The "TUMOUR GRADING DETERMINATION" panel box covered all the information needed to determine the grade of a tumor. This information consisted of the total number of "hot-spot" images that were already being analyzed, the total number of positive and negative Ki67 cells, and the total percentage of the Ki67 index.

The next component tool used in this CAD software was the editable box. This editable box is a tool that allows users to choose whether to insert data information in text or numeric style. This box has two general properties. First, the developer can design this box to be editable, which means the user can edit the information inside the box at any time. Second, the developer can design this box to be uneditable where the editing process is restricted, which means the user can only read the information displayed in the box. Figure 11.1 shows there were eight editable boxes placed in the CAD software. The first editable box was the "Input image" box. This box displayed the filename of the selected image for analysis. Three editable boxes in the "IMAGE ANALYSIS" panel box were used to display the percentage of the Ki67 index and the counting results of positive and negative Ki67 cells for the current image. The remaining four boxes were located in the "TUMOUR GRADING DETERMINATION" panel box. In this panel box, one box was used to display the total number of "hot-spot" images used in the analysis. Two boxes were used to display the total numbers of positive and negative Ki67 cells, while the remaining box was used to show the total percentage of the Ki67 index.

11.3.2 System Operation of the CAD Software

Figure 11.2 illustrates the the grading of a tumor based on Ki67 cell counting. The process of determining an astrocytoma tumor grading by the CAD software consisted of six major stages: (a) image enhancement, (b) color deconvolution, (c) feature extraction, (d) counting the number of positive and negative Ki67 cells, (e) calculating the percentage of the Ki67 index, and (f) determining the astrocytoma tumor grading.

First, the user must clear the memory to ensure all the previous data and other troublesome issues that may disrupt the performance of the software were removed. The next stage was to select the input image, which was the "hot-spot" image of

```
                    ┌──────────┐                              ┌───┐
                    │  Start   │                              │ A │
                    └────┬─────┘                              └─┬─┘
                         ▼                                      ▼
                ┌─────────────────┐                 ┌──────────────────────┐
                │ Clear all previous │               │ Calculate Percentage │
                │      data       │                 │    of Ki67 Index     │
                └────────┬────────┘                 └──────────┬───────────┘
                         ▼                                      ▼
                  Select                                  No. of 'Hot-Spot'      No
                 'Hot-Spot'  ◄──────────────            Images == 3  ──────────┐
                   image                                       │ Yes            │
                         ▼                                      ▼               │
                ┌─────────────────┐                 ┌──────────────────────┐   │
                │     Image       │                 │   Determine Tumour   │   │
                │  Enhancement    │                 │      Grading         │   │
                └────────┬────────┘                 └──────────┬───────────┘   │
                         ▼                                      ▼               │
                ┌─────────────────┐                    Display results:        │
                │     Colour      │                    i)  Total no. of positive │
                │  Deconvolution  │                        Ki67 cells           │
                └───┬─────────┬───┘                    ii) Total no. of negative │
                    │         │                            Ki67 cells           │
         ┌──────────┘         └──────────┐             iii) Total percentage of  │
         ▼                               ▼                  Ki67 index           │
  ┌─────────────┐              ┌──────────────┐        iv) Grade of a tumour     │
  │ DAB Staining │              │ Haematoxylin │                                │
  │Concentration │              │   Staining   │                  ▼             │
  └──────┬──────┘              │Concentration │             ┌──────────┐        │
         ▼                     └──────┬───────┘             │   End    │        │
  ┌─────────────┐                     ▼                     └──────────┘        │
  │   Feature   │              ┌──────────────┐                                 │
  │ Extraction  │              │   Feature    │                                 │
  └──────┬──────┘              │ Extraction   │                                 │
         ▼                     └──────┬───────┘                                 │
  ┌─────────────┐                     ▼                                         │
  │Count Positive│             ┌──────────────┐                                 │
  │ Ki67 Cells  │              │    Count     │                                 │
  └──────┬──────┘              │  Negative    │                                 │
         │                     │  Ki67 Cells  │                                 │
         └──────────┬──────────┘──────┬───────┘                                 │
                    ▼                                                            │
                  ┌───┐                                                          │
                  │ A │                                                          │
                  └───┘                                                          │
```

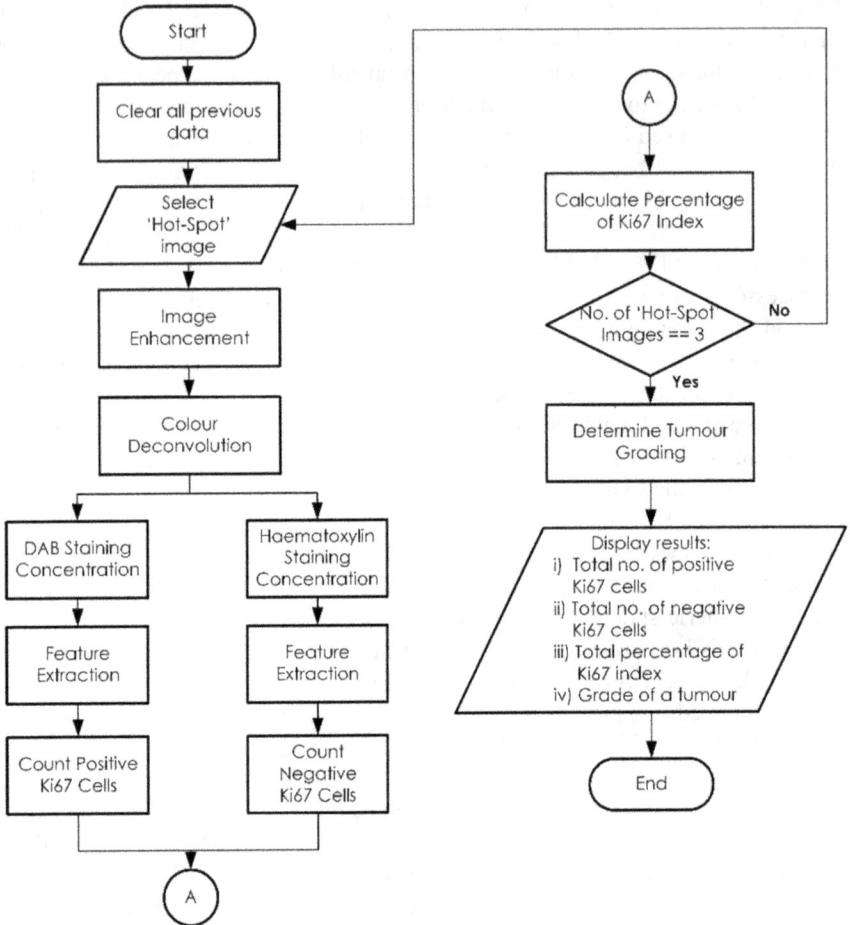

FIGURE 11.2   Process flowchart of CAD software.

the astrocytoma. After selecting the input image, the software performed the image enhancement technique. The image enhancement was applied to simplify the analysis process by enhancing the visual appearance of targeted features in the input image. Later, the color deconvolution was used to separate between the DAB and haematoxylin stain concentrations. Afterward, the software extracted the related features that would be useful for differentiating between the Ki67 cells with other unwanted objects. Then the software counted the number of positive and negative Ki67 cells. After obtaining the counting results, the system calculated the percentage of the Ki67 index. Typically, pathologists need at least three "hot-spot" images of astrocytoma before determining the tumor grade. Thus, the analysis process repeated until the user obtained the results for three "hot-spot" images. The last stage was to determine the grade of astrocytoma according to the WHO classification standard.

11.3.3 IMAGE ACQUISITION

A total of five histopathological slides were selected from the Department of Pathology, Hospital Universiti Sains Malaysia (HUSM). These slides were prepared between 2016 and 2018. Among these five slides, 15 "hot-spot" area images were captured under 40× magnification using an Olympus BX51 microscope and Cell^F software that worked as an interface with the digital camera, which was attached to the microscope. The "hot-spot" area refers to a selected area of malignant cells with a high proliferative activity potentially associated with a more aggressive biological behavior than that represented in the whole of the tumor [16]. Pathologists assess at least three "hot-spot" areas for each slide before determining the grade of a tumor. Approximately, the pathologists need to count a minimum of 400 cells in each "hot-spot" area to calculate the Ki67 index.

For this study, the pathologists used IHC staining to stain the tissue specimens. The positive Ki67 cells were stained using diaminobenzidine (DAB) while the negative Ki67 cells were stained by haematoxylin. The resulting of stains made the positive Ki67 cells appear a granular brown color, whereas the negative Ki67 cells were presented as a diffuse blue color. The sample "hot-spot" area images were then saved in a (*.jpg) format with a resolution of 4140 × 3096 pixels and 24-bit RGB.

11.3.4 IMAGE ENHANCEMENT

Sometimes, there was a lack of contrast and brightness in the images captured from the microscope. This was, perhaps, due to the illumination conditions while capturing the image. Due to this problem, it was necessary to implement an image enhancement technique to improve the quality of captured images. For this study, the goal of applying image enhancement was to improve the quality of the input images by enhancing the contrast and brightness of the images. Before enhancing the appearance of the input images, the system resized the input image into new pixel dimensions. Since the captured images came with high resolutions and large sizes, the CAD system needed to resize the images to increase the processing speed and save storage space. Thus, the proposed system resized the input image into 1360 × 1024 pixels. This size was the lowest resolution without compromising the useful features of the Ki67 cells.

The next process was to convert the color space of the image from a RGB to L*a*b* color space. This color space was selected due to the exact color representation and its device-independent color model. Then, the contrast enhancement technique was applied to the luminosity, "L" channel while the $a*$ and $b*$ channels remained unchanged. In this study, Contrast-Limited Adaptive Histogram Equalization was used to enhance the contrast of the astrocytoma images. The CLAHE technique is a variation of an adaptive histogram equalization which reduces noise amplification by limiting contrast amplification [17]. Instead of using the whole image, this technique was performed on small regions called "tiles." The contrast of each tiles was enhanced, resulting in the histogram of the output region approximately matching the histogram specified by the desired histogram value [18]. Equation 11.1 demonstrates the central equation for enhancing the image:

$$I(x,y) = p(x,y)q(x,y) * \max luminosity \qquad (11.1)$$

where $q(x,y)$ can be calculated using the expression:

$$q(x,y) = J(x,y) / \max luminosity \qquad (11.2)$$

$I(x,y)$ is the output of the enhanced luminosity channel, $p(x,y)$ represents the new luminance value for pixel (x,y) after applying the contrast enhancement technique, which was the CLAHE technique, $q(x,y)$ is the processed luminance image, and $J(x,y)$ is the luminosity value at the luminance channel. For this study, the luminosity values were scaled to the range of [0, 1]. The issue raised in this study was the selection of a maximum luminosity value. This is because the images taken had various illumination conditions. Thus, the output images were affected if the luminosity value was constant. Based on the observation of the captured images, it was shown that low-quality images required a high luminosity value to enhance the contrast, whereas good quality images just needed a low luminosity value to enhance the contrast. Hence, to solve this issue, the system calculated the maximum luminosity value automatically using Equation 11.3. This value was then inserted into the enhancement algorithm in Equation 11.1. The equation to calculate the luminosity value is as follows:

$$max\ luminosity = avg(luminance) / \left[max(luminance) - min(luminance) \right] \qquad (11.3)$$

where the $avg(luminance)$ is the average of all the luminosity values in the $L*$ channel. The $max(luminance)$ is the maximum value in the luminosity channel, and the $min(luminance)$ is the minimum value in the luminosity channel. Then, the resultant contrast-enhanced image was converted back to a RGB color space for visualization purposes.

11.3.5 COLOR DECONVOLUTION

In 2001, Ruifrok and Johnson [19] designed a technique based on color image analysis for separating slide concentrations, which was used for quantifying cells with immunohistochemical staining. This technique is called color deconvolution. The color deconvolution technique is extensively recognized, especially in digital pathology applications. This technique separates the multiple stained color images into images representing the stain concentrations. Various types of microorganisms with different types of staining can be found in a biological sample stained slide. With the implementation of the color deconvolution technique, it can differentiate the staining attached to the cell nuclei, cytoplasm, specific protein, and other organelles. The outcome of the image can be used for a number of applications such as texture analysis on each stain, densitometry analysis, and the intensity and area calculations for each stain [19].

The basic operation of the color deconvolution is from Lambert-Beer's Law, where it measures the intensity of light after passing through a material and the amount of stain with an absorption factor corresponding to the specific R, G, and B channels. At first, the system converts the RGB values in the image into optical density (OD) values, in which these OD values are linear with the concentration of absorbance material. The OD values can be achieved by calculating the relative absorption for each RGB channel in slides stained with a single stain. At this moment, every single stain will have three OD values that represent each R, G, and B channels, respectively. These values can be represented in a matrix form:

$$\begin{bmatrix} p11 & p12 & p13 \\ p21 & p22 & p23 \\ p31 & p32 & p33 \end{bmatrix} \tag{11.4}$$

where every row represents a specific stain used in the color image, and every column indicates the OD values for each RGB channel in a single stain.

After obtaining the OD values, the next process is to perform an ortho-normal transformation on optic density image. The purpose of this transformation is to acquire independent information on the individual stains used. The transformation needs to be normalized in order to achieve the correct balancing of the absorption factor for each stain. The normalization process can be executed by dividing each OD vector values with its total length. For this study, the color deconvolution was used to separate between the DAB and haematoxylin concentrations. Figure 11.3 shows an example of the resultant image after applying color deconvolution with a given color vector. As mentioned in Section 11.3.3, the DAB concentration resulted in the appearance of positive Ki67 cells in a brown color, while the haematoxylin concentration caused the negative Ki67 cells to appear in blue. Figure 11.3 shows that all the objects in the image were successfully separated according to a specific stain.

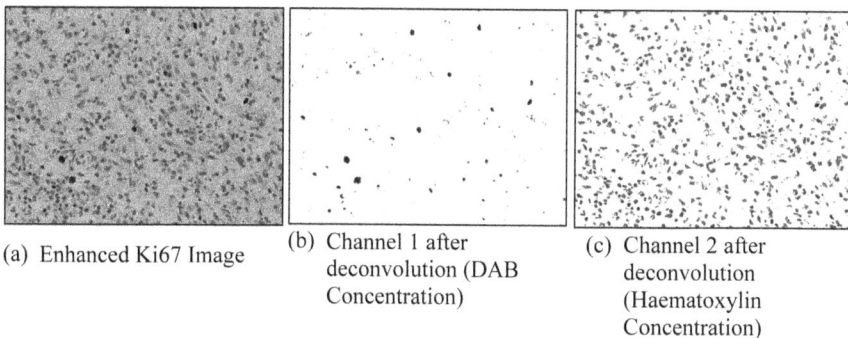

(a) Enhanced Ki67 Image

(b) Channel 1 after deconvolution (DAB Concentration)

(c) Channel 2 after deconvolution (Haematoxylin Concentration)

FIGURE 11.3 An example of a resultant image after applying the color deconvolution technique.

11.3.6 Feature Extraction

Feature extraction is one of the favorable techniques used for image processing. The purpose of this technique was to select relevant and informative features, which would be useful in object recognition. This part was crucial because it can affect the efficiency of the recognition system. A feature is usually related to the input variable, and the selection of an appropriate feature needs to be done carefully since the result can influence the performance of the system, especially in counting the Ki67 cells. Features such as color, texture, shape, and size are some of the examples used to represent an object. The extracted features will provide useful information for identifying the real Ki67 cell and removing unnecessary objects like stained artifacts. One of the most important characteristics for distinguishing between positive and negative Ki67 cells was by determining the color intensity. However, this issue was already addressed by using a color deconvolution method, as discussed in Section 11.3.5. This section will highlight analyzing the shape and the properties of the Ki67 cells, where some of the features that characterize the appearance of a Ki67 cell will be measured to differentiate between the Ki67 cells and unwanted objects.

Three main features of the Ki67 cells were selected, namely, area, circularity, and solidity based features. In image processing, the area refers to the number of pixels in a shape. For this study, the area represented the number of pixels of each object in the image, whether it was a Ki67 cell or an unwanted object. The area of the cell can be calculated by using Equation 11.5 [20]

$$Area\ of\ cell,\ A = \sum_{x=1}^{X}\sum_{y=1}^{Y} I(x,y) \tag{11.5}$$

where $I(x,y)$ is referred to as the image pixel value. X and Y are the sizes of each dimension (row and columns) of image I. A shape feature, usually known as a shape descriptor, is a representation of 2D or 3D shape that contains a set of numbers to describe the specific characteristics concerning the geometry of a given shape [21]. In this study, the circularity feature was used to distinguish between positive and negative Ki67 cells with the unwanted objects. The shape of the Ki67 cells is very subjective. Thus, there was no limit range in defining the shape of the Ki67 cells. Value 1 indicated that the object is a circle. The higher the circularity value indicated the object was less circular. The circularity was identified by first calculating the object areas with Equation 11.5. Next was to calculate the perimeter using Equation 11.6 [22]. The perimeter describes the number of pixels in the boundary of the cell.

$$Perimeter = \sum_{i=1}^{N-1} d_i = \sum_{i=1}^{N-1} |x_i - x_{i+1}| \tag{11.6}$$

where d is the distance measurements between two specified pixels. The x_i is the boundary of the object. Then, using Equation 11.7, calculate the circularity of every object in the image.

$$Circularity = \frac{(Perimeter)^2}{4\pi *Area\ of\ cell} \tag{11.7}$$

The third feature used in this analysis was solidity. Solidity is used to measure the density of the cell. Solidity was determined by dividing the area within the cell with the convex area. The convex area was the area enclosed with a convex hull. A value of 1 indicated a solid object, while a value of less than 1 defined an irregular object or one containing holes. Equation 11.8 presents the formula used to calculate the solidity.

$$Solidity = \frac{|Area\ of\ cell|}{|Convex\ area|} \tag{11.8}$$

Next, the system calculated the average and the standard deviation for every feature used. Then the average values were subtracted with the standard deviation values to obtain the threshold values for every feature. These values were used as inputs to the system for extracting between the Ki67 cells and the unwanted objects. Any unwanted objects were discarded from the proposed system if the value of the objects were smaller than the threshold values.

11.3.7 COUNTING KI67 CELLS AND TUMOR GRADING DETERMINATION

In this stage, the binary image from the output image of the feature extraction process was used to count the Ki67 cells. The binary image had only two possible values, 0 or 1 for each pixel. The value 0 represented the background pixels of the image, while value 1 described the pixels of the object, which was the Ki67 cell. The counting process was done by using the MATLAB function, *bwlabel*. The *bwlabel* function first identified all the connected components in the image. All the pixels related to the first connected component were labeled with a value of 1, while all the pixels related to the second connected component are labeled with a value of 2 and so forth. This process repeated continuously until the system received the number of positive and negative Ki67 cells. The next stage was to calculate the percentage of the Ki67 index. The value of the Ki67 index was obtained by dividing the number of positive Ki67 cells with the total number of Ki67 cells in the image. Equation 11.9 presents the formula used to calculate the Ki67 index.

$$Ki67\ Index = \frac{No.\ of\ positive\ Ki67\ Cells}{No.\ positive\ Ki67\ Cells + No.\ of\ negative\ Ki67\ Cells} \times 100\% \tag{11.9}$$

After acquiring the percentage value, the last stage was to determine the astrocytoma tumor grading. According to the 2016 WHO Classification of Tumours of the

TABLE 11.1
Astrocytoma Grading Classification Based on Ki67 Index Value

Grades	Types of Tumor	Range of Ki67 Index
I	Pilocytic Astrocytoma	0%–1%
II	Diffuse Astrocytoma	2%–4%
III	Anaplastic Astrocytoma	5%–14%
IV	Glioblastoma	15%–20%

Central Nervous System, there are four stages of astrocytoma, which were already clarified in Section 1.1.3. Table 11.1 displays the grading classification of astrocytoma based on the range values of the Ki67 index and the type of tumor. The system generated the grading results based on the Ki67 value that fell within the ranges as given in Table 11.1.

11.4 RESULT AND ANALYSIS

The CAD software was designed using MATLAB version 2018. A personal computer running on Intel Core i7-5500U, 2.4 GHz, and 16 GB RAM was used to develop the system. From five sample slides, there were 15 "hot-spot" images of Ki67 selected for this study.

11.4.1 IMAGE ENHANCEMENT ANALYSIS

The enhanced images were compared with the original images for assessing the performance and accuracy of the system in improving the quality of original images. All enhanced images were saved in the (*.jpg) format. Figure 11.4 shows an example of a comparison between the original image and output image after applying an image enhancement technique. Figure 11.4 also provides a comparison of RGB histograms between the original image and the output image. An image histogram refers to a bar graph that represents the frequency of the intensity values that occur in an image. Each bar indicates one intensity level. The horizontal axis refers to the intensities of the image, which can be a color or grayscale intensities. The vertical axis explains the frequency of the intensity values.

Figure 11.4(a) illustrates the original Ki67 image of the astrocytoma. From this image, it is shown that most of the positive Ki67 cells are visible. However, some negative Ki67 cells appear to have low contrast and dull color. Thus, it can affect the performance of the system as these cells are difficult to identify. After implementing the image enhancement technique, the contrast of each cell was improved, as shown in Figure 11.4(c). Besides that, morphological, texture, size, and shape of the Ki67 cells were also clearly revealed.

Contrast refers to the difference between dark and light areas of a scene. In an image histogram, contrast indicates the range of intensity values effectively used

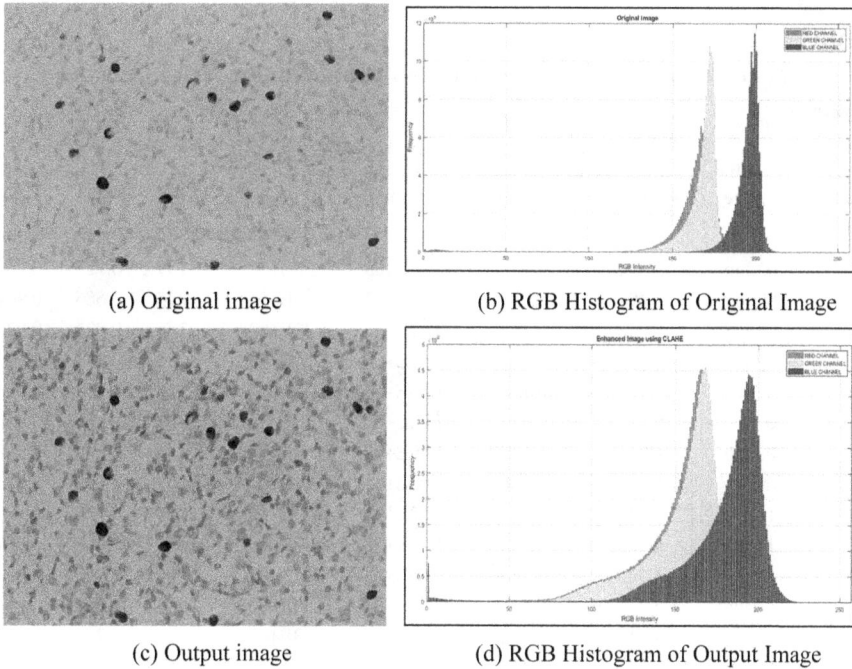

(a) Original image

(b) RGB Histogram of Original Image

(c) Output image

(d) RGB Histogram of Output Image

FIGURE 11.4 Comparison between original image and output image after applying an image enhancement technique.

within an image. A broad histogram displays an image with good contrast, while a narrow histogram reflects an image with low contrast that may appear flat or full. Figure 11.4(b) displays the RGB histogram for the original image. This figure shows that there are approximately no values between 20 and 120. The intensity for all channels was mostly concentrated in the middle range of the histogram. Following the implementation of the image enhancement technique, the range of the histogram has been expanded, as shown in Figure 11.4(d). The intensity values for all RGB channels were well-spread out and mostly filled the entire intensity range.

Table 11.2 presents the detailed results of the performance analysis of the proposed system of enhancing the input images. Five quantitative measures were carried out in this study. The measurements consisted of the Tenengrad criterion (TEN), entropy measurement (H), absolute mean brightness error (AMBE), structural similarity index (SSIM), and universal image quality index (UIQI). The TEN analysis was used to identify whether the sharpness of an enhanced image had been improved or not. The TEN value of the enhanced image was expected to be higher than the original image. A higher value of TEN demonstrated sharper edges. Entropy analysis (H) is a method used to calculate the amount of information contained within an image. The entropy of the enhanced image should not be lower than the entropy of the original image. Lower entropy conveys a loss of some image details while higher entropy values specify that the image is rich in detail. For AMBE measurement,

TABLE 11.2

Performance Analysis of the Proposed System in Enhancing 15 Original Images

Slide	Original Image	H		TEN		AMBE	SSIM	UIQI
		Original	Enhanced	Original	Enhanced			
1	01.jpg	5.282	6.679	1.86E+03	1.07E+04	14.376	0.713	0.980
	02.jpg	5.444	6.768	2.25E+03	1.15E+04	23.799	0.713	0.966
	03.jpg	5.496	6.859	3.02E+03	1.71E+04	20.825	0.686	0.967
2	04.jpg	5.462	6.542	2.62E+03	6.22E+03	17.755	0.824	0.975
	05.jpg	5.552	5.961	3.33E+03	6.94E+03	11.181	0.884	0.989
	06.jpg	4.975	6.150	1.85E+03	4.45E+03	15.592	0.831	0.984
3	07.jpg	5.512	6.842	2.04E+03	9.14E+03	18.349	0.736	0.966
	08.jpg	5.213	6.615	1.69E+03	8.22E+03	14.841	0.748	0.978
	09.jpg	5.097	6.356	1.63E+03	6.09E+03	18.892	0.797	0.974
4	10.jpg	4.936	6.371	1.53E+03	7.45E+03	17.959	0.725	0.983
	11.jpg	4.803	6.205	1.25E+03	5.99E+03	20.056	0.748	0.981
	12.jpg	5.145	6.577	1.08E+03	5.97E+03	14.378	0.743	0.980
5	13.jpg	4.942	6.307	1.92E+03	7.34E+03	14.136	0.768	0.984
	14.jpg	4.817	6.165	1.58E+03	5.63E+03	13.418	0.787	0.985
	15.jpg	5.361	6.712	2.36E+03	1.23E+04	22.688	0.713	0.967
AVERAGE						**17.216**	**0.761**	**0.977**

the system calculates the difference between the mean brightness of the original image and the enhanced image. A good brightness preserving method will have a low AMBE value. SSIM is a measurement indicator that measures the similarity between the original and enhanced images. The SSIM measure combines luminance, contrast, and structural comparisons. The resulting performance index takes values between –1 to 1. The maximum value of 1 is achieved when x and y are identical. SSIM is another analysis that measures the image distortion between the original and output image. The values vary from –1 to 1. The greater the similarity between the two images, the closer the UIQI value to one.

Overall, the results for all selected component measurements in Table 11.2 shows an improvement. As shown in Table 11.2, all entropy values for enhanced images were higher than those obtained from the original images. This suggests that the visibility and detail of the information for Ki67 cells was improved. These results were necessary for the feature extraction process, where the system extracted several features to define the ideal Ki67 cell. The values achieved from the enhanced image were also higher than the original images for TEN analysis. This signified the objects in the enhanced images had maximum edge sharpness, and the structural information in the images was also improved. Table 11.2 also shows acceptable AMBE results with an average of 17.216. This value defines that the brightness of the enhanced images was still well-preserved. Following the SSIM results, the system was able to get a reasonable value with an average of 0.761. This value expressed that the enhanced images were still identical to the original image. According to the UIQI results, the proposed system was successful in improving the quality of the original images, since the average value was high with 0.977.

11.4.2 Cell Counting and Ki67 Labeling Index Analysis

This section discusses the system performance in counting both positive and negative Ki67 cells. The counting results obtained from the CAD system were compared with manual counting, which was performed by a pathologist. The cells were counted individually, and the results of counting with the percentages of Ki67 index were noted for comparison purposes. For the evaluation process, several quantitative analyses were used to measure the accuracy of the proposed system in counting the cells. Three analyses were conducted to evaluate the inter-observer agreement of the counting methods. The first analysis was to calculate the relative accuracy of that system. The purpose of this analysis was to find how close a measured value was to a standard value on relative terms. The relative accuracy could be calculated by finding the relative error between the measured and exact values. The expression for calculating the relative error was as follows [23]:

$$\varepsilon_r = \left| \frac{V_A - V_E}{V_E} \right| \tag{11.10}$$

where ε_r is the relative error, V_A is the results of counting from the proposed system, and V_E is the value of manual counting cells. The relative accuracy was calculated according to Equation 11.11 [24]. Table 11.3 shows the analysis results of counting the positive Ki67 cells.

$$Relative\ Accuracy = 1 - \varepsilon_r \tag{11.11}$$

The results in Table 11.3 show moderate values of relative error and relative accuracy. The average relative error and relative accuracy were 0.5. Generally, the proposed system was able to give acceptable results for counting, although four images produced a relative error of more than 0.5. Several factors were identified, which influenced the sensitivity result of the counting. First, it could be due to the complex nature of the cells, which has a heterogeneous shape. Second, it was possibly due to the quality of the captured images. Before the analysis process, some of the objects or cells were difficult for the system to identify. After applying an image enhancement technique, the contrast of those cells or objects were improved. Hence, the system was able to identify and count each of the cells in the image. For these reasons, it affected the counting result produced by the system. The following table will show the analysis results of counting the negative Ki67 cells.

According to the results in Table 11.4, it was clearly displayed that the proposed system was able to detect and count the negative Ki67 cells. The relative error was

TABLE 11.3

Comparison of the Counting Performance for Positive Ki67 Cells between Manual and Automated Counting

Image	Positive Ki67 Cells Counting		Relative Error	Relative Accuracy
	Manual Counting	Automated Counting		
01.jpg	6	7	0.17	0.83
02.jpg	15	10	0.33	0.67
03.jpg	5	4	0.20	0.80
04.jpg	16	16	0.00	1.00
05.jpg	9	12	0.33	0.67
06.jpg	12	11	0.08	0.92
07.jpg	14	22	0.57	0.43
08.jpg	8	10	0.25	0.75
09.jpg	15	16	0.07	0.93
10.jpg	11	8	0.27	0.73
11.jpg	3	7	1.33	-0.33
12.jpg	4	14	2.50	-1.50
13.jpg	34	16	0.53	0.47
14.jpg	29	16	0.45	0.55
15.jpg	25	16	0.36	0.64
Average			**0.50**	**0.50**

obtained at an average of 0.11. For relative accuracy, the proposed system was able to achieve an average of 0.89. The second analysis was to calculate the Pearson correlation coefficient (PCC). This statistical test was commonly used in linear regression, and was used to measure how strong the correlation was between manual and automated cell counts. A p-value of less than 0.05 was considered statistically significant. The Pearson correlation coefficient (r) can be calculated as [25]:

$$r_{xy} = \frac{Cov(X,Y)}{\sigma_x \sigma_y} \tag{11.12}$$

where $Cov(X,Y)$ was the covariance of two variables X and Y. The σ_x was the standard deviation of X, and σ_y was the standard deviation of Y. $Cov(X,Y)$ can be expressed as [25]:

$$Cov(X,Y) = \Sigma\left[(X - EX)(Y - EY)\right] \tag{11.13}$$

where EX was the mean of X and EY was the mean of Y. A scatter plot with a regression line was plotted to find the relation between the manual and automated counting. Figure 11.5 presents the correlation of the Ki67 cell count with manual and automated counting.

TABLE 11.4

Comparison of the Counting Performance for Negative Ki67 Cells between Manual and Automated Counting

Image	Manual Counting	Automated Counting	Relative Error	Relative Accuracy
		Negative Ki67 Cells Counting		
01.jpg	206	194	0.06	0.94
02.jpg	238	213	0.11	0.89
03.jpg	240	211	0.12	0.88
04.jpg	304	247	0.19	0.81
05.jpg	242	218	0.10	0.90
06.jpg	201	197	0.02	0.98
07.jpg	480	405	0.16	0.84
08.jpg	329	278	0.16	0.84
09.jpg	353	296	0.16	0.84
10.jpg	209	206	0.01	0.99
11.jpg	183	162	0.11	0.89
12.jpg	248	221	0.11	0.89
13.jpg	432	417	0.03	0.97
14.jpg	394	330	0.16	0.84
15.jpg	512	468	0.09	0.91
Average			**0.11**	**0.89**

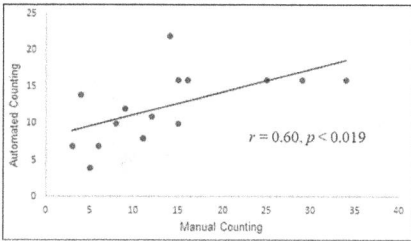

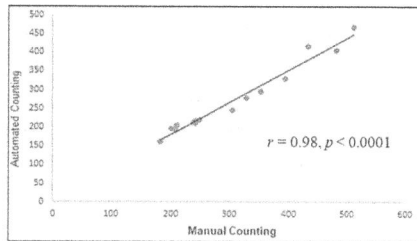

(a) Correlation result of counting positive Ki67 cells between manual and automated counting

(b) Correlation result of counting negative Ki67 cells between manual and automated counting

FIGURE 11.5 Correlation of Ki67 cells counting by manual and automated counting.

Figure 11.5(a) shows a moderate positive correlation in counting positive Ki67 cells between manual and automated counting. From that figure, only a few data points were close to the regression line. The Pearson correlation coefficient between these techniques was ($r = 0.60$, $p < 0.0019$). Meanwhile, in Figure 11.5(b), most of the points were close to the straight line. The PCC test showed there was a strong

TABLE 11.5

Comparison of Ki67 Index Results between Manual and Automated Counting

Image	Manual Counting (%)	Ki67 Index (%) Automated Counting (%)	Absolute Error (%)
01.jpg	2.83	3.48	0.65
02.jpg	5.92	4.48	1.44
03.jpg	2.04	1.86	0.18
04.jpg	5.00	6.08	1.08
05.jpg	3.59	5.22	1.63
06.jpg	5.63	5.29	0.34
07.jpg	2.83	5.15	2.32
08.jpg	2.37	3.47	1.10
09.jpg	4.08	5.13	1.05
10.jpg	5.00	3.74	1.26
11.jpg	1.61	4.14	2.53
12.jpg	1.59	5.95	4.36
13.jpg	7.30	3.70	3.60
14.jpg	6.86	4.48	2.38
15.jpg	4.66	3.31	1.35
Average (%)			**1.68**

positive correlation between these two techniques in counting negative Ki67 cells with ($r = 0.98$, $p < 0.0001$).

The third analysis was to calculate the absolute error between the two proposed automated counting techniques and the manual counting technique. The absolute error was calculated based on the difference of Ki67 index results between the automated counting and manual counting technique. The absolute error can be calculated as [26]:

$$\Delta x = \left| x_0 - x \right| \tag{11.14}$$

where Δx was the absolute error, x_0 was the Ki67 index result from the proposed automated counting, and x was the Ki67 index result from manual counting technique. Table 11.5 presents the comparison of the percentage results for the Ki67 index between the manual and automated counting systems. The absolute error results obtained by the automated counting system produce encouraging results, with an average of 1.68%.

11.4.3 TUMOR GRADING ANALYSIS

The last step was to determine brain tumor grading. There was a difference between the manual counting technique and the CAD system in deciding the tumor grading.

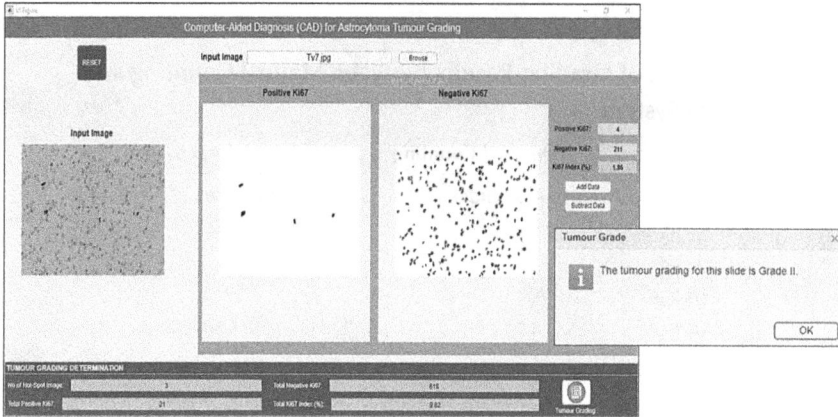

FIGURE 11.6   The final result of the CAD system.

For the manual counting technique, the Ki67 index was calculated for each of "hot-spots" area images. Then, the pathologists calculated the total average by adding all the Ki67 index results and dividing them with the number of "hot-spots" area images. Thus, the grading was determined based on the total average value. Comparatively, the CAD system calculated the Ki67 index by finding the total number of positive Ki67 cells for all "hot-spots" area images. Then, the results were divided with the total number of positive and negative Ki67 cells for all "hot-spots" area images. The resultant Ki67 value was used to determine brain tumor grading. As mentioned in Section 3.2, a pop-up window appeared and displayed the grading result after obtaining the Ki67 index value. Figure 11.6 illustrates the final process of the CAD system, which was to display the result of tumor grading.

As shown in Figure 11.6, once the users obtained all the values required in the "TUMOUR GRADING DETERMINATION" panel, the user clicks on the "Tumour Grading" icon to display the grading result. For the grading analysis, Table 11.6 compares the grading results between the manual counting and the CAD system.

Based on Table 11.6, the proposed techniques were able to provide an accurate grading result. Most of the grading results obtained from the proposed techniques had the same result as the manual counting technique. Interestingly for slide 4, the grading result obtained by the CAD system had the same result, although the Ki67 index result was higher than the average Ki67 index obtained by the manual counting. Compared with slide 2 results, the difference between Ki67 index values was small. However, the grading achieved by the CAD system was different from the manual counting result. This occurred when the Ki67 index values did not fall within the grading range and consequently led to different grading results.

11.5   CONCLUSION

This chapter introduced the development of a CAD system for astrocytoma histo-pathological images. The objective of developing the CAD system was to help the

TABLE 11.6

Comparison of Grading Results between Manual Counting and the CAD System

Slide	Image	Manual Counting		CAD System	
		Average Ki67 Index (%)	Grade	Ki67 Index (%)	Grade
1	01.jpg	3.60	2	3.29	2
	02.jpg				
	03.jpg				
2	04.jpg	4.74	2	5.56	3
	05.jpg				
	06.jpg				
3	07.jpg	3.09	2	4.67	2
	08.jpg				
	09.jpg				
4	10.jpg	2.73	2	4.69	2
	11.jpg				
	12.jpg				
5	13.jpg	6.27	3	3.80	2
	14.jpg				
	15.jpg				

pathologists reduce their workload, particularly during the counting process for deciding astrocytoma tumor grading. The approach taken for determining the tumor grading was based on Ki67 cell counting. Previous studies have shown that Ki67 is known as a predictive indicator for measuring the proliferation of cancer cells. Thus, Ki67 can become an excellent marker for identifying abnormal cells in tumor cell populations.

Generally, the development of a CAD system comprises of six stages: image enhancement, color deconvolution, feature extraction, counting positive and negative Ki67 cells, calculating the percentage of Ki67 index, and determination of tumor grading. Image enhancement was used to improve the quality of the captured images for further analysis. A color deconvolution technique was applied to separate the color intensities between the DAB and haematoxylin staining, which were used to stain the histopathological slides. The feature extraction technique was used to extract the important features, which were related to Ki67 cells. The next step was to count the number of positive and negative Ki67 cells. After obtaining the counting results, the system calculated the percentage of the Ki67 index. This percentage value was then used to determine the tumor's grade based on the classification provided by the WHO. Overall, the CAD system was able to produce good results in identifying and counting the Ki67 cells. The average relative accuracy for counting positive Ki67 cells was 0.5, while for counting negative Ki67 cells, the result was 0.89.

ACKNOWLEDGMENT

Special thanks to the pathologists from the Department of Pathology, Health Campus Universiti Sains Malaysia for helping and contributing to this study.

REFERENCES

1. Brownstein K and Stevenson E. 2004. *The Essential Guide to Brain Tumors.* Edited by E Vassall. San Francisco, CA: National Brain Tumor Foundation.
2. Li LT, Jiang G, Chen Q and Zheng JN. 2015. Ki67 is a promising molecular target in the diagnosis of cancer (Review). *Mol. Med. Rep.*, **11**, 1566–1572.
3. Starborg M, Gell K, Brundell E and Höög C. 1996. The murine Ki-67 cell proliferation antigen accumulates in the nucleolar and heterochromatic regions of interphase cells and at the periphery of the mitotic chromosomes in a process essential for cell cycle progression. *J. Cell Sci.*, **109**, 143–153.
4. Rosai J and Ackerman LV. 1979. The pathology of tumors, Part III: Grading, staging & classification. *CA, Cancer J. Clin.*, **29**, 66–77.
5. Damjanov I. 2011. Grading of tumors. In *Encyclopedia of Cancer*, edited by M Schwab. Springer Berlin Heidelberg, pp. 1591–1593.
6. Louis DN, Perry A, Reifenberger G, von Deimling A, Figarella-Branger D, Cavenee WK, Ohgaki H, Wiestler OD, Kleihues P and Ellison DW. 2016. The 2016 World Health Organization classification of tumors of the central nervous system: a summary. *Acta Neuropathol.*, **131**, 803–820.
7. Abdelzaher E. 2017. WHO grading of astrocytoma. *Pathol. Outl.* https://www.patholog youtlines.com/topic/cnstumorwhograding.html (accessed February 7, 2020).
8. Collins VP, Jones DTW and Giannini C. 2015. Pilocytic astrocytoma: pathology, molecular mechanisms and markers. *Acta Neuropathol.*, **129**, 775–788.
9. Palys V. 2020. Astrocytomas. *Viktor Notes Neurosurg. Resid.* http://www.neurosurger yresident.net › Onc10. Astrocytomas.pdf (accessed February 6, 2020)
10. D'Alessio A, Proietti G, Sica G and Scicchitano BM. 2019. Pathological and molecular features of glioblastoma and its peritumoral tissue. *Cancers*, **11**, 1–19.
11. Ştefănescu D, Streba C, Cârţână ET, Săftoiu A, Gruionu G and Gruionu LG. 2016. Computer aided diagnosis for confocal laser endomicroscopy in advanced colorectal adenocarcinoma. *PLoS One*, **11**, 1–9.
12. Misawa M, Kudo SE, Mori Y, Nakamura H, Kataoka S, Maeda Y, Kudo T, Hayashi T, Wakamura K, Miyachi H, Katagiri A, Baba T, Ishida F, Inoue H, Nimura Y and Mori K. 2016. Characterization of colorectal lesions using a computer-aided diagnostic system for narrow-band imaging endocytoscopy. *Gastroenterology*, **150**, 1531–1532.
13. Supriyanti R, Chrisanty A, Ramadhani Y and Siswandari W. 2018. Computer aided diagnosis for screening the shape and size of leukocyte cell nucleus based on morphological image. *Int. J. Electr. Comput. Eng.*, **8**, 150–158.
14. Win KY, Choomchuay S, Hamamoto K, Raveesunthornkiat M, Rangsirattanakul L and Pongsawat S. 2018. Computer aided diagnosis system for detection of cancer cells on cytological pleural effusion images. *Biomed Res. Int.*, **2018**, 1–21.
15. Lo CM, Wu YH, (Jack) Li YC and Lee CC. 2020. Computer-aided bacillus detection in whole-slide pathological images using a deep convolutional neural network. *Appl. Sci.*, **10**, 1–12.
16. Fulawka L and Halon A. 2017. Ki-67 evaluation in breast cancer: The daily diagnostic practice. *Indian J. Pathol. Microbiol.*, **60**, 177–184.
17. Pisano ED, Zong S, Hemminger BM, DeLuca M, Johnston RE, Muller K, Braeuning MP and Pizer SM. 1998. Contrast limited adaptive histogram equalization image

processing to improve the detection of simulated spiculations in dense mammograms. *J. Digit. Imaging*, **11**, 193–200.

18. Mathworks. 2020. Adapthisteq. *Mathworks Doc.* https://www.mathworks.com/help/images/ref/adapthisteq.html (accessed August 21, 2020).

19. Ruifrok AC and Johnston DA. 2001. Quantification of histochemical staining by color deconvolution. *Anal. Quant. Cytol. Histol.*, **23**, 291–299.

20. Gonzalez RC and Woods RE. 2007. Representation and description. In *Digital Image Processing*. Upper Saddle River, NJ: Prentice Hall, pp. 817–864.

21. Kazmi IK, You L and Zhang JJ. 2013. A survey of 2D and 3D shape descriptors. In *2013* 10th International Conference Computer Graphics, Imaging, and Visualization. pp. 1–10.

22. Writh MA. 2004. Shape analysis and measurement. *Purdue Univ. Cytom. Lab.* http://www.cyto.purdue.edu/cdroms/micro2/content/education/wirth10.pdf (accessed April 15, 2019).

23. Purkait P, Biswas B, Das S and Koley C. 2013. Measurement of errors. In *Electrical and Electronics Measurements and Instrumentation*. New Delhi: McGraw-Hill Education.

24. Zhang Y, Wang H, Yang Z and Li J. 2014. Relative accuracy evaluation. *PLoS One*, **9**, 1–13.

25. Wälder O. 2008. *Mathematical methods for engineers and geoscientists.* Berlin, Heidelberg: Springer Science & Business Media.

26. Abramovitz M and Stegun IA. 1972. *Handbook of Mathematical Functions with Formulas, Graphs and Mathematical Tables.* New York: Dover.

12 Improved Classification Techniques for the Diagnosis and Prognosis of Cancer

Pankaj Dadheech, Ankit Kumar, S. R. Dogiwal,
Vipin Jain, Vijander Singh, and Linesh Raja

CONTENTS

12.1   INTRODUCTION

The healthcare market in India is one of the largest and fastest-growing industries in the world, it consumes nearly 10% of the GDP of a developed or developing nation; the healthcare industry contributes a significant amount to the country's economy. The Indian healthcare sector provides new and existing players with special opportunities for achieving and performing innovative research. Healthcare in India was also awarded "polio-free" status by the World Health Organization (WHO). According to research by McKinsey & Company, in the next decennary, consumer awareness and demand for better services and facilities will increase, and in India the healthcare industry will become the third-largest service sector employer. The latest innovation in healthcare data mining is the "big data analytics" revolution. In the healthcare industry, big data consists of electronic health datasets or flat-file data which are disordered, complex, and so large that they are nearly impossible to manage with the available tools or traditional hardware and software techniques. For the healthcare data/information, there is a very large amount of data available for understanding the patterns and trends; hence, big data analytics has the potential to improve healthcare services and provide cost reductions. This chapter explores data mining applications, and shows the difference these can make to the patients and their daily lives; it will also provide suggestions for future work/directions in healthcare. The hospital-based survey also provides benefits for various data mining techniques, as it can show results in different ways, such as clustering, association rules, and classification in the healthcare domain. This chapter also defines cancer and morphology patterns among various patients in Haryana and the surrounding state with the help of the above-defined different data mining techniques [1].

12.1.1 MEDICAL SERVICES IN INDIA

If the Indian economy grows faster than the economies of developed nations and the education rate keeps on increasing, then much of the Indian will be middle class by 2025, and the middle class can afford quality healthcare. According to a CII study, India needs 50 billion dollars annually to fulfill its healthcare requirement for the next 15 years, until 2040; India needs 2 million beds, and requires an immediate investment of 82 billion dollars. According to the PWC, 60% of patients are outpatients in the private sector. Nearly 40% of hospital beds are in the private sector. Around 30% of the medical market is covered by this economic segment. Now, in the Indian market, hospitals are realizing that IT can be effective and efficient for hospital growth. Indian healthcare services are fast-growing according to a CII-McKinsey study; Indian hospitals are the first choice of foreign tourists for health diagnosis. India is gaining a significant reputation for medical tourism from Gulf countries. Figure 12.1 shows the various shares of healthcare spending in India. Data mining techniques are used to diagnose different cancers, such as the early detection of breast cancer, which is one of the leading cancers in Asian countries.

12.1.2 DATA MINING IN FIELD OF HEALTHCARE

Data mining explores the hidden patterns or information in a data warehouse. This knowledge, i.e., information extracted from a vast dataset, is presented in an understandable form. Later on, HMIS (Healthcare Management Information System) is in the healthcare domain and fake cases are found. Healthcare and commercial databases are growing at an unpredicted rate. To handle these huge data sets, we need mature data mining algorithms which can be combined with older statistical methods. Table 12.1 shows the data to knowledge evolution.

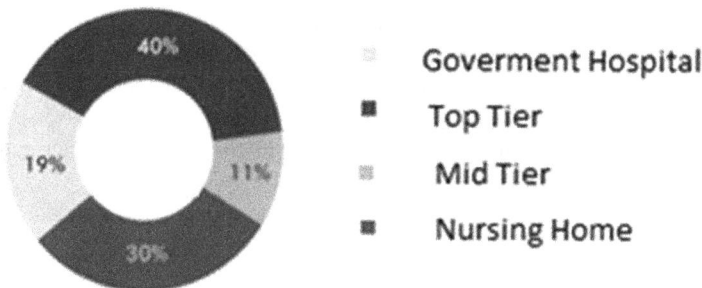

FIGURE 12.1   Shares in healthcare spending in India.

TABLE 12.1
Data to Knowledge Evolution

Evolutionary Step Year Wise	Questions From Business perspective	Technologies Used	Product Providers
Data Collection (1960–1980)	"What was the total revenue for the hospital in the last five years"	Computers, tapes, disks	CDC, IBM
Accessing Data (1980–1990)	"Total number of patient available department wise in the hospital"	Structured query language (SQL), relational databases (RDBMS)	IBM, Microsoft, Informix, Oracle, Sybase
Data Warehousing and Decision Support (1990s)	"Total number of patient available department wise in the hospital. Drill down to a single patient"	Data warehouses, multi-dimensional databases, OLAP	Micro strategy, Arbor, Cognos, Comshare, Pilot
Data Mining (Emerging Technology)	"How many patients will require cosmetics this month? Why?"	Massive databases, advanced algorithms, multiprocessor computers	SGI, IBM, Pilot startups, Lockheed

12.1.3 Architecture for Data Mining

Data mining tools and techniques operate by extracting, importing, and analyzing the data. A hospital data warehouse contains a combination of patient details and hospital details, i.e., patient beds and medicines. This warehouse information can be collected in variety of relational database systems, such as Sybase, Oracle, and MySQL, in an optimized manner so that it can be easy and fast to access. Now the OLAP (online analytical processing) database comes into the picture. With the help of facts and dimensions, a multi-dimensional structure is created, which helps users to analyze data for business purposes. The data mining techniques must be merged with the data warehouse and the OLAP server to produce new predictions and results.

Figure 12.2 shows the knowledge discovery process.

i. **Data cleaning** removes the inconsistent data.
ii. **Data integration** combines data from various sources.
iii. **Data selection and transformation** extracts the relevant data which is used for further analytics.
iv. **Data mining** contains intelligent methods (algorithms) to extract patterns out of the data (pattern evolution).
v. **Knowledge discovery** contains the overall visualization of the data.

12.1.4 Data Mining in Healthcare

Generally, healthcare data is available in flat-files, relational databases, or in advanced database systems, such as images which are collected from different data sources,

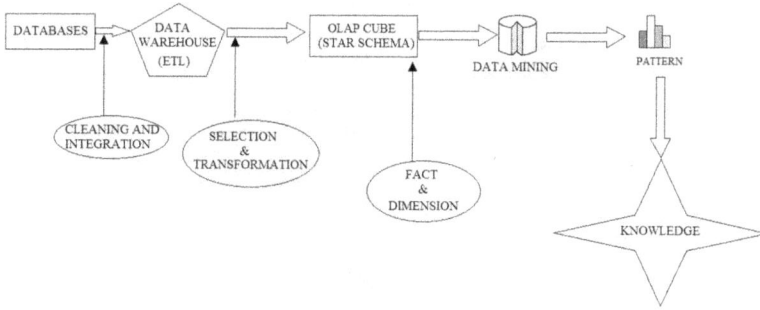

FIGURE 12.2 Knowledge discovery process.

including OPD, laboratory information systems, operation theater modules, radio-therapy modules, chemotherapy modules, blood banks, and the drug store. Various types of healthcare data are represented in Table 12.2.

12.1.4.1 Nature of Healthcare Data

Healthcare data is very specific. To mine healthcare data, all the information needs to be changed into numeric values. The methods for this task are described in medical textbooks which usually come from a certain range, and it is beyond the scope of this chapter. With the help of numeric data it is easy to apply mining algorithms to extract knowledge out of the data.

12.1.4.2 Patient Data Set

Data mining determines knowledge from a patient's dataset, and this dataset is a collection of different data sources, as mentioned above. In this research work, we collected data from the RCCR (Regional Cancer Center Rohtak) and prepared and analyzed the datasets for the different regions of Haryana and its surrounding states; then we applied a data mining algorithm to extract hidden patterns. This helped us to determine the cancer site and morphological patterns of various patients, and it also helped to develop a suggestive management information system to improve cancer treatment.

12.1.4.3 Preliminary Analysis of Dataset

Preliminary analysis of a dataset is an essential step for transforming unstructured data or row data into a format suitable for applying data mining techniques and improving the quality of the data. Preliminary analysis is to identify hospital and government needs, perform economic, technical analysis and green analytics, perform a cost–benefit analysis, and create a suggestive management information system. There should be enough expertise available for database queries and data mining algorithms and software for doing the analysis.

12.1.5 MEDICAL DATA SELECTION AND PREPARATION

Medical data selection and preparation (MDP) is a crucial and very time-consuming process for data mining. It takes around 45% to 55% of the total time

TABLE 12.2
Types of Healthcare Data

ECG	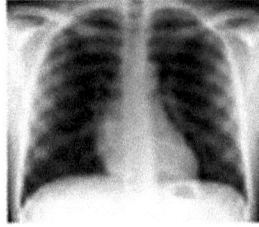
EEG	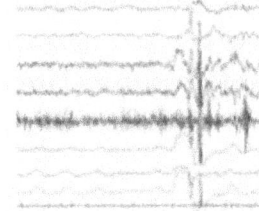
RTG	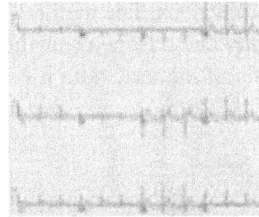

to prepare a dataset for the data mining process. Healthcare data selection and processing aims to establish a data warehouse for data mining algorithms and information sharing. MDSP is connected to various data sources, such as laboratories, operation theaters, blood banks, drug stores, and therapy modules, which hold valuable data, such as patient information, including their workflow and past prescriptions. There are two processes involved in MDSP, namely data selection and the processing. The medical data selection process selects data from different sources, as mentioned above. And all the relevant data is updated at a centralized data warehouse according to the patient details. Then, the MDS process selects whether the data is useful or not and then passes it to the research department. Medical data preparation involves examining or logging the data and verifying the data for accuracy and exactness. This is because some selected data may be missing or in different formats. At this stage, all necessary information is extracted from the data selection process to apply to further data mining processes, the age of the patient should be "81" but recorded as "18" so this comes under the human error. Data cleaning can reduce the missing, inconsistent, and noisy data that affect the results.

12.1.6 Issues and Challenges

i. **Domain expertise:**
   Domain knowledge related to healthcare data is very important; it helps in finding out different patterns from the database.
ii. **Visualization of healthcare data mining result:**
   After discovering the knowledge, the results should be easily understood and directly usable by humans.
iii. **Handling incomplete, noisy, and diverse data:**
   Healthcare data contains exceptional cases and incomplete data objects. Therefore the accuracy of the discovered patterns will be low. For exceptional cases, we have to apply data cleaning and data analysis methods.
iv. **Pattern discovery:**
   It is very difficult to cover thousands of patterns; hence, many patterns are undiscovered, and many of the patterns are uninteresting to the user.

12.1.7 Cancer Treatments Using Decision Support System

To design a DSS, the two main factors involve stakeholder involvement and the type of decision they want to make. A healthcare DSS should be well organized, including procedures for decision-making, strategic planning, and structures according to the government regulations should be kept in mind. The effectiveness of a DSS depends on the methodology used to design the system. We can divide this into three approaches:

i. Clinical algorithms
ii. Heuristics approaches
iii. Mathematical approaches

Data scientists have revealed that neural network systems can provide good planning for patient care, length of patient stay, and mortality rate. However, to implement a DSS with this methodology, extensive research and resources are required. Another very popular approach is data mining techniques, which help to identify rules and patterns concerning various problems. Data mining DSS is built based on the data, and it is effective for cost reduction and improving quality of care. As the nurses and doctors have to deal with various complex diagnoses, it becomes time consuming for them to adopt a new technology or a system. That is why DSSs are not widely accepted, due to time complexity and other constraints, but they are perceived as effective and efficient to use. Figure 12.3 shows the simulation model of a decision support system and Table 12.3 shows the various uses of data mining algorithms.

12.2 REVIEW OF LITERATURE

In the last section, we discussed various data mining fields, including medical services in India, data mining in a healthcare context, issues and challenges in

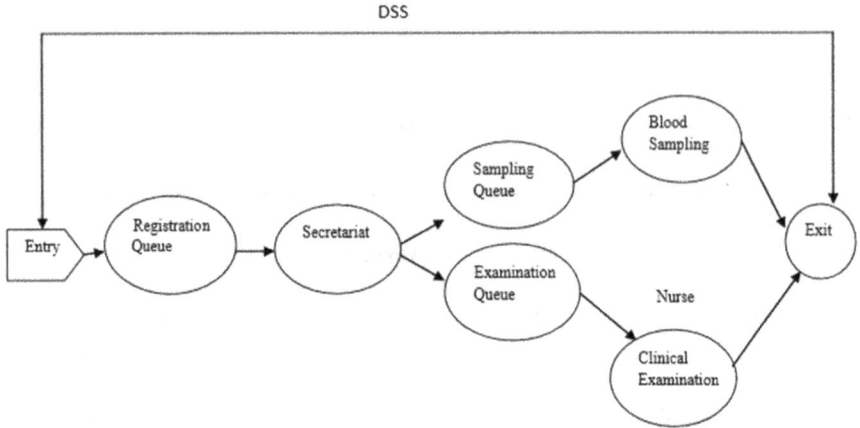

FIGURE 12.3   Simulation model of DSS.

TABLE 12.3
Uses of Data Mining Algorithm

Example of Task	Algorithm to Use
Predicting a discrete value	Naïve Bayes algorithm,
Categorization of a patient according to outcomes and explores related factors.	Decision trees algorithm, Neural network algorithm,
Calculating the probability of server failure within the next four months.	Clustering algorithm.
Predicting an attribute	Linear regression algorithm,
Predict next year sales.	Time-series algorithm,
Calculating or generating a risk factor or score on a given demographics.	Decision trees algorithm.
Sequence prediction	Sequence clustering
Capturing and analyzing the sequential activities of outpatient visit, to formulate common activities.	algorithm.
Searching similar items in transactions:	Association algorithm,
Use of market basket analysis.	Decision trees algorithm.
Suggesting additional product to buyer for purchase.	

healthcare data mining, and data mining applications in cancer treatments. Now, we will discuss research works on data mining in the healthcare industry. In the first section of the chapter, the review process is described. This covers a range of papers that I have studied and about how I conducted my review. The next section consists of a summary of all the papers that I researched to perform the current study and achieve the objective. This section is further divided into various categories based on the solutions in the field of our objective. The third section covers solutions given by various researchers. Then in Section 4, there is a discussion on the strengths and weaknesses which I have observed after studying various papers. The last section covers a summary of all the papers.

12.2.1 Review Process Adapted

Data mining is a term that has been used since the last century in various aspects. We see a new way of using data mining for the benefit of the humanities. If we continue to research data mining, it'll take years. But in the healthcare field, it is a newly emerging area of research. Many healthcare centers are now taking advantage of data mining applications. To perform this research work, we adopted a hierarchical concept of generalization to specialization. After reviewing the healthcare analytics of the world, from the country to the state, we came to Haryana to start experiments with the healthcare data in Haryana and its surroundings.

In this review process, all the research papers were classified based on their solutions.

We reviewed three main categories:

i. **Algorithms used in data mining:** This category involved the review of those papers which were based on the various algorithms used in the data mining process; how researchers used those algorithms and what solutions were provided by them.

ii. **Cancer causes and treatments:** This category involved papers based on cancer, its causes, symptoms, prevention, and cures, which are used by doctors across the world. It also included papers on the various solutions used by medical scientists to cure cancer patients.

iii. **Data mining in the field of healthcare:** This category consisted of review papers based on the application of data mining in the healthcare domain. This showed the devices, therapies, and procedures used by doctors and the works of medical scientists in the development of new equipment to help prevent this disease.

12.2.2 Categorical Review of Literature

12.2.2.1 Literature Review on Algorithm Classification

In 2012 Patil [2] launched an advanced wireless sensor network that aimed to provide online health forecasts through the real-time monitoring of critical body signals. For integrating and executing historical patient data, they implemented cluster algorithms (graph theoretical, k-means). The comparative tests of vital signals from clustered algorithms added additional measurements to the hazard warnings and made it more accurate for the doctor to diagnose.

A regression model was created by Carel et al. [3] for asthma drug use using a KDD approach to time-series datasets for historical asthma medication. The clustering and decision tree algorithms were used on the geographic patient sample. The results showed that 274 asthma patients received 9319 approved drugs; the classification also showed that corticosteroid drugs were the most significant indicator of the trend.

In 2014 Reyes et al. [4] were working to build an integrated method for the study of primary healthcare, using clustering methods (partitioned algorithms).

They used Java 12.6 as a language Eclipse 3.4 and JBoss 4.2 as a server to build the solution.

In 1999, Goil et al. [5] discussed a multi-dimensional device scalability framework for OLAP as well as OLAP data mining integration. It generated massive datasets on parallel computers with distributed memory.

12.2.2.2   Literature Review on Cancer Causes and Treatments

In 2013 Kawsar Ahmed et al. [6] gathered data of 400 cancer and non-cancer patients from numerous diagnostic centers, and applied a k-means clustering algorithm for the detection of important and irrelevant results. Finally, they established a test for lung cancer that was very effective in identifying a person's lung cancer predisposition.

In 2013 they also discussed the use of classification based technologies, such as artificial neural network guidelines, naïve Bayes, and decision trees, concerning healthcare results (also see Krishnaiah et al. [7]). They focused on common pulmonary symptoms, such as weight loss, pain in the legs and arms or chest, and short-sightedness. They introduced an early warning approach to help save lives.

12.2.2.3   Literature Review on Data Mining in Health Care

In the year 2013, a novel method of data mining was implemented by Akay, Dragomir, and Erlandsson [8], which tracked the experience of diabetic patients with drugs and medical devices. The paper explained how forums were turned into the vectors of search patterns in response to the patients' feedback on their prescriptions and computers. It also offered the impression it could be success for opioid patients.

Subhashet al. in 2013 [9] researched the prevention of infant hunger in developed countries using data processing methods and strategies. They included a selection of literature and used a decision tree methodology. They concluded that information obtained from surveys could be used to more efficiently mitigate infant malnutrition and could also be utilized for potential forecasts.

In 2012 Durairaj, Sivagowry, and Persia [10] addressed methods of data processing for the useful compilation of information from heart disease treatment systems. The study contrasted the efficiency of decision-making, naïve Bayes, neural network, and k-mean strategies for cardiac diagnosis extraction. The tools that are used for grouping, clustering, and membership purposes. Like text mining, medical data prediction can be further strengthened.

In 2009, Bellazzi [11] suggested providing remote medical centers with clinical decision support systems (CDSS). This artificial intelligence platform develops frameworks focused on information and data processing. It also defines FOCL hybrid algorithms which are used in the decision support framework and provide a highly efficient, modified version of FOCL.

In 2012, Xylogiannopoulos et al. [12] introduced a middleware model internet inter-orb protocol (IIOP) which was used in a distributed system and resulted in efficient response times between client and server. This model could be used in health clinics for cost reduction and performance efficiency. It could further be applied in medical referral systems and electronic consultation systems.

In 2010, Santhi et al. [13] suggested methodological problems for assessment of the data mining model, translation bioinformatics and bioinformatics aspects of genetic epidemiology on data collection, and data-driven approaches in the field of medical computing, data aggregation, and convergence.

In 2010, Sung Ho Ha et al. [14] stressed the current healthcare issues and their applications; they discussed how healthcare data mining applications were growing in number and producing better healthcare services and policy, and detecting diseases and preventing deaths in hospitals. It also discussed fraudulent insurance claims. This paper gave a broad idea of how to extract knowledge from a database.

12.2.2.4 Issue Wise Solution Approach

Table 12.4 shows various issues with solution approaches used in data mining.

12.3 PROBLEM STATEMENT AND OBJECTIVES

12.3.1 Problem Statement

"Improvisation of Data Mining Techniques in Cancer Site among Various Patients using Market Basket Analysis Algorithm."

Cancer is one of the most deadly diseases worldwide, and many people are currently suffering from this disease. There is no such treatment that provides 100% successful cure of this disease. With the increase of technologies in this area, doctors can explore this disease more and more. One of the best fields of IT, i.e., data mining, is showing very good results in the healthcare domain.

Data mining mainly includes three phases that help to find patterns among database artifacts and examine cancer sites. We study data or medical records by collecting them from any number of hospitals and then analyze that data to find the patients suffering from cancer. There are many types of cancer. From these datasets, we can extract the type of cancer, no of patients suffering from each type of cancer, and the precautions and medicines provided to them in Haryana and its surrounding areas.

To examine this, software vendors were developed for integration, analysis, and reporting services. We used Microsoft tools to generate reports for cancer sites. This study could be continued further and would be helpful for curing large numbers of cancer patients.

12.3.2 Objectives

Four objectives were set up to critically analyze patient data, and the objectives of the research work were as follows:

a. To study the successful data mining techniques and tools to improve the diagnosis of health diseases.
b. To study the various types of algorithms in data mining.
c. To study healthcare data and discover patterns to develop a suggestive management information system to improve cancer treatment.
d. Reducing the mortality rate due to cancer and increasing health awareness.

TABLE 12.4

Issue Wise Solution Approaches Used in Data Mining

Author	Publication Year	Approaches	Accuracy
Yan et al. [15]	2003	Multilayer perceptron	63.61%
Andreeva, P. [16]	2006	Kernel density	84.45%
		Neural network	82.78%
		Decision tree	75.71%
		Naïve Baye	78.58%
Hara et al. [17]	2008	Immune multi-agent neural network	82.33%
		Automatically defined groups	67.81%
Sitar-Taut et al. [18]	2009	Decision tree	60.42%
		Naïve Bayes	62.13%
Chang et al. [19]	2009	Decision tree with sensitivity analysis	86.88%
		Decision tree	90.86%
		Artificial neural network	92.61%
Rajkumar et al. [20]	2010	k-NN	45.67%
		Decision tree	52.04%
		Naïve Bayes	52.33%
Srinivas et al. [21]	2010	One dependency augmented Naïve Bayes classifier	80.46%
		Naïve Bayes	84.14%
Anbarasi et al. [22]	2010	Genetic with classification via clustering	88.35%
		Genetic with naïve Bayes	96.53%
		Genetic with decision tree	99.21%
Kangwanariyakul et al. [23]	2010	Bayesian neural network	78.41%
		RBF-kernel support vector machine	60.76%
		Polynomial support vector machine	70.55%
		Linear support vector machine	74.53%
		Probabilistic neural network	70.57%
		Back-propagation neural network	78.42%
Osareh et al. [24]	2010	SVM-POLY	95.29%
		PNN	92.83%
		SVM-RBF	95.44%
		k-NN	94.16%
Abdi et al. [25]	2013	AR_MLP	97.25%
		AR_PSO-SVM	98.93%
		SVM	94.57%

12.3.3 TOOLS AND TECHNOLOGIES USED

12.3.3.1 SQL Server Integration Services (SSIS)

SQL server integration services developed by Microsoft are tools or platforms for performing ETL operations. Table 12.5 shows the features of the SQL server integration services, and Table 12.6 shows the features of a data warehouse.

TABLE 12.5
Features of SSIS

Feature	Datacenter	Standard	Enterprise
Import and export wizard for SQL server	Yes	Yes	Yes
In-build data source connector	Yes	Yes	Yes
Runtime and designer for run time	Yes	Yes	Yes
Task used by import and export wizard	Yes	Yes	Yes
Logging and log provider	Yes	Yes	Yes
Data profiling tools	Yes	Yes	Yes
Extensibility of programmable object	Yes	Yes	Yes

TABLE 12.6
Features of a Data Warehouse

Feature	Datacenter	Standard	Enterprise
Data warehousing and auto-staging	Yes	Yes	Yes
Change in captured data	Yes	Yes	—
Compression in data	Yes	Yes	—
Query optimization with the help of star join	Yes	Yes	—
Automatic view by on query optimizer	Yes	Yes	—
Cubes partition	Yes	Yes	—

12.3.3.2 SQL Server Analysis Services (SSAS)

Table 12.7 shows the features of SQL server analysis services, and Table 12.8 shows the data mining features in SQL server analysis services.

12.4 METHODOLOGIES/ALGORITHMS USED

12.4.1 NAÏVE BAYES ALGORITHM

The naïve Bayes is a Bayes-based classification algorithm. It is used to model forecasts. Quite often, we use this algorithm to rapidly produce a mining model to find relationships between inputs and predetermined columns [26]. This model can be used for initial data scans. The naïve Bayes algorithm output screen can be seen in Figure 12.4.

Data required for naïve Bayes models:

- Key column: Each model should contain one primary column.
- Input columns: Input columns must be either discrete or continuous, and they should be independent of each other.
- Predictable column: There should be one predictable attribute which must contain continuous or discrete values.

TABLE 12.7
Features of SSAS

Feature	Datacenter	Standard	Enterprise
Backup facility	Yes	Yes	Yes
Dimension and cube design	Yes	Yes	Yes
Power Pivot	Yes	Yes	Yes
Distributed and Partition cubes	Yes	Yes	—

TABLE 12.8
Data Mining Features in SSAS

Feature	Datacenter	Standard	Enterprise
Bunch of comprehensive data mining algorithm	Yes	Yes	Yes
Integrated tools: Editors, model query builder, wizards, viewers	Yes	Yes	—
Tuning optimization for algorithm	Yes	Yes	—
Pipeline and text mining	Yes	Yes	—
Sequence prediction	Yes	Yes	—

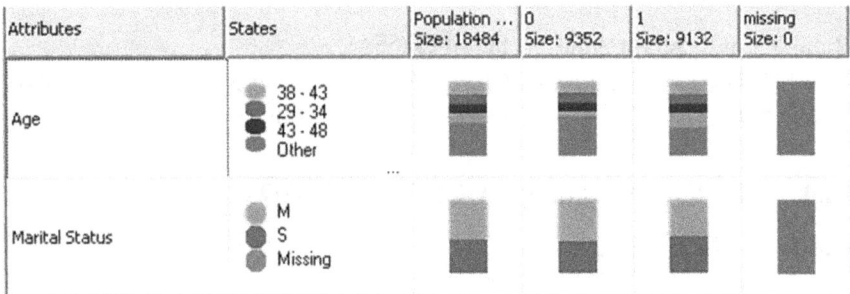

FIGURE 12.4   Naïve Bayes algorithm output screen.

12.4.2   CLUSTERING ALGORITHM

A clustering algorithm is a segmentation algorithm which uses iterative techniques to make clusters which contain similar type of characteristics. These clusters are useful for finding similar objects in the data for prediction. Figure 12.5 depicts how these clusters look.

A clustering algorithm primarily identifies the relations in the dataset and generates clusters based on that relationship. We can visualize the grouped data with a scatter plotter. Each scatter plot represents cases in a data set. After defining the

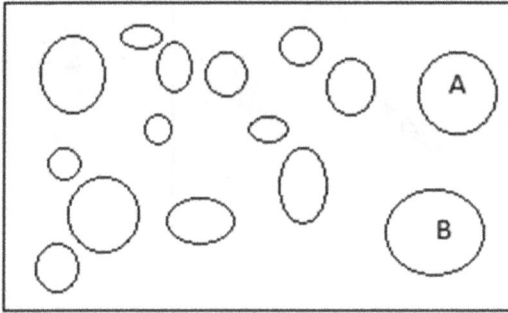

FIGURE 12.5   How cluster looks like.

cluster, it will again calculate how well the cluster represents the grouping of the point, and it will redefine the cluster to better represent the data [27].

Data required for a clustering algorithm:

- A single key column: The model should contain the primary key.
- Input columns: Each model should contain one input column.
- Predictable column: This column is an optional field in the model.

12.4.3   NEURAL NETWORK ALGORITHM

In the mining, model networks are dependent on the number of states (input columns and predictable columns) present in them. The neural network algorithm contains three layers of neurons in a network which is created by this algorithm. These three layers contain input–output layers and an optional hidden layer [28].

The input layer determines all the values of the input properties and the data mining process probabilities.

The hidden layer receives feedback from the neuron origin and produces neuron output. This layer includes the weights for the different inputs. The input to the hidden neuron is defined by weight. Input importance depends on the weight; the greater the weight, the more important its value will be. Weight can be negative, and in this case input will be neglected.

The output layer represents the predictable attribute values for the mining model.

12.4.4   TIME-SERIES ALGORITHM

This algorithm uses a regression technique which is optimized for the forecasting of continuous values. This algorithm does not require an additional column for predicting trends like a decision tree.

Historical information represents the data which is used to create the model, and it is represented on the left of the vertical in Figure 12.6. Predicted information represents the forecasting of the model, and it appears at the right of the vertical.

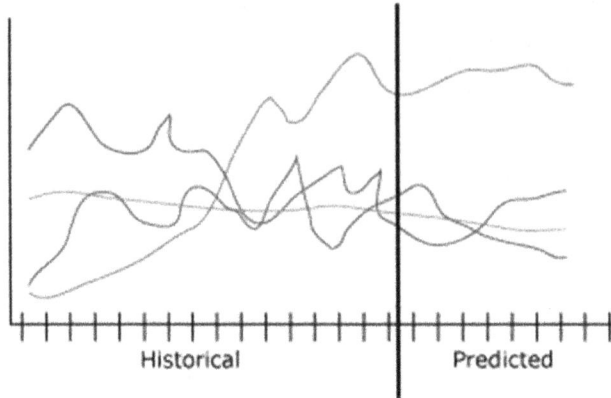

FIGURE 12.6   Time-series trends.

12.4.5   DECISION TREE ALGORITHMS

This is an algorithm based on classification and regression, which provides both continuous and discrete predictive modeling. It makes predictions based on the input relation between the columns in the dataset for discrete attributes [29]. For continuous attributes, it will determine where the decision tree splits by using a linear regression technique. If the model includes more than one column, a different decision tree will be formed for each repetitive column. The algorithm trees construct the mining structures by constructing a sequence of divisions in the tree. Splits are viewed as nodes. Every algorithm adds a new node to the column if it predicts a correlation between the input columns. The decision tree algorithm uses a feature selection technique to select attributes which will improve the quality and performance of the mining algorithm. If we use a more predictable attribute, then the model will take a very long time to process, or it will show an out of memory error [30].

12.4.6   ASSOCIATION ALGORITHM

This algorithm is useful for engines. A suggestion engine recommends items to the consumer or shows the need. The algorithm association [31] is also useful to evaluate the market.

12.5   EXPERIMENTAL ALGORITHM AND RESULTS

In the previous section, we discussed the design specification for performing this research work. The steps involved in the design specification are the basic steps followed in the experimental analysis. In this chapter, we will discuss the various experiments carried out and results obtained from these experiments.

12.5.1 Details of Experiment Carried Out

The experiments carried out to achieve the above-mentioned objectives are described in various stages. From data analysis to integration to reporting, the following experiments were performed.

12.5.1.1 Stage 1: Environmental Setup

The above-mentioned hardware and software requirements were fulfilled, and all the software was downloaded. To establish the environment for the experiment, the steps below were followed.

Windows 7 OS was installed on the system.

The visual studio was installed on the machine.

The SQL server 2008 r2 was installed.

We installed SQL server 2008 r2, which included integration, analytical, and reporting services,

12.5.1.2 Stage 2: Create OLAP Cubes

The data present in the OLAP system was present in a multi-dimensional structure, and it was created with the help of facts and dimensions [32]. The dimensions had a granularity of viewing data. Therefore, day, month, and year was a TIME dimension hierarchy which specified various aggregation levels. Another dimension which we used was in/outpatient data: Patient age, cancer disease group, sex, diagnosis. We used an OLAP [33] cube creation with the help of facts and dimensions, but it was difficult to find trends and patterns in large OLAP dimensions; therefore, we used data mining techniques.

12.5.1.3 Stage 3: Applying Market Basket Analysis

We applied algorithms on generated OLAP cubes to extract useful data from the large datasets.

```
D = The whole database
s = Choose a random sample from database D
S = Large_patientsitemsets in s
F = patientsitemsets having >= min_chance_of_cancer
Report if Error return(F)
```

12.5.2 Sampling Algorithm

```
P = Partition Patients_Database(n) n = The number of
Partitions
for (i = 1 ; i ≤ n ; i + +) {
Read from Partition(pi ∈ P)
Li = gen _large_patientsitemsets(pi)
}
```

```
/* Merge Phase */
for (i = 2 ; Lij != Ø , j = 1, 2, . . . , n ; i + +)
{
CGi = Uj=1,2,...,nLij
}
/* 2nd Phase */
for (i = 1 ; i ≤ n ; i + +) {
Read from Partition(pi ∈ P)
forall candidates c ∈ CGgen count(c, pi) }
LG = {c ∈ CG|c.count ≥ min_chance_of_cancer} return(LG);
```

```
/* Processing Step */
The processor searches its partitions to find locally wide
supports for patients.
```

```
Compute L1 and calculate C2 = Apriori_Gen(L1)
```

```
Virtually Prune C2
```

```
Initialize the common portion of the rest of the patients
```

```
Configure to create a uniform network.
/* Parallel Step: Every processor i runs this in its partition
   Di */
```

```
while(some processor has not finished counting the items on
the shared part)
{
  while( processor i has not finished counting the
patientsitemsets in the shared part)
  {
Scan the next interval on Di and count the patients item sets
in the shared part
```

```
    Find the locally large patients item sets among the ones
    in the shared part
    Generate new patients from these locally large
    patientsitemsets
    Perform virtual partition pruning and put shared part
    Remove globally small patientsitemsets in the shared part
  }
}
```

```
Generate Parallel Patients dataset of different diseases
procedurePatientDieases
  {
          let Patients set L = Ø;
    let People set F = {Ø};
    while (F != Ø)
```

```
                                 {
                                     let Male set C = Ø;
        forall database tuples t
        {
                forall patientsitemsets f in F
                    {
            if (t contains f)   {
                                        let Cf =Male
                                        patientsitemsets that are
                                        extensions of f and
                                        contained in t forall
                                        patientsitemsets cf in Cf
                    {
                      if(Cf C)    cf .count + +
                        else {
                                cf .count = 0
                                C = C + cf
                                }
                    }
                                        }
                        }
        }
let F = Ø
forall patientsitemsets c in C {
if count(c)/|D| > min_chance_of_cancer
L = L + c
 if c should be used as a People in the next pass F = F + c
}
Find large Patientsitemset or similar data set
```

12.5.3   RESULTS AND DISCUSSION

In the previous section, we described the various experiments performed to achieve
the objective.

The objective of this chapter consisted of two main parts.

- To determine the cancer site and the morphology patterns between various
  patients.
- To explore data mining applications and challenges in healthcare.

We chose SQL server 2008 integration, analytical, reporting services. We were able
to provide a dependency network for cancer patients. In our research, we explored
data mining applications and challenges in healthcare.

Market Basket Analysis
Market basket analysis showed the dependency of the attributes which were strongly
correlated. Figure 12.7 shows the market basket analysis of patient data. Here we

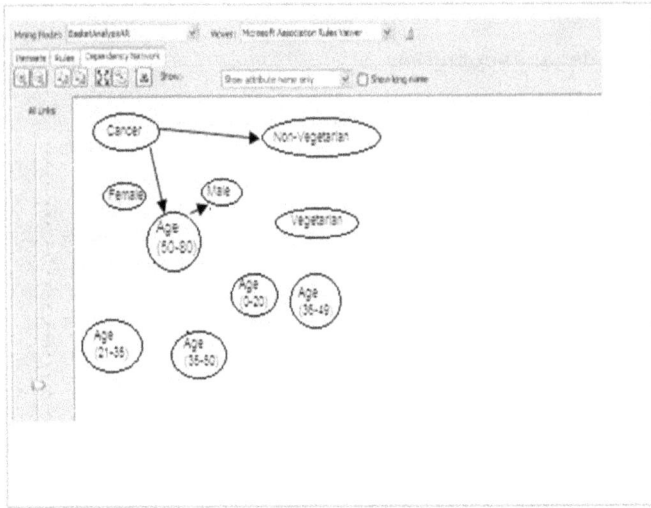

FIGURE 12.7   Market basket analysis on patient data.

found some very interesting patterns in the data. Initially, we defined the attributes for patient analysis, such as age, gender, vegetarian, non-vegetarian. From the above analysis, we observed that patients in the age group of 35–40, male, and non-vegetarian occurred more frequently in the cancer treatment category.

The graph depicted in Figure 12.8 shows the no. of cancer patients in different age groups. The above analytical graph is calculated on eight years of patient data. It

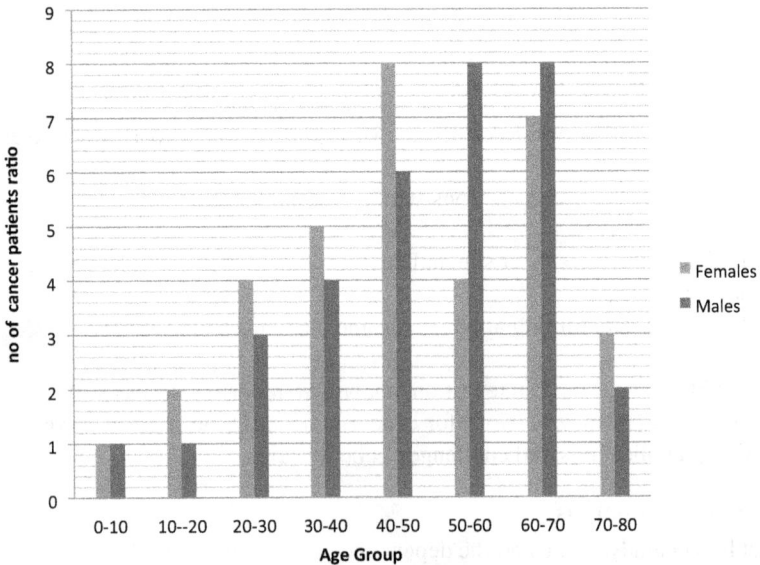

FIGURE 12.8   Graph showing no. of cancer patients in different age groups.

TABLE 12.9

Naïve Bayes Algorithm V/s Market Basket Analysis

Naïve Bayes Algorithm

By using a naïve Bayes algorithm we can use predictive modeling. Naïve Bayes algorithm discovers the relation between input and predictable column.

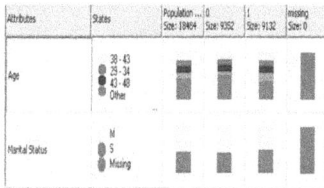

Output screen of naïve Bayes algorithm.

Clustering Algorithm

It is a segmentation algorithm used to make clusters which contain similar type of characteristics. Clustering algorithms are used for finding a similar object in data for prediction.

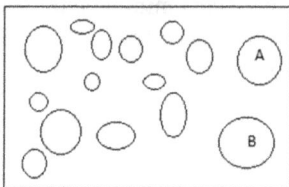

Output screen for a cluster.

Time-Series Algorithm

Time-series algorithm uses regression technique which is optimized for the forecasting of continuous value. This algorithm required historical data of patients sets.

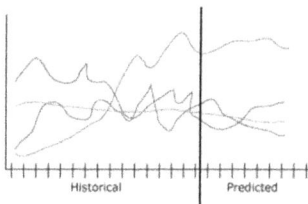

Output screen of time series.

Market Basket Analysis

Market basket analysis show dependency on the attributes which are strongest correlated. In my research work, we find very interesting patterns out of different patient's dataset. Firstly we defined the attribute for patient's analysis such as age, gender, veg, and non-veg. From the above analysis, we can see that patients in the age group of 35–40, male, and non-vegetarians are more in the cancer predict. So the result shows people belong this category check their health timely for predict cancer diseases.

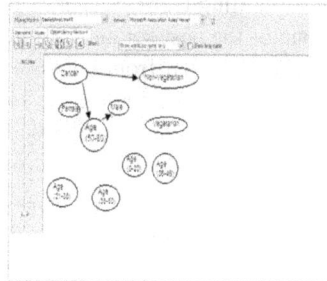

Output screen for market basket analysis.

shows very interesting graphs over cancer patients. Most of the cancer patients present between the 40–70 years of age group, and most of them are count of the female patient is more than male patients.

12.5.4 COMPARISON STUDY

Table 12.9 shows the comparative study of a naïve Bayes algorithm V/s market basket analysis.

12.6 CONCLUSION AND FUTURE SCOPE

12.6.1 CONCLUSION

Health issues are arising faster than ever before and scientists are constantly running behind in finding technologies to provide solutions to the new diseases. Data mining is one of the best solutions to treat patients with the help of past experience and knowledge extracted from data collected over years. Data mining has a significant impact in the field of healthcare. Healthcare industries are improving their outputs with the use of various techniques and equipment developed by medical scientists. In this chapter, we proposed to carry out a study of cancer site and morphology patterns among various patients in the context of complex data, such as text, images, sounds and videos. In our approach, we combined OLAP with data mining which resulted in a high level of analysis, and helped us to discover hidden patterns in the data and provide visualization of the information.

12.6.2 FUTURE SCOPE

This research work provided an association between OLAP and data mining (OLAP mining) for analysis. For future work, we found various issues to address in the future. The first issue was to provide an association between OLAP and predictive analytics using a data mining algorithm. The second was to provide a facility from which doctors could query the data cube on aspects of a business problem and translate this problem into a MDX (multi-dimension expression) query automatically.

REFERENCES

1. Shweta Kharya discussed. 2012. Using Data Mining Techniques for Diagnosis and Prognosis of Cancer Disease. *International Journal of Computer Science, Engineering and Information Technology (IJCSEIT)*, 2(2), pp. 55–66.
2. Patil, D., and Wadhai, V. 2012. Dynamic Data Mining Approach to WMRHM. In *Proceedings of 7th IEEE Conference on Industrial Electronics and Applications (ICIEA)*, Singapore, pp. 1978–1983.
3. Last, M., Carel, R., and Barak, D. 2007. Utilization of Data-Mining Techniques for Evaluation of Patterns of Asthma Drugs Use by Ambulatory Patients in a Large Health Maintenance Organization. In *Seventh IEEE International Conference on Data Mining Workshops (ICDMW 2007)*, Omaha, NE, pp. 169–174. doi: 10.1109/ICDMW.2007.50.

4. Reyes, A. J. O., Garcia, A. O., and Mue, Y. L. 2014. System for Processing and Analysis of Information Using Clustering Technique. *Latin America Transactions, IEEE (Revista IEEE America Latina)* 12(2), 364–371.

5. Goil, S., and Choudhary, A. 1999. A Parallel Scalable Infrastructure for OLAP and Data Mining. In *International Symposium Proceedings on Database Engineering and Applications, IDEAS '99*, Montreal, QC, pp. 178–186.

6. Ahmed, K., Emran, A., Jesmin, T., Mukti, R. F., Zamilur Rahman, Md., and Ahmed, F. 2013. Early Detection of Lung Cancer Risk Using Data Mining. *Asian Pacific Journal of Cancer Prevention: APJCP*, 14(1), 595–598.

7. Krishnaiah, V., Narsimha, G., and Chandra, N. S. 2013. Diagnosis of Lung Cancer Prediction System Using Data Mining Classification Techniques. *International Journal of Computer Science and Information Technologies* 4(1), 39–45.

8. Akay, A., Dragomir, A., and Erlandsson, B.-E. 2013. A Novel Data-Mining Approach Leveraging Social Media to Monitor and Respond to Outcomes of Diabetes Drugs and Treatment. *IEEE Point-of-Care Healthcare Technologies (PHT)* 2013, 264–266.

9. Ariyadasa, S. N., et al. 2013. Knowledge Extraction to Mitigate Child Malnutrition in Developing Countries (Sri Lankan Context). In *4th International Conference on Intelligent Systems, Modelling and Simulation*, IEEE Computer Society, Washington, DC, pp. 321–326.

10. Durairaj, M., et al. 2012. An Empirical Study on Applying Data Mining Techniques for the Analysis and Prediction of Heart Disease. In *International Conference on Information Communication and Embedded Systems (ICICES)*, IEEE, Chennai, pp. 265–270.

11. Bellazzi, R., Sacchi, L., and Concaro, S. 2009. Methods and Tools for Mining Multivariate Temporal Data in Clinical and Biomedical Applications. In *Annual International Conference of the IEEE Engineering in Medicine and Biology Society*, pp. 5629–5632.

12. Xylogiannopoulos, F. 2012. Developing an Efficient Health Clinical Application: IIOP Distributed Objects Framework. In *IEEE/ACM International Conference on Advances in Social Networks Analysis and Mining*, Istanbul, pp. 759–764.

13. Santhi, P., and Murali Bhaskaran, V. 2010. Performance of Clustering Algorithms in Healthcare Database. *International Journal for Advances in Computer Science* 2(1), 26–31.

14. Ha, S. H., and Joo, S. H. 2010. A Hybrid Data Mining Method for the Medical Classification of Chest Pain. *World Academy of Science, Engineering and Technology, Open Science Index 37, International Journal of Computer and Information Engineering* 4(1), 99–104.

15. Yan, H., Zheng, J., Jiang, Y., Peng, C., and Li, Q.. 2003. Development of a Decision Support System for Heart Disease Diagnosis Using Multilayer Perceptron. In *Proceedings of the 2003 International Symposium on Circuits and Systems (ISCAS) '03*. 5, pp. V–V.

16. Andreeva, P. 2006. Data Modelling and Specific Rule Generation via Data Mining Techniques. In *International Conference on Computer Systems and Technologies—CompSysTech*, pp. IIIA.17-1–IIIA.17-6.

17. Hara, A., and Ichimura, T. 2008. Data Mining by Soft Computing Methods for the Coronary Heart Disease Database. In *IEEE Fourth International Workshop on Computational Intelligence & Application*, SMC Hiroshima Chapter, Hiroshima University, Japan, 10-11.

18. Sitar-Taut, V. A. 2009. Using Machine Learning Algorithms in Cardiovascular Disease Risk Evaluation. *Journal of Applied Computer Science & Mathematics* 3(1), 29–32.

19. Chang, C. L., and Chen, C. H. 2009. Applying Decision Tree and Neural Network to Increase Quality of Dermatologic Diagnosis. *Expert Systems with Applications, Elsevier* 36, 4035–4041.

20. Rajkumar, A. and Reena, G. S. 2010. Diagnosis of Heart Disease Using Data mining Algorithm. *Global Journal of Computer Science and Technology* 10(10), pp 38-43.

21. Srinivas, K., Rani, B. K., and Govrdhan, A. 2010. Applications of Data Mining Techniques in Healthcare and Prediction of Heart Attacks. *International Journal on Computer Science and Engineering (IJCSE)* 02(02), 250–255.

22. Anbarasi, M., Anupriya, E., and Iyengar, N. 2010. Enhanced Prediction of Heart Disease with Feature Subset Selection Using Genetic Algorithm. *International Journal of Engineering, Science and Technology* 2(10), 5370–5376.

23. Kangwanariyakul, Y., Nantasenamat, C., Tantimongcolwat, T., and Naenna, T. 2010. Data Mining of Magnetocardiograms for Prediction of Ischemic Heart Disease. *EXCLI Journal*, 9, 82–95, 2010.

24. Osareh, A., and Shadgar, B. 2010. Machine Learning Techniques to Diagnose Breast Cancer. In *5th International Symposium on Health Informatics and Bioinformatics*, pp. 114–120.

25. Abdi, M. J., and Giveki, D. 2013. Automatic Detection of Erythemato-Squamous Diseases Using PSO–SVM Based on Association Rules. *Engineering Applications of Artificial Intelligence* 26, 603–608.

26. Kumar, A., Goyal, D., and Dadheech, P. 2018. A Novel Framework for Per-formance Optimization of Routing Protocol in VANET Network. *Journal of Advanced Research in Dynamical & Control Systems* 10, 2110–2121, ISSN: 1943-023X.

27. Dadheech, P., Goyal, D., Srivastava, S., and Kumar, A. 2018. A Scalable Data Processing Using Hadoop & MapReduce for Big Data. *Journal of Advanced Re-search in Dynamical & Control Systems* 10, 2099–2109, ISSN: 1943-023X.

28. Dadheech, P., Goyal, D., Srivastava, S., and Choudhary, C. M.. 2018. An Efficient Approach for Big Data Processing Using Spatial Boolean Queries. *Journal of Statistics and Management Systems (JSMS)* 21(4), 583–591.

29. Kumar, A. and Sinha, M. 2014. Overview on Vehicular Ad Hoc Network and Its Security Issues. In *International Conference on Computing for Sustainable Global Development (INDIACom)*, pp. 792–797. doi: 10.1109/IndiaCom.2014.68280712.

30. Dadheech, P., Kumar, A., Choudhary, C., Beniwal, M. K., Dogiwal, S. R., and Agarwal, B. 2019. An Enhanced 4-Way Technique Using Cookies for Robust Authentication Process in Wireless Network. *Journal of Statistics and Management Systems* 22(4), 773–782. doi: 10.1080/09720510.2019.1609557.

31. Kumar, A., Dadheech, P., Singh, V., Raja, L., and Poonia, R. C. 2019. An Enhanced Quantum Key Distribution Protocol for Security Authentication. *Journal of Discrete Mathematical Sciences and Cryptography* 22(4), 499–507. doi: 10.1080/09720529.2019.1637154.

32. Kumar, A., Dadheech, P., Singh, V., Poonia, R. C., and Raja, L. 2019. An Improved Quantum Key Distribution Protocol for Verification. *Journal of Discrete Mathematical Sciences and Cryptography* 22(4), 491–498. doi: 10.1080/09720529.2019.1637153.

33. Kumar, A., and Sinha, M. 2019. Design and Analysis of an Improved AODV Protocol for Black Hole and Flooding Attack in Vehicular Ad-Hoc Network (VANET). *Journal of Discrete Mathematical Sciences and Cryptography* 22(4), 453–463. doi: 10.1080/09720529.2019.16371512.

13 Discovery of Thyroid Disease Using Different Ensemble Methods with Reduced Error Pruning Technique

Dhyan Chandra Yadav and Saurabh Pal

CONTENTS

13.1 INTRODUCTION

Barrea et al. [1] established that thyroid hormones can have a serious effect on human health, due to this problem many functions of the body, such as metabolism, do not work correctly when the thyroid is not able to produce enough hormones. Thyroid disorders are a more widespread problem in women compared to men. Thyroid disease directly affects women's menstrual cycle and ovulation system. If the ovulation system does not properly work, then women cannot conceive. Thyroid disorders create problems in pregnant women for both the woman and her baby. Thyroid disorders can affect human life in various ways, such as:

1. Hypothyroidism
2. Hyperthyroid
3. Thyroiditis
4. Goiter
5. Thyroid nodules
6. Thyroid cancer

13.1.1 HYPOTHYROIDISM

Ejtahed et al. [2] identified a hormone disorder in which thyroid hormones were not produced in sufficient quantities by the thyroid, due to this problem many functions of the body such as metabolism did not work properly.

Symptoms of a hypothyroid:

Some symptoms of a hypothyroid are described as:

- Feeling excessively cold.
- Constipation.
- Muscles are weak.
- Weight gain.
- Always remain depressed.
- Excess feeling of fatigue.
- Skin is yellow and dry.
- Dry and thin hair.
- Slow heart rate.
- Sweating less than an average person.
- Swollen face. Hoarseness.
- Bleeding more than normal during menstruation.
- Hypothyroid patients may suffer from bad cholesterol.

13.1.2 HYPERTHYROIDISM

Hyperthyroidism is a type of disorder in which hormones are produced more than usual or more than the human body needs. This disorder affects the immune system of the body.

Some of its symptoms are described as:

- More weight gain than normal people while consuming a normal food intake.
- Patients eat more food compared to normal people.
- Heartbeat is higher than normal people.
- An increased feeling of nervousness and restlessness.
- Burning sensations.
- Low sleepiness.
- Tingling in the toes.
- Sweating more than normal people.
- Muscle weakness.
- Having more bowel movements than normal people.
- Menstrual periods shorter and lighter than normal.
- There are different types of changes in the eyes.
- Fear of weakening of bone.

13.1.3 THYROIDITIS

Thyroiditis is an inflammation of the thyroid in the body. It is produced when the body's immune system produces antibodies, and the thyroid is damaged.

The main reasons for this disease are as follows:

- Diseases like diabetes and rheumatic fever are covered under autoimmune diseases.
- Genetics.
- Having a viral or bacterial infection.

It is produced by some types of special medicines; it is mainly of two types:

- Hashimoto's disease
- Postpartum thyroiditis

13.1.4 GOITER

A goiter is a type of disease in which the thyroid gland is larger than average in size. It is also more common in women than in men.

The main reasons for this disease are as follows:

- Hashimoto's disease
- Thyroid nodule disease found.
- Swelling of the throat also occurs due to gland cancer.
- Thyroid nodules cause inflammation on one side of the gland.

13.1.5 Thyroid Nodules

Thyroid nodules are a liquid or blood-filled solid. Most thyroid nodules don't cause symptoms. The majority of thyroid nodules don't require treatment.

13.1.6 Thyroid Cancer

Thyroid cancer is caused by cancer cells from the tissues of the thyroid gland. Most thyroid cancers occur only to those who have thyroid nodules. A survey in 2020 showed that thyroid cancer cases in women were higher than in men.

13.1.7 Reduced Error Pruning

Ting et al. [3] introduced a reduced error pruning method based on an idea by Quinlan; this method observed each node in the decision tree as a candidate for pruning to remove subtrees rooted at that nodes to prepare them as leaf nodes (Figure 13.1 and Table 13.1).

13.1.8 AdaBoost M1

Wu et al. [4] introduced a binary classifier as an adaptive boosting algorithm, and it used for weak learners. The training data trained the weak learners with weighted samples. The main work of boosting was to take decisions for weak learners and convert into strong learners (Figure 13.2).

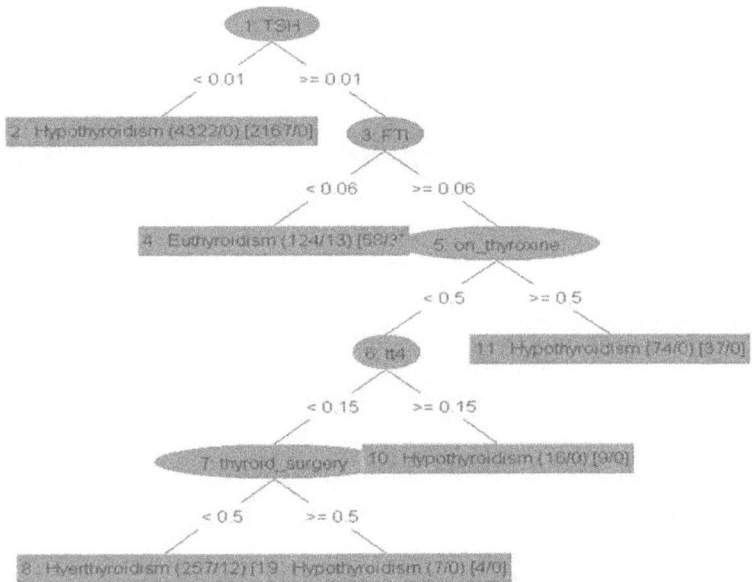

FIGURE 13.1 Representation of decision reduced pruning tree for thyroid disease.

TABLE 13.1

Representation of the Size of a Reduced Pruning Tree for Thyroid Disease

TSH < 0.01 : Hypothyroidism (4322/0) [2167/0]

TSH >= 0.01

| FTI < 0.06 : Euthyroidism (124/13) [58/3]

| FTI >= 0.06

| | on_thyroxine < 0.5

| | | tt4 < 0.15

| | | | thyroid_surgery < 0.5 : Hyerthyroidism (257/12) [125/2]

| | | | thyroid_surgery >= 0.5 : Hypothyroidism (7/0) [4/0]

| | | tt4 >= 0.15 : Hypothyroidism (16/0) [9/0]

| | on_thyroxine >= 0.5 : Hypothyroidism (74/0) [37/0]

Size of the tree: 11

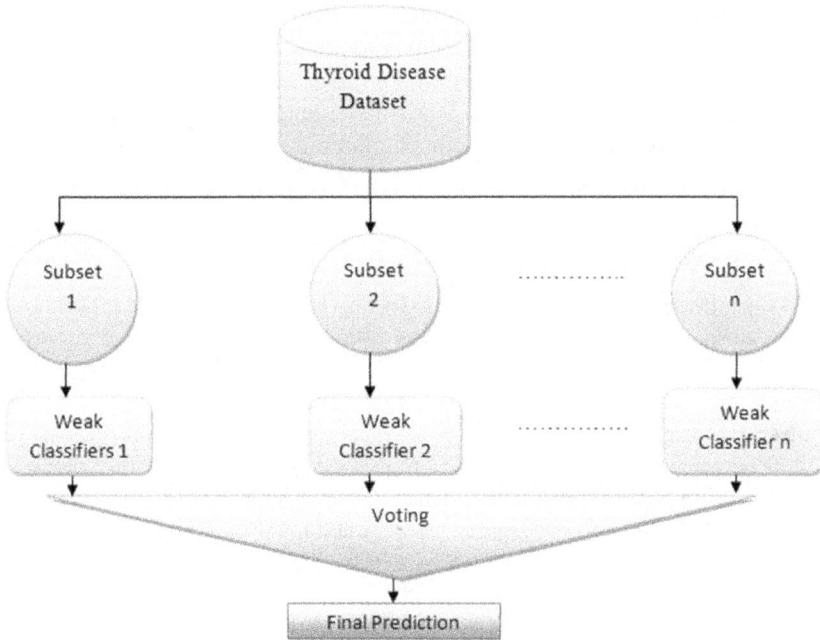

FIGURE 13.2 Representation of AdaBoostM1 ensemble method for thyroid disease.

The boosting predictions were generated by a weighted majority vote to produce the final prediction.

$$F_m(x) = F_{m-1}(x) + y_m h_m(x) \tag{13.1}$$

$$F_m(x) = F_{m-1}(x) + argmin_h \sum_{i=1}^{n} L(y_i, F_{m-1}(x_i) + (h(x_i))) \tag{13.2}$$

At each stage, the decision tree $h_m(x)$ was chosen to minimize a loss function L given the current model $F_{m-1}(x)$ (Table 13.2).

13.1.9 BAGGING

Hong et al. [5] introduced a bootstrap aggregation and reduced the variance between multiple estimators as bagging. The bagging algorithm trained multiple different trees on different subsets of the data by selecting and replacing and analyzing the ensemble:

$$f(n) = 1/x \sum_{z=1}^{X} f_x(n) \tag{13.3}$$

Where, x is the number of subsamples, and bagging was used to prepare the training set for base learners. The base learners used a voting algorithm and provided output as a classification (Figure 13.3).

The bootstrap aggregation stops overfitting when combined with several base learners. It is a variance machine learning algorithm for decision trees. The subset created a model with a sample of observations and features, and this provided the best split for the training dataset. These processes rapidly calculated the previous to the last sample subset and created many parallel models. The final prediction was made on the aggregation of the prediction by all the models, as shown on Table 13.3.

13.1.10 RANDOM SUBSPACE

Cimen [6] introduced an ensemble method using a random subspace algorithm. A random subspace ensemble algorithm was used for classification and regression subsamples. A random subspace ensemble algorithm was different from the other ensemble methods because it trained the subsample model on randomly selected samples of features, contrasting all the feature sets. This algorithm reduced correlations between predictors and the first regression of the subset feature space was running in T training subsets. Then, base regression was applied to each of these subsets, and a final decision was based on weighted majority voting (Figure 13.4).

In this process, a REP tree worked as a base. The base classifier performed its work based on majority voting, and the best model was selected as the final outcome (Table 13.4).

13.2 RELATED WORK

Guo et al. [7], discussed thyroid cancer and predicated high new case numbers and estimated deaths. They used a support vector machine (SVM) to differentiate between a benign thyroid and malignant thyroid. The authors classified a benign with a malignant thyroid dataset and found an accuracy of 0.8185 due to the area under the receiver operating a characteristic curve of 0.8376.

TABLE 13.2

Representation of an AdaBoostM1 Ensemble Tree Model for Thyroid Disease

TSH < 0.01 : Hypothyroidism (4322/0) [2167/0]
TSH >= 0.01
| FTI < 0.06 : Euthyroidism (124/13) [58/3]
| FTI >= 0.06
| | on_thyroxine < 0.5
| | | tt4 < 0.15
| | | | thyroid_surgery < 0.5 : Hyerthyroidism (257/12) [125/2]
| | | | thyroid_surgery >= 0.5 : Hypothyroidism (7/0) [4/0]
| | | tt4 >= 0.15 : Hypothyroidism (16/0) [9/0]
| | on_thyroxine >= 0.5 : Hypothyroidism (74/0) [37/0]
Size of the tree : 11
Weight: 5.48

…………………………..
…………………………..
…………………………..

FTI < 0.06
| t3 < 0.02
| | TSH < 0.15
| | | t4U < 0.1 : Euthyroidism (18.16/1.38) [57.38/0.55]
| | | t4U >= 0.1
| | | | FTI < 0.02 : Euthyroidism (10.25/0.14) [4.06/0]
| | | | FTI >= 0.02
| | | | | FTI < 0.03 : Hypothyroidism (153.86/0) [8.24/1.73]
| | | | | FTI >= 0.03
| | | | | | | TSH < 0.04
| | | | | | | | age < 0.78
| | | | | | | | | query_hypothyroid < 0.5 : Hypothyroidism (206.15/1.91) [190.76/189.72]
| | | | | | | | | query_hypothyroid >= 0.5 : Euthyroidism (3.86/0) [0.31/0]
| | | | | | | | age >= 0.78 : Euthyroidism (6.28/0.07) [0/0]
| | | | | | | TSH >= 0.04 : Euthyroidism (27.06/0.82) [19.31/0]
| | TSH >= 0.15 : Euthyroidism (24.92/0) [13.77/0]
| t3 >= 0.02
| | TSH < 0.01 : Hypothyroidism (27.51/0) [0.63/0]
| | TSH >= 0.01
| | | t3 < 0.03 : Euthyroidism (110.67/0) [23.16/0]
| | | t3 >= 0.03 : Hypothyroidism (13.02/0) [99.22/0]
FTI >= 0.06 : Hypothyroidism (1666.1/793.45) [4515.33/1834.99]

Size of the tree : 23
Weight: 0.44
Number of performed Iterations: 10

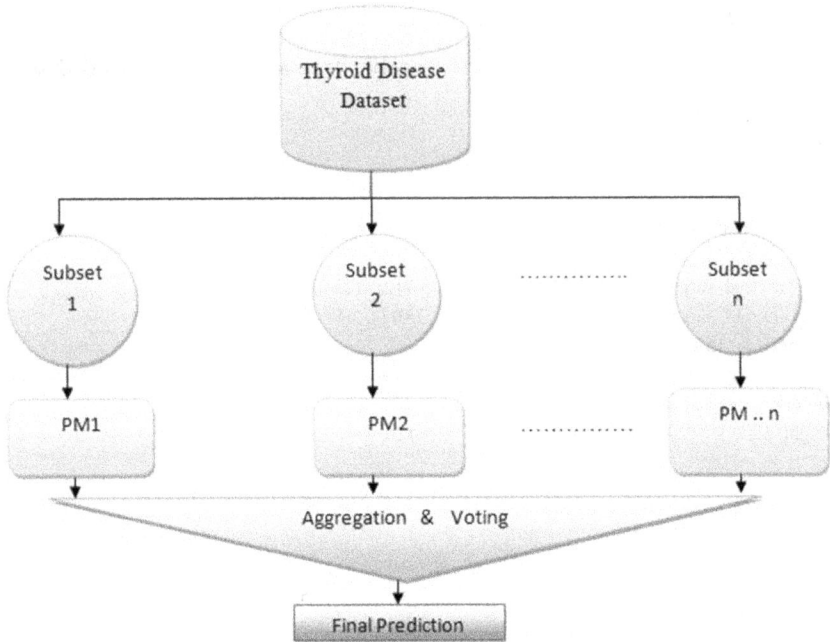

FIGURE 13.3   Representation of Bagging ensemble method for thyroid disease.

Razia et al. [8], identified patterns of disease accuracy using a machine learning model. They predicted that inputs correlated based on high-dimensional and multi-dimensional data for the best fit of a set of observations using decision trees, k-NNs, k-means, and artificial neural networks (ANNs). Chaubey et al. [9] created predictions using different machine learning techniques, such as logistic regression, decision trees and k-nearest neighbor (k-NN) algorithms. They evaluated the algorithmic performance using accuracy, and the authors used 807 patients in Kashmir. The prediction and classification make better and comfort disease diagnosis.

Dharamkar et al. [10] discussed medical diagnosis and extracting patterns. They used a thyroid classification, feature selection, C4.5, random forest, multilayer perceptron and Bayesian net for thyroid, diabetes, and cancer. The authors found that machine learning fusing C4.5 and a random forest classification technique provided better results compared to other classifiers.

Shin et al. [11] evaluated the diagnostic performance of machine learning with differentiating follicular adenoma. They calculated and compared an ANN and SVM for both classifier models. The measurement of sensitivity, specificity, and accuracy of the ANN was 32.3%, 90.1%, and 74.1% and the SVM based models were 41.7%, 79.4%, and 69.0%, respectively.

Shankar et al. [12] classified a thyroid dataset using optimal feature selection and a kernel-based classifier process. They found there was an improvement in the classification process when using gray wolf optimization. The authors evaluated the accuracy, sensitivity, and specificity: 97.49%, 99.05%, and 94.5%, respectively. With

TABLE 13.3
Representation of a Bagging Ensemble Tree Model for Thyroid Disease

TSH < 0.01 : Hypothyroidism (4336/0) [2183/0]
TSH >= 0.01
| FTI < 0.06 : Euthyroidism (112/11) [53/3]
| FTI >= 0.06
| | on_thyroxine < 0.5
| | | tt4 < 0.15
| | | | thyroid_surgery < 0.5
| | | | | t3 < 0.03
| | | | | | tt4 < 0.05 : Hypothyroidism (5/1) [3/1]
| | | | | | tt4 >= 0.05 : Hyerthyroidism (228/0) [112/1]
| | | | | t3 >= 0.03
| | | | | | t4U < 0.12 : Hypothyroidism (5/0) [3/1]
| | | | | | t4U >= 0.12 : Hyerthyroidism (4/0) [4/0]
| | | | thyroid_surgery >= 0.5 : Hypothyroidism (11/0) [3/0]
| | | tt4 >= 0.15 : Hypothyroidism (17/0) [10/0]
| | on_thyroxine >= 0.5 : Hypothyroidism (82/0) [29/0]
Size of the tree: 17
....
....
....
TSH < 0.01 : Hypothyroidism (4317/0) [2171/0]
TSH >= 0.01
| FTI < 0.06
| | t3 < 0.03 : Euthyroidism (117/9) [60/6]
| | t3 >= 0.03 : Hypothyroidism (4/0) [1/0]
| FTI >= 0.06
| | on_thyroxine < 0.5
| | | tt4 < 0.15
| | | | thyroid_surgery < 0.5 : Hyerthyroidism (256/9) [124/1]
| | | | thyroid_surgery >= 0.5 : Hypothyroidism (12/0) [6/0]
| | | tt4 >= 0.15 : Hypothyroidism (18/0) [9/0]
| | on_thyroxine >= 0.5 : Hypothyroidism (76/0) [29/0]

Size of the tree : 13
Time taken to build the model: 1.04 seconds

a high computation time, classifiers obtained a good performance for predicting thyroid illnesses.

Ma et al. [13] diagnosed thyroid cancer and treatment using an ultrasound test dataset. They used machine learning, a clustering DNN, a k-NN, a logistic regression and a support vector machine. They found that on average cross-validation had an accuracy of 0.87 in 10-sec. The authors measured four image distances, including the Euclidean distance, the Manhattan distance, the Chebyshev distance, and the Minkowski distance, and found the Chebyshev distance to be the best.

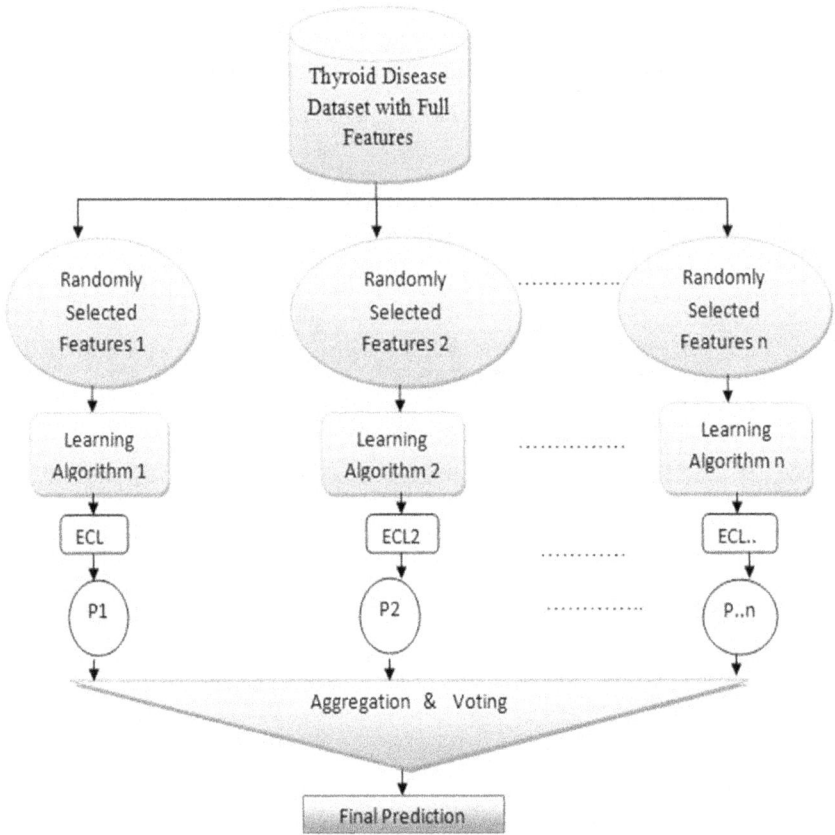

FIGURE 13.4   Representation of Random Subsample ensemble method for thyroid disease.

Wang et al. [14] investigated the capabilities of imaging radiomics using machine learning-based multiparametric magnetic resonance. They measured the performance of the machine learning for PTC aggressiveness as a receiver operating on characteristic curve values as 0.92. This technique could be helpful for future treatment strategies of aggressive PTC.

Duggal and Shukla [15] discussed several methods of feature selection and classification for thyroid disease diagnosis. The authors used a naïve Bayes, support vector machine, and random forest, and found that the support vector machine was the most accurate technique for the hypothyroid, hyperthyroid, sick euthyroid and euthyroid (negative) classes.

Yadav and Pal [16] discussed using a decision tree for a very complex and large dataset. Decision tree techniques classified the hyperthyroidism, hypothyroidism, and euthyroidism classes. The authors used the J48, random tree and Hoeffding algorithms for accuracies of 99.12%, 97.59%, and 92.37%, respectively. Finally, they found a better accuracy of 99.2% with a sensitivity of 99.36% for the ensemble model.

TABLE 13.4
Representation of a Random Subsample Ensemble Tree Model for Thyroid Disease

TSH < 0.01 : Hypothyroidism (4322/0) [2167/0]
TSH >= 0.01
| TSH < 0.03
| | t3 < 0.01
| | | TSH < 0.01 : Hyerthyroidism (30/13) [12/3]
| | | TSH >= 0.01 : Euthyroidism (18/4) [13/4]
| | t3 >= 0.01
| | | t3 < 0.02
| | | | t3 < 0.02
| | | | | t4U < 0.1 : Hyerthyroidism (51/8) [24/8]
| | | | | t4U >= 0.1
| | | | | | t4U < 0.1 : Hypothyroidism (5/0) [1/0]
| | | | | | t4U >= 0.1 : Hyerthyroidism (25/11) [16/6]
| | | | t3 >= 0.02 : Hyerthyroidism (96/6) [48/6]
| | | t3 >= 0.02
| | | | t3 < 0.02
| | | | | t3 < 0.02 : Hyerthyroidism (18/7) [7/2]
| | | | | t3 >= 0.02 : Hypothyroidism (35/0) [11/0]
| | | | t3 >= 0.02
| | | | | t3 < 0.03
| | | | | | t3 < 0.02
| | | | | | | TSH < 0.01
| | | | | | | | t4U < 0.09 : Hypothyroidism (2/0) [1/0]
| | | | | | | | t4U >= 0.09
| | | | | | | | | t4U < 0.11 : Hyerthyroidism (22/1) [6/0]
| | | | | | | | | t4U >= 0.11
| | | | | | | | | | t4U < 0.11 : Hypothyroidism (6/1) [1/0]
| | | | | | | | | | t4U >= 0.11 : Hyerthyroidism (12/1) [4/3]
| | | | | | | TSH >= 0.01 : Hyerthyroidism (11/4) [5/2]
| | | | | | t3 >= 0.02 : Hyerthyroidism (11/2) [6/1]
| | | | | t3 >= 0.03
| | | | | | t3 < 0.04
| | | | | | | TSH < 0.01 : Hyerthyroidism (6/2) [5/1]
| | | | | | | TSH >= 0.01
| | | | | | | | TSH < 0.02 : Hypothyroidism (7/0) [2/2]
| | | | | | | | TSH >= 0.02 : Hyerthyroidism (3/1) [3/1]
| | | | | | t3 >= 0.04 : Hypothyroidism (6/0) [2/0]
| TSH >= 0.03
| | t3 < 0.01 : Euthyroidism (67/3) [32/5]
| | t3 >= 0.01
| | | TSH < 0.09
| | | | t4U < 0.1 : Hyerthyroidism (7/2) [8/3]
| | | | t4U >= 0.1

(*Continued*)

TABLE 13.4 (CONTINUED)

Representation of a Random Subsample Ensemble Tree Model for Thyroid Disease

| | | | | t3 < 0.01 : Hypothyroidism (6/1) [6/4]
| | | | | t3 >= 0.01 : Euthyroidism (24/15) [13/8]
| | | TSH >= 0.09 : Euthyroidism (10/3) [7/1]
Size of the tree: 47

....

....

....

t3 < 0.01
| t4U < 0.1
| | t3 < 0.01 : Hypothyroidism (227/37) [96/17]
| | t3 >= 0.01
| | | t3 < 0.01 : Euthyroidism (11/0) [3/0]
| | | t3 >= 0.01 : Hypothyroidism (39/4) [35/4]
| t4U >= 0.1
| | t3 < 0.01 : Euthyroidism (73/26) [33/11]
| | t3 >= 0.01 : Hypothyroidism (11/2) [7/1]
t3 >= 0.01
| t3 < 0.02
| | t3 < 0.02 : Hypothyroidism (982/128) [466/71]
| | t3 >= 0.02 : Hyerthyroidism (23/0) [11/0]
| t3 >= 0.02 : Hypothyroidism (3434/101) [1749/46]

Size of the tree : 15
Time taken to build model: 1.46 seconds

Rao et al. [17] examined the capacities of the thyroid to find related illnesses. They tested T3, T4, and thyroid stimulating hormone (TSH) level values in a blood report dataset. The authors classified hyperthyroidism and hypothyroidism using a naïve Bayes machine learning classifier and found valuable information that could support thyroid diagnosis.

Chen et al. [18] considered serum samples from 199 patients with thyroid dysfunction. They used a multi-feature fusion convolutional neural network, support vector machine, and decision tree (DT) for classification. The authors found classification accuracies of 94.01%, 91.91%, and 90.34%, respectively. Finally, the authors found that MCNN had a good diagnostic effect for identifying thyroid dysfunction.

Halicek et al. [19] considered 216 tissue samples of the thyroid from 82 patients to evaluate the performance of HSI for tumor detection. They developed a CNN model for tumor detection and measured AUC scores of 0.90. The authors used various levels of CNN for better HSI and found that salivary glands, label-free HSI, and autofluorescence may offer the best performance for tumor detection.

Gupta et al. [20] considered thyroid disease and function of the thyroid gland using machine learning algorithms. They used a random forest, k-NN, and decision tree for thyroid disease diagnosis. Authors used a MALO features selection method for an accuracy of 95.94% with a random forest classifier, and 95.66% and 92.51% was achieved using a decision tree classifier and a k-NN classifier, respectively.

Yadav and Pal [21] discussed thyroid disease in women over the age of 30 years. They used a decision tree, random forest, and regression tree classification (CART) and enhanced the results using a bagging ensemble technique. The authors used a decision tree, random forest tree, and extra tree to provide accuracies of 98%, 99%, and 93%, respectively. The authors developed a model by combining the three basic tree classifiers and found better accuracy of 100% with a seed value of 35 and a num-fold value of 10.

13.3 METHODOLOGY

In this chapter, we used previous work and brought together older research information along with the clinical attributes from several databases. We collected 7200 instances and 22 attributes with multiple class variables.

13.3.1 DATA DESCRIPTION

The dataset was collected from the UCI repository, and the sample size of the class are follows:

Euthyroidism	166
Hyperthyroidism	368
Hypothyroidism	6666

In hyperthyroidism, hormone production of the thyroid gland will always be high by as the body can not manage metabolism. In hypothyroidism, hormone production by the thyroid gland will be low, as the body cannot balance the energy system. Euthyroidism is a normal condition in the body, and the thyroid gland produces a normal amount of hormones (Table 13.5).

Adilah et al. [22] introduced a boxplot and demonstrated its importance. The whisker plot with a boxplot graphically depicts groups of numerical data through their quartiles. The line in the boxplot assigns variability outside the upper and lower quartiles and the variable boundaries within the box (Figure 13.5).

Shakeel et al. [23] introduced a histogram and demonstrated its importance at different levels. A histogram represents data with bars at different levels. The bars represent various range shape by continuous sample data. A histogram represents a vertical bar graph and summarizes continuous data (Figure 13.6).

13.3.2 PEARSON CORRELATION

Liu et al. [24] created a Pearson correlation coefficient in combination with other variables. The Pearson correlation coefficient generates correlations between variables. It is also referred to as a product-moment correlation coefficient, and it measures the linear correlation between two variables X and Y, which varies between +1

TABLE 13.5

Thyroid Dataset Variables Representation

Attributes	Domain Range
Age	[0.01, 0.97](Year 1–97)
Sex	[0, 1](0 = Male, 1 = Female)
On_thyroxine	[0, 1](0 = False, 1 = True)
Query_on_thyroxine	[0, 1](0 = False, 1 = True)
On_antithyroid_medication	[0, 1](0 = False, 1 = True)
Sick	[0, 1](0 = False, 1 = True)
Pregnant	[0, 1](0 = False, 1 = True)
Thyroid_surgery	[0, 1](0 = False, 1 = True)
I131_treatment(Patients Receiving Radioiodine)	[0, 1](0 = False, 1 = True)
Query_hypothyroid	[0, 1](0 = False, 1 = True)
Query_hyperthyroid	[0, 1](0 = False, 1 = True)
Lithium	[0, 1](0 = False, 1 = True)
Goiter	[0, 1](0 = False, 1 = True)
Tumor	[0, 1](0 = False, 1 = True)
Hypopituitary	[0, 1](0 = False, 1 = True)
Psych	[0, 1](0 = False, 1 = True)
TSH (thyroid stimulating hormone)	[0.0, 0.53]
T3(triiodothyronine)	[0.0005, 0.18]
TT4(total thyroxine)	[0.0020, 0.6]
T4U(Thyroxine utilization rates)	[0.017, 0.233]
FTI(Thyroid Function Tests)	[0.0020, 0.642]
Class	[Hypothyroidism, Hyperthyroidism, Euthyroidism]

and −1. The variable tends to be positive, negative, or have no tendency, depending on the correlation intensity within the range (Figure 13.7).

13.3.3 FORMULA REPRESENTATION

In this chapter, we used a statistical formula for the analysis of thyroid disease. Negi and Kanda; Yadav and Pal; Kuthirummal et al.; Kumar et al.; Ashrafian et al. introduced a statistical formula (4–12) with an error prediction [25–31].

In statistics, the mean absolute error (MAE) calculates the errors between observations and expressive phenomenon. The formula measures comparisons of predicted and observed values as:

$$Mean\,Absolute\,Error = \frac{1}{m}\sum_{k=1}^{m}|p_i - p| \tag{13.4}$$

Where, m is the number of errors and $|p_i - p|$ represents the errors.

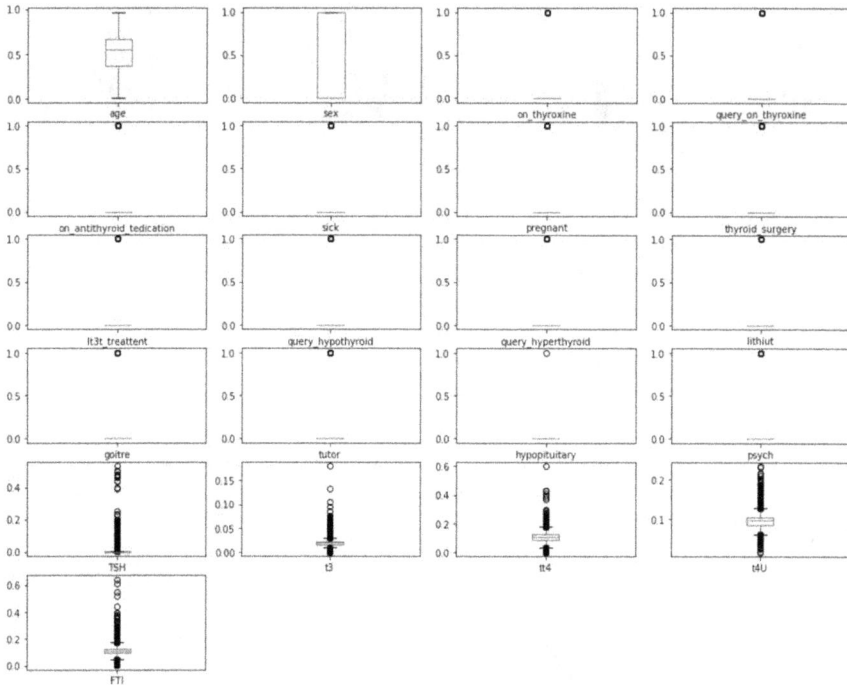

FIGURE 13.5 Representation of a whisker plot with a boxplot for thyroid disease.

The root mean square error measures the difference between the predicted values and values observed by a model.

$$Root\ Mean\ Square\ Error = \sqrt{\frac{1}{m}\sum_{k=1}^{m} p_i - \hat{p}} \qquad (13.5)$$

Where $(m, p_i,$ and $\hat{p})$ represents variables of the sample, forecasts, and observed values, respectively for $(k = 1...m)$.

The relative absolute error generates a ratio between the mean error and errors produced by a trivial model.

$$Relative\ Absolute\ Error = \frac{\sum_{k=1}^{m} |p_i - a_i|}{\sum_{k-1}^{m} |\bar{a} - a_i|} \qquad (13.6)$$

Where, p_i Calculate predicted value and a_i calculates the actual value for $(k = 1...m)$ samples.

The absolute error provides the difference between the exact value and the approximation values, and evaluates the magnitude of difference in term of the exact values.

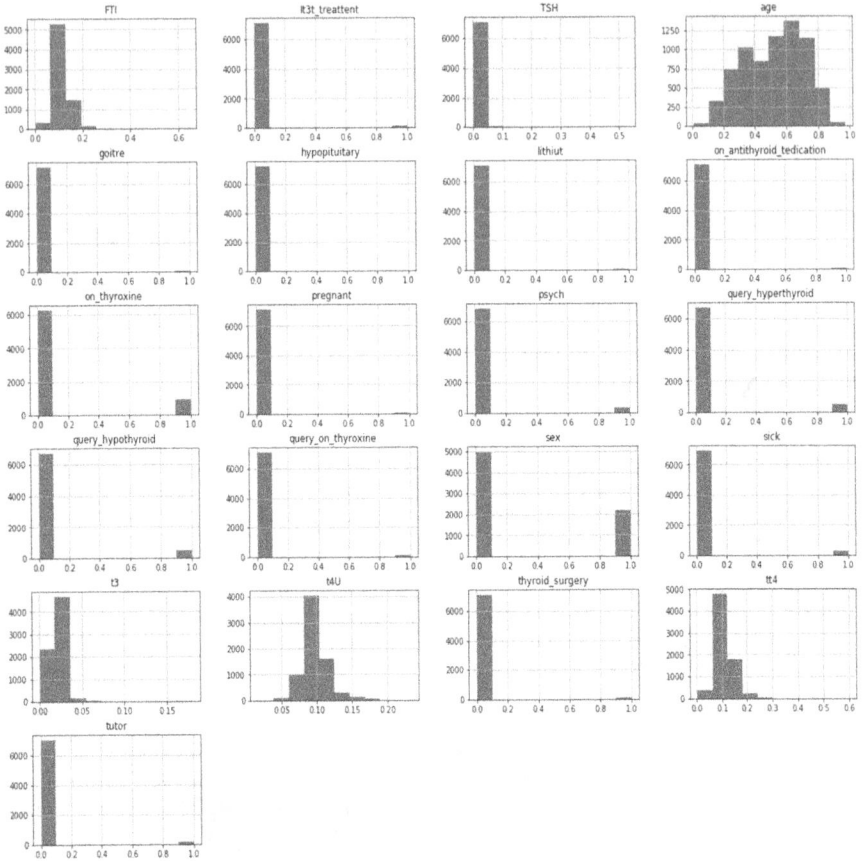

FIGURE 13.6 Representation of a histogram for thyroid disease.

The root relative squared error (RRSE) ratio normalizes the total squared error and total squared error using predictors. RRSE reduces the error with the root of the same dimensional value as the final prediction.

$$Root\ Relative\ Square\ Error\ \left(RRSE\right) = \frac{\sum_{j=1}^{n}\left(V_{ij}\_T_{ij}\right)^2}{\sum_{j=1}^{n}\left(T_{j}\_\overline{T}\right)^2} \tag{13.7}$$

Where, $V_{(ij)}$ is the value predicted by the individual program, i for sample case j (out of n sample cases); T_j is the target value for sample case j; $\overline{T}$ represents perfect fit in the sample cases.

$$\overline{T} = \sum_{j=1}^{n}\left(T_{j} - T\right) \tag{13.8}$$

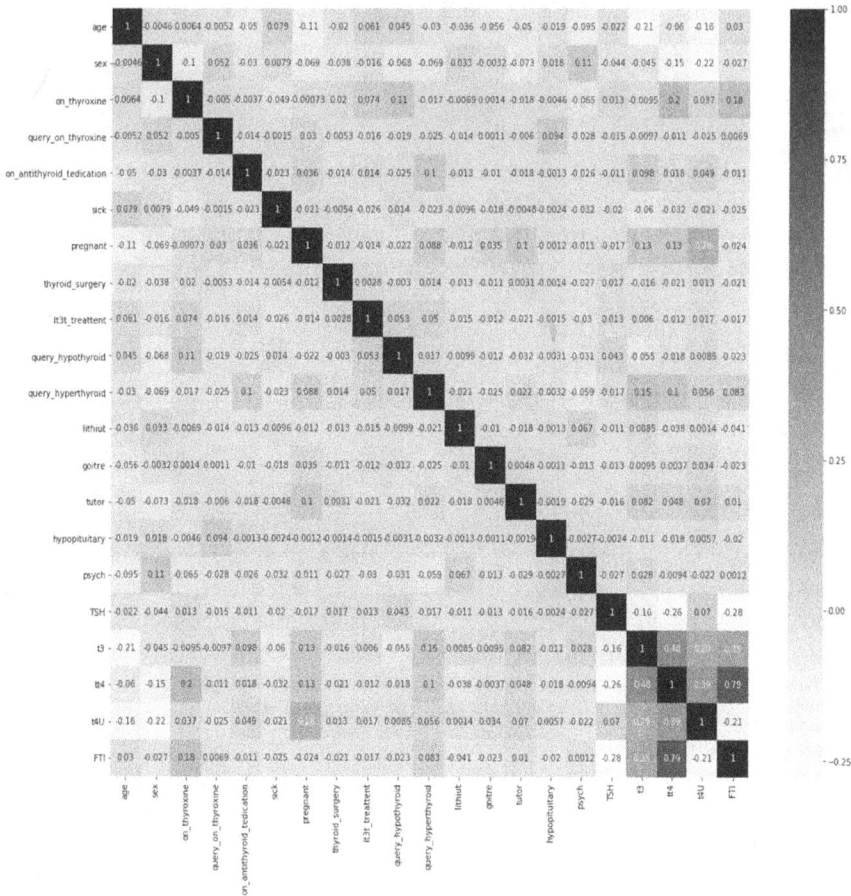

FIGURE 13.7 Representation of a Pearson correlation for thyroid disease attributes.

The classification accuracy evaluates the number of records correctly classified as:

$$\text{Classification Accuracy} = (TP + TN) / (TN + FN + TP + FP) \qquad (13.9)$$

The sensitivity evaluates a positive test result of an individual model as:

$$\text{Sensitivity} = TP / (TP + FN) \qquad (13.10)$$

The specificity evaluates a negative test of an individual model as:

$$\text{Specificity} = TN / (TN + FP) \qquad (13.11)$$

A receiver operating characteristic curve visualizes a graphical plot and connects it between sensitivity and specificity for each combination of tests. It provides the benefit of using the test as the *x*-axis and *y*-axis shows (1 − specificity) and sensitivity, respectively.

13.3.4 YOUDEN'S J

Youden's J is known as informedness, and its likelihood generates a single unit using sensitivity and specificity. The single unit has a range of 0–1 for a positive and perfect test. The values move from 0–1 that will move in test perfection means 1 indicates no false-positive or false-negative value then the test will be perfect.

$$J = \text{Sensitivity} + \text{Specificity} - 1 \tag{13.12}$$

The J values are assigned with a vertical line and have maximum values in the ROC index curve.

13.4 PROPOSED MODEL

The reduced error pruning decision tree generates a candidate subsample for pruning and formating in a leaf node, but reduced error pruning has a drawback in that it cannot manage error complexity and accuracy as an ensemble model. In this chapter, we tested a reduced error pruning tree in three different environments as an AdaBosstM1, bagging, and random subsample. The meta classifiers ensemble methods AdaBoostM1 with reduced error pruning were applied to the training set and it took a decision tree model for the weak learner to generate strong learners to fit into the training sample. The bagging meta classifiers ensemble model with reduced error pruning reduced variance and stopped overfitting by aggregating weak learners and generating a decision model in a tree format. The random subsample meta classifier ensemble method with reduced error pruning generated training of a subsample randomly for the unbalanced feature set and generated a decision tree model (Figure 13.8).

13.5 RESULTS

In this chapter, we used 7200 instances thyroid disease with 22 attributes, and Table 13.6 represents the evaluated values of the weighted average of the true positive rate (0.92), false-positive rate (0.63), precision value (0.87), recall value (0.92), F-measure (0.90), and receiver operating characteristic (0.98) for three different classes: hypothyroid, hyperthyroid, and euthyroid.

Table 13.7 represents the bagging algorithm detailed for receiver operating characteristic values for three different classes. The bagging ensemble method evaluated a true positive rate (0.99), false-positive rate (0.007), precision value (0.99), recall value (0.99), F-measure (0.99), and receiver operating characteristic (1).

In Table 13.8 a random subspace ensemble model with reduced error pruning calculate the true positive rate (0.97), false-positive rate (0.34), precision value (0.97), recall value (0.77), F-measure (0.96), and receiver operating characteristic (0.99).

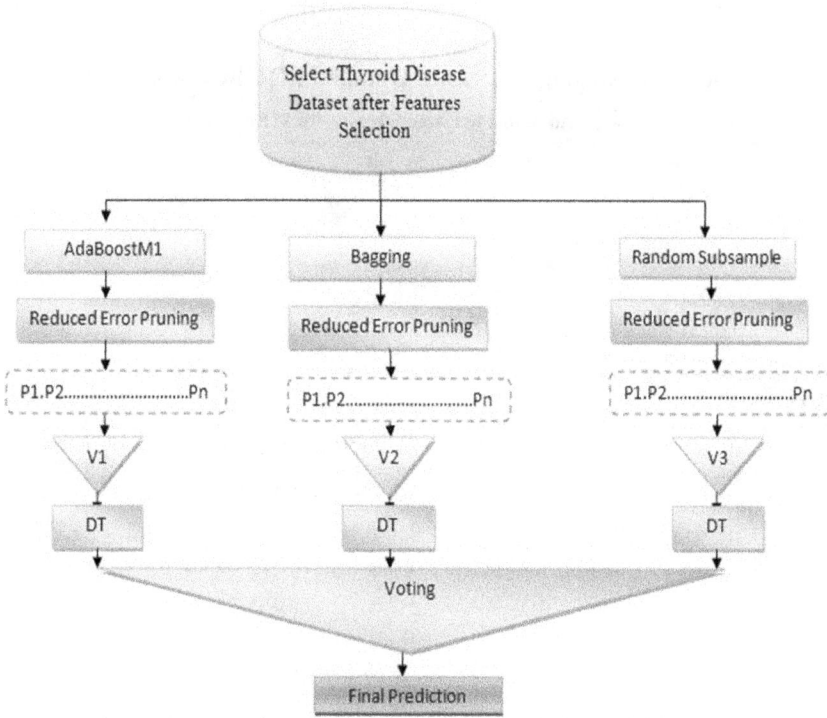

FIGURE 13.8 Representation Proposed Model for thyroid disease attributes.

TABLE 13.6

Representation of an AdaBoost M1 Algorithm Detailed ROC by Class

	TP Rate	FP Rate	Precision	Recall	F- Measure	ROC	Class
				Bagging Algorithm (With REP)			
	0.996	0.007	0.999	0.996	0.998	1	Hypothyroidism
	0.989	0.001	0.976	0.986	0.982	1	Hyperthyroidism
	1	0.002	0.917	1	0.957	0.999	Euthyroidism
Wgt. Avg.	0.996	0.007	0.996	0.996	0.996	1	

Table 13.9 shows a confusion matrix for the three different ensemble methods with a reduced error pruning decision tree which calculated the true positive, false-positive, true negative, and false negative for the three classes of thyroid disease.

Figure 13.9, Figure 13.10, and Figure 13.11 represent a stratified error summary, receiver operating characteristic, and accuracy details for an AdaBoostM1, bagging, and random subspace three ensemble model with a reduced error pruning decision tree model.

TABLE 13.7

Representation of a Bagging Algorithm Detailed ROC by Class

AdaBoost M1 Algorithm (With REP)

	TP Rate	FP Rate	Precision	Recall	F- Measure	ROC	Class
	0.998	0.689	0.948	0.998	0.972	0.982	Hypothyroidism
	0	0.027	0	0	0	0.961	Hyperthyroidism
	0	0	0	0	0	0.998	Euthyroidism
Wgt. Avg.	0.924	0.639	0.877	0.924	0.9	0.988	

TABLE 13.8

Representation of a Random Subspace Algorithm Detailed ROC by Class

Random Subspace Algorithm (With REP)

	TP Rate	FP Rate	Precision	Recall	F- Measure	ROC	Class
	0.998	0.376	0.971	0.998	0.984	0.996	Hypothyroidism
	0.508	0	0.984	0.508	0.67	0.998	Hyperthyroidism
	0.88	0.002	0.924	0.88	0.901	0.999	Euthyroidism
Wgt. Avg.	0.97	0.349	0.97	0.77	0.966	0.997	

TABLE 13.9

Representation of a Confusion Matrix

Algorithms	Confusion Matrix
AdaBoostM1	a b c <-- classified as
	6650 16 0 | a = Hypothyroidism
	368 0 0 | b = Hyerthyroidism
	0 166 0 | c = Euthyroidism
Bagging	a b c <-- classified as
	6642 9 15 | a = Hypothyroidism
	4 364 0 | b = Hyerthyroidism
	0 0 166 | c = Euthyroidism
Random Subspace	a b c <-- classified as
	6651 3 12 | a = Hypothyroidism
	181 187 0 | b = Hyerthyroidism
	20 0 146 | c = Euthyroidism

13.6 DISCUSSION

After the analysis, we found that the bagging ensemble method with a reduced error pruning evaluated high values for the true false rate, precision, recall, F-measure, and receiver operating characteristic as 0.99, 0.99, 0.99, 0.99, and 1, respectively, and

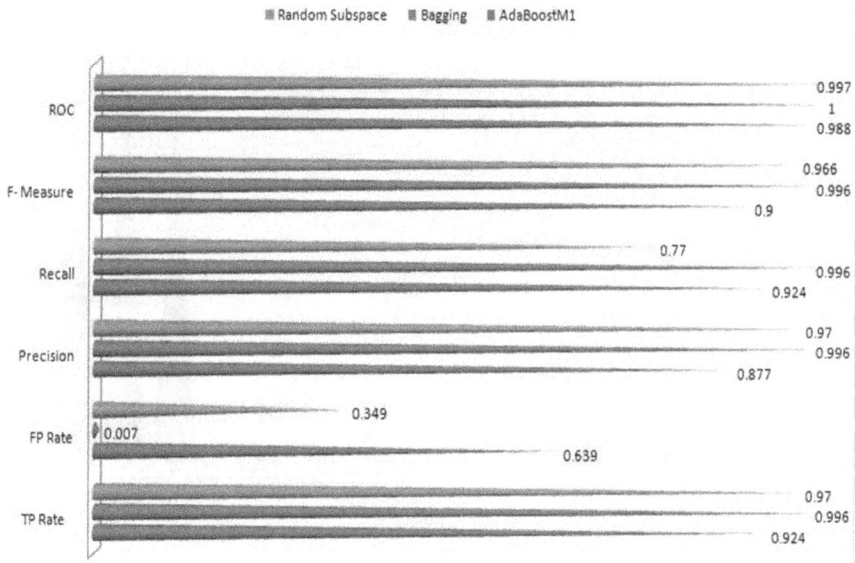

FIGURE 13.9 Representation of an AdaBoost M1, bagging, and random subspace algorithm detailed ROC.

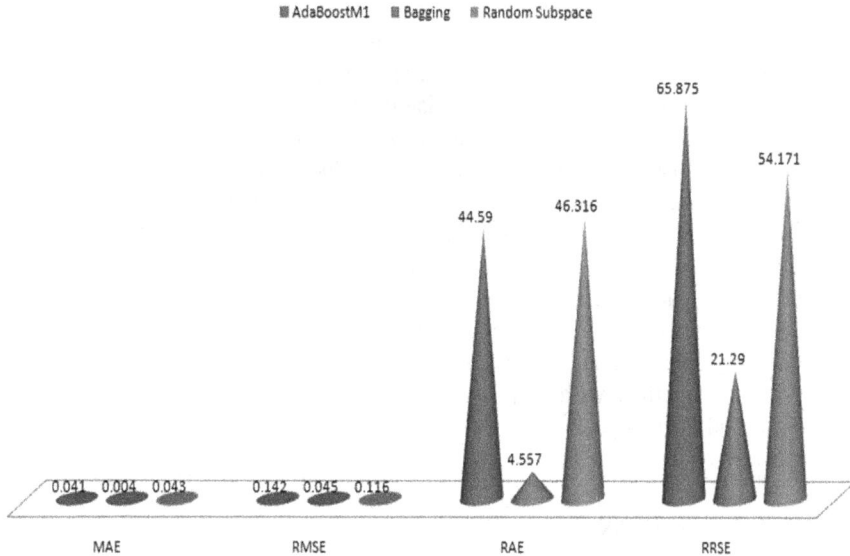

FIGURE 13.10 Comparison of a stratified error summary.

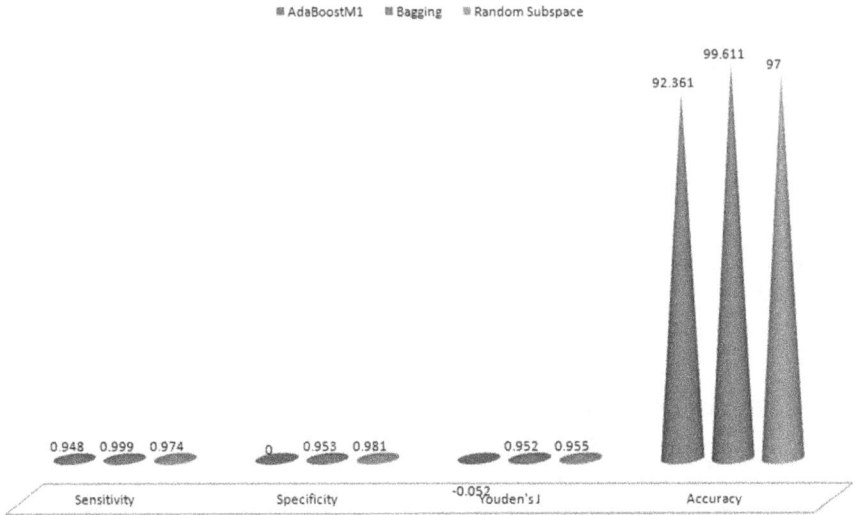

FIGURE 13.11 Comparison of a detailed accuracy model for AdaBoostM1, bagging, and random subspace.

AdaBoostM1 and random subspace ensemble methods provided low values. The false-positive rate of the bagging algorithms was low compared to other algorithms.

With the results, as shown in Figure 13.10, the bagging ensemble method with a reduced error pruning tree calculated low mean absolute values, root mean square values, and relative absolute error values with root relative squared error values; other ensemble methods with error pruning decision trees calculated high error values.

From the Figure 13.11, we found the sensitivity, Youden's J, and classification accuracy values of the bagging ensemble with reduced error pruning were always high compare to AdaBoostM1 and random subspace ensemble methods with reduced error pruning. The classification accuracy was 99.61% obtained with the bagging ensemble method and the values 92.36% and 97.00% calculated by AdaBoostM1 and random subspace ensemble methods, respectively.

On the basis of previous performance on different datasets, the authors, Yadav, Pal, Chaurasia, Verma, and Kumar [32–36], calculated a high accuracy on the basis of an ensemble method with majority voting. In this chapter, we calculated high Youden's *J* and accuracy, and a low error rate using the bagging ensemble method with majority voting.

13.7 CONCLUSION

In this chapter, we used a thyroid dataset of three different classes: hypothyroid, hyperthyroid, and euthyroid with 7200 instances and 22 attributes. The dataset was collected from the UCI repository, and contained only two illnesses: hypothyroid and hyperthyroid. These two thyroid illnesses depend on three different basic thyroid gland hormones in the bloodstream: triiodothyronine (T3), thyroxin (T4), and TSH levels. We used three different machine learning ensemble methods with reduced error pruning: AdaBoostM1, bagging and random subspace meta classifier algorithms. With the results, we analyzed all the algorithms for metrics of the true false rate, precision, recall, F-measure, receiver operating characteristic, mean absolute values, root mean square values, relative absolute error values, root relative squared error, sensitivity, specificity, Youden's J, accuracy, and found that a bagging meta classifier ensemble method with reduced error pruning calculated better values compared to the AdaBoostM1 and random subspace meta classifier ensemble methods with reduced error pruning. In future, we will calculate meta classifiers ensemble methods using pessimistic error pruning with error complex pruning and generate a new decision tree model for complex, substantial medical datasets.

CONFLICT OF INTEREST

Authors have no conflict of interest.

FUNDING

This study was not funded.

ACKNOWLEDGMENTS

The author is grateful to Veer Bahadur Singh Purvanchal University Jaunpur, Uttar Pradesh, for providing financial support to work as Post-Doctoral Research Fellowship.

REFERENCES

1. Barrea, L., Gallo, M., Ruggeri, R.M., Giacinto, P.D., Sesti, F., Prinzi, N., Adinolfi, V., Barucca, V., Renzelli, V., Muscogiuri, G. and Colao, A. 2020. Nutritional status and follicular-derived thyroid cancer: An update. *Critical Reviews in Food Science and Nutrition*, pp. 1–35.
2. Ejtahed, H.S., Angoorani, P., Soroush, A.R., Siadat, S.D., Shirzad, N., Hasani-Ranjbar, S. and Larijani, B. 2020. Our little friends with big roles: Alterations of the gut microbiota in thyroid disorders. *Endocrine, Metabolic & Immune Disorders-Drug Targets* (Formerly Current Drug Targets-Immune, Endocrine & Metabolic Disorders), 20(3), pp. 344–350.
3. Ting, W.C., Lu, Y.C.A., Ho, W.C., Cheewakriangkrai, C., Chang, H.R. and Lin, C.L. 2020. Machine learning in prediction of second primary cancer and recurrence in colorectal cancer. *International Journal of Medical Sciences*, 17(3), p. 280.
4. Wu, Y., Ke, Y., Chen, Z., Liang, S., Zhao, H. and Hong, H. 2020. Application of alternating decision tree with AdaBoost and bagging ensembles for landslide susceptibility mapping. *Catena*, 187, p. 104396.
5. Hong, H., Liu, J. and Zhu, A.X. 2020. Modeling landslide susceptibility using LogitBoost alternating decision trees and forest by penalizing attributes with the bagging ensemble. *Science of the Total Environment*, 718, p. 137231.
6. Cimen, E. 2020. A random subspace based conic functions ensemble classifier. *Turkish Journal of Electrical Engineering & Computer Sciences*, 28(4), pp. 2165–2182.

7. Guo, B.J., He, X., Wang, T., Lei, Y., Curran, W.J., Liu, T., Zhang, L.J. and Yang, X. 2020. Benign and malignant thyroid classification using computed tomography radiomics. In *Medical Imaging 2020: Computer-Aided Diagnosis* (Vol. 11314, p. 1131440). International Society for Optics and Photonics.

8. Razia, S., Kumar, P.S. and Rao, A.S. 2020. Machine learning techniques for Thyroid disease diagnosis: A systematic review. In *Modern Approaches in Machine Learning and Cognitive Science: A Walkthrough* (pp. 203–212). Cham: Springer.

9. Chaubey, G., Bisen, D., Arjaria, S. and Yadav, V. 2020. Thyroid disease prediction using machine learning approaches. *National Academy Science Letters-India.*

10. Dharamkar, B., Saurabh, P., Prasad, R. and Mewada, P. 2020. An ensemble approach for classification of thyroid using machine learning. In *Progress in Computing, Analytics and Networking* (pp. 13–22). Singapore: Springer.

11. Shin, I., Kim, Y.J., Han, K., Lee, E., Kim, H.J., Shin, J.H., Moon, H.J., Youk, J.H., Kim, K.G. and Kwak, J.Y. 2020. Application of machine learning to ultrasound images to differentiate follicular neoplasms of the thyroid gland. *Ultrasonography*, 39(3), p. 257.

12. Shankar, K., Lakshmanaprabu, S.K., Gupta, D., Maseleno, A. and De Albuquerque, V.H.C. 2020. Optimal feature-based multi-kernel SVM approach for thyroid disease classification. *The Journal of Supercomputing*, 76(2), pp. 1128–1143.

13. Ma, X., Xi, B., Zhang, Y., Zhu, L., Sui, X., Tian, G. and Yang, J. 2020. A machine learning-based diagnosis of thyroid cancer using thyroid nodules ultrasound images. *Current Bioinformatics*, 15(4), pp. 349–358.

14. Wang, H., Song, B., Ye, N., Ren, J., Sun, X., Dai, Z., Zhang, Y. and Chen, B.T. 2020. Machine learning-based multiparametric MRI radiomics for predicting the aggressiveness of papillary thyroid carcinoma. *European Journal of Radiology*, 122, p. 108755.

15. Duggal, P. and Shukla, S. 2020. Prediction of thyroid disorders using advanced machine learning techniques. In *2020 10th International Conference on Cloud Computing, Data Science & Engineering (Confluence)* (pp. 670–675). IEEE.

16. Yadav, D.C. and Pal, S. 2020. Discovery of hidden pattern in thyroid disease by machine learning algorithms. *Indian Journal of Public Health Research & Development*, 11(1), pp. 61–66.

17. Rao, P.S., Hussain, M. and Abhigna, C. 2020. Finding risk factors in thyroid and cardiovascular system using Naive Bayesian (NB) machine learning technique. *Research & Reviews: A Journal of Embedded System & Applications*, 8(1), pp. 1–5.

18. Chen, H., Cheng, C., Wang, H., Chen, C., Guo, Z., Tong, D., Li, H., Li, H., Si, R., Lai, H. and Lv, X. 2020. Serum Raman spectroscopy combined with a multi-feature fusion convolutional neural network diagnosing thyroid dysfunction. *Optik*, p. 164961.

19. Halicek, M., Dormer, J.D., Little, J.V., Chen, A.Y. and Fei, B. 2020. Tumor detection of the thyroid and salivary glands using hyperspectral imaging and deep learning. *Biomedical Optics Express*, 11(3), pp. 1383–1400.

20. Gupta, N., Jain, R., Gupta, D., Khanna, A. and Khamparia, A. 2020. Modified ant lion optimization algorithm for improved diagnosis of thyroid disease. In *Cognitive Informatics and Soft Computing* (pp. 599–610). Singapore: Springer.

21. Yadav, D.C. and Pal, S. 2020. Prediction of thyroid disease using decision tree ensemble method. *Human-Intelligent Systems Integration*, pp. 1–7.

22. Adilah, A.N., Zarif, M.M. and Idris, A.M. 2020. Rainfall trend analysis using box plot method: Case study UMP campus Gambang and Pekan. In *IOP Conference Series: Materials Science and Engineering* (Vol. 712, No. 1, p. 012021). IOP Publishing.

23. Shakeel, P.M., Desa, M.I. and Burhanuddin, M.A. 2020. Improved watershed histogram thresholding with probabilistic neural networks for lung cancer diagnosis for CBMIR systems. *Multimedia tools and applications*, 79(23), pp. 17115–17133.

24. Liu, Y., Mu, Y., Chen, K., Li, Y. and Guo, J. 2020. Daily activity feature selection in smart homes based on pearson correlation coefficient. *Neural Processing Letters*, pp. 1–17.
25. Negi, H.S. and Kanda, N. 2020. An appraisal of spatio-temporal characteristics of temperature and precipitation using gridded datasets over NW-Himalaya. In *Climate Change and the White World* (pp. 219–238). Cham: Springer.
26. Yadav, D.C. and Pal, S. 2019. To generate an ensemble model for women thyroid prediction using data mining techniques. *Asian Pacific Journal of Cancer Prevention: APJCP*, 20(4), p. 1275.
27. Yadav, D.C. and Pal, S. 2020. Prediction of heart disease using feature selection and random forest ensemble method. *International Journal of Pharmaceutical Research*, 12(4), pp. 56–66.
28. Liu, Y., Mu, Y., Chen, K., Li, Y. and Guo, J. 2020. Daily activity feature selection in smart homes based on pearson correlation coefficient. *Neural Processing Letters*, pp. 1–17.
29. Kuthirummal, N., Vanathi, M., Mukhija, R., Gupta, N., Meel, R., Saxena, R. and Tandon, R. 2020. Evaluation of Barrett universal II formula for intraocular lens power calculation in Asian Indian population. *Indian Journal of Ophthalmology*, 68(1), p. 59.
30. Kumar, V., Mishra, B.K., Mazzara, M., Thanh, D.N. and Verma, A. 2020. Prediction of malignant and benign breast cancer: A data mining approach in healthcare applications. In *Advances in Data Science and Management* (pp. 435–442). Singapore: Springer.
31. Ashrafian, A., Gandomi, A.H., Rezaie-Balf, M. and Emadi, M. 2020. An evolutionary approach to formulate the compressive strength of roller compacted concrete pavement. *Measurement*, 152, p. 107309.
32. Yadav, D.C. and Pal, S. 2019. Thyroid prediction using ensemble data mining techniques. *International Journal of Information Technology*, pp. 1–11.
33. Yadav, D.C. and Pal, S. 2019. Calculating diagnose odd ratio for thyroid patients using different data mining classifiers and ensemble techniques. *International Journal of Advanced Trends in Computer Science and Engineering*, 9(4), pp. 5463–5470.
34. Chaurasia, V. and Pal, S. 2020. Applications of machine learning techniques to predict diagnostic breast cancer. *SN Computer Science*, 1(5), pp. 1–11.
35. Verma, A.K., Pal, S. and Kumar, S. Prediction of different classes of skin disease using machine learning techniques. In *Smart Innovations in Communication and Computational Sciences* (pp. 91–100). Springer, Singapore.
36. Verma, A.K., Pal, S. and Kumar, S. 2020. Prediction of skin disease using ensemble data mining techniques and feature selection method—a comparative study. *Applied biochemistry and biotechnology*, 190(2), pp. 341–359.

14 Reliable Diagnosis and Prognosis of COVID-19

*Marjan Mansourian, Hamid Reza Marateb,
Maja von Cube, Sadaf Khademi,
Mislav Jordanic, Miguel Ángel Mañanas,
Harald Binder, and Martin Wolkewitz*

CONTENTS

14.1   INTRODUCTION

COVID-19, also known as severe acute respiratory syndrome coronavirus (SARS-CoV-2), has caused a disaster in 2020 [1]. Since December 2019, when pneumonia cases of an unknown cause occurred in Wuhan, Hubei, China, the virus has been spread worldwide. Compared to its two ancestors, viruses causing SARS and the Middle East respiratory syndrome (MERS) [2, 3], COVID-19 causes pneumonia in humans at a higher pace and lower mortality [1, 4]. More than 29.3 million confirmed global COVID-19 cases and 931,000 deaths, on September 15, 2020 [5], are just some of the consequences of this virus, while the related socio-economic and long-term adverse effects are devastating [6]. The number of confirmed cases per 100,000 of the population, for the top ten countries with the highest absolute daily number of deaths, is shown in Figure 14.1. This plot shows the increasing trend of COVID-19. Also, the number of cumulative deaths in the six WHO regions is shown in Figure 14.2.

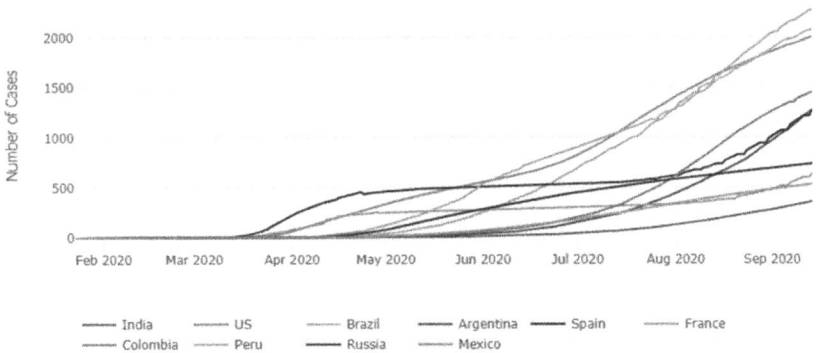

FIGURE 14.1 The number of confirmed cases per 100,000 population in the top ten countries with the highest absolute daily number of deaths, based on COVID-19 Dashboard by the Center for Systems Science and Engineering at Johns Hopkins University (JHU) [5].

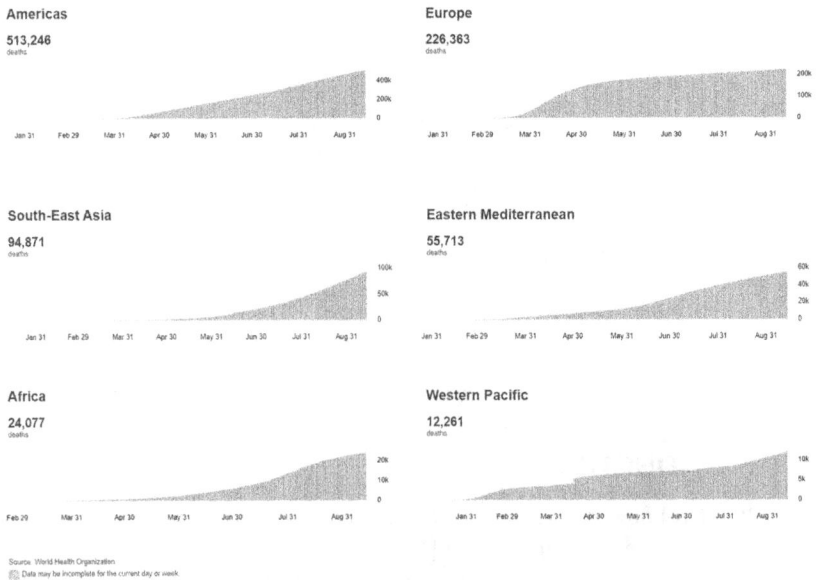

FIGURE 14.2 The cumulative number of deaths for six WHO regions, based on WHO Coronavirus Disease (COVID-19) Dashboard. (Accessed on September 15, 2020.)

The medical research community has been responding to the pandemic at an incredible speed. Many papers were published in different parts of the world about the descriptive characteristics of the COVID-19 patients, some preliminary or advanced statistical analysis in the journals, or even as preprints [7]. The number of published COVID-19 papers and preprints was estimated as 36,970 by September 15, 2020, based on the COVID-19 related publications on Publons. Many COVID-19 papers received many citations [1], regardless of the validity of the experimental design or statistical analysis. Then, some journals started to re-examine the published papers, resulting in the retraction of some published works, e.g., in the *LANCET* [8]. The pace of COVID-19 publications was also observed in preprints that are supposed to provide early access to results [9]. This raises an important question of whether such a high number of studies could help fight COVID-19 as a data science issue [10] when there are problems with the study design, statistical analysis, or even flaws in the calculation of performance indices [11, 12]. "Fighting panic with information" [13] implies using proper statistical models to deduce the risk of bias [11]. Focusing on the diagnosis and prognosis of COVID-19 patients [12, 14], we aimed to review some of the published papers in this domain and provide some guidelines for proper statistical and data mining analysis of COVID-19 studies.

14.2 COVID-19 DIAGNOSIS

Either molecular tests (viral RNA) or serological tests (anti-SARS-CoV-2 immuno-globulins) are used for COVID-19 diagnosis. The former includes the reverse tran-scriptase-polymerase chain reaction (RT-PCR) as a reference standard for COVID-19 diagnosis [15]. However, false-positive results occur when using the RT-PCR test [16, 17]. Moreover, the results may change over time during the disease period [18], and the test material is not entirely available globally [19]. Serological tests are alter-natives to the RT-PCR tests that are often cheaper and more accessible at the point of care. Their testing includes chemiluminescent immunoassays (CLIAs), lateral flow immunoassays (LFIAs), or enzyme-linked immunosorbent assays (ELISAs) [15, 20]. According to the systematic review and meta-analysis performed by Lisboa Bastos et al. [15], the pooled sensitivity of CLIA, ELISA, and LFIA methods were 98%, 84%, and 66%, while their specificity was higher than or equal to 97%. Serologically based antibody tests are, thus, additions to molecularly based tests [21]. The human antibody response to SARS-CoV-2 infection is shown in Figure 14.3. Plasma cells produce their SARS-CoV-2-specific receptors (IgM, IgG, or IgA antibodies) in the fifth stage that could be detected by serological tests [21].

Chest CT-scans were also proposed as a primary tool for COVID-19 studies [22, 23], but its discrimination of COVID-19 or non-COVID-19 patients is not accept-able. Using RT-PCR in the computer-aided diagnosis as the gold standard is prob-lematic since it is not 100% accurate [24–26]. Moreover, immunity status against SARS-CoV-2 cannot be assessed by RT-PCR tests [27]. A combination of the diag-nostic tests could thus be used as the gold standard (e.g., the confirmation of the chest CT-scan by the radiologists and the RT-PCR [28]), or composite reference standards could be proposed [29] or the agreement rate (e.g., Kappa agreement rates [30]) must

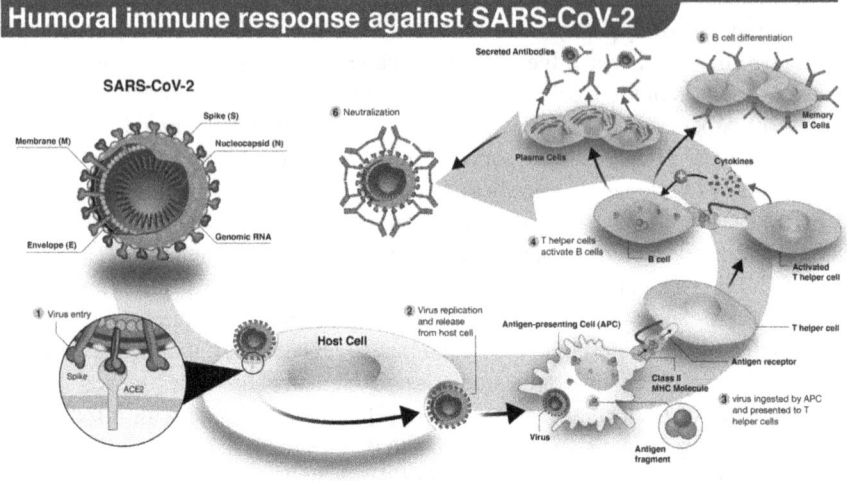

FIGURE 14.3 The human antibody response to SARS-CoV-2 infection. (Reproduced with permission from [21].)

be provided instead of the diagnostic accuracy measures [31]. Some of the COVID-19 diagnosis methods are reviewed in supplementary Table A.1 (https://doi.org/10.6 084/m9.figshare.12990707).

14.2.1 GUIDELINES

TRIPOD, transparent reporting of a multivariable prediction model for individual prognosis or diagnosis, is one of the most comprehensive checklists for clinical prediction models [32]. The checklist is shown in Figure 14.4. Its entire 22 items are required to provide a rigorous and reliable model presentation and validation. There are few studies in which such items are entirely taken into account in biomedical engineering, while such a checklist is mandatory in many highly prestigious medical journals. Moreover, as the TRIPOD [32] and STARD [33] checklists were not initially designed for machine learning and artificial intelligence methods, although many concepts are applicable, their extensions, namely TRIPOD-ML [34] and STARD-AI [35], are being developed. We aimed to highlight some of the critical issues that were not taken into account in the COVID-19 diagnosis studies to underline possible flaws and biases they could create. Note, that based on our previous discussion, using RT-PCR as the gold standard is problematic, and the agreement rate must be provided instead of the traditional signal detection theory parameters. Even if such a gold standard is available, the following issues must be taken into account in COVID-19 diagnostic studies:

14.2.2 THE CONFIDENCE INTERVAL

It is necessary to provide the 95% confidence interval (CI) of the performance indices based on the STARD and TRIPOD (Figure 14.4; item 16). This directly shows

TRIPOD Checklist: Prediction Model Development and Validation

TR⚕POD

Section/Topic	Item		Checklist Item	Page
Title and abstract				
Title	1	D;V	Identify the study as developing and/or validating a multivariable prediction model, the target population, and the outcome to be predicted.	
Abstract	2	D;V	Provide a summary of objectives, study design, setting, participants, sample size, predictors, outcome, statistical analysis, results, and conclusions.	
Introduction				
Background and objectives	3a	D;V	Explain the medical context (including whether diagnostic or prognostic) and rationale for developing or validating the multivariable prediction model, including references to existing models.	
	3b	D;V	Specify the objectives, including whether the study describes the development or validation of the model or both.	
Methods				
Source of data	4a	D;V	Describe the study design or source of data (e.g., randomized trial, cohort, or registry data), separately for the development and validation data sets, if applicable.	
	4b	D;V	Specify the key study dates, including start of accrual; end of accrual; and, if applicable, end of follow-up.	
Participants	5a	D;V	Specify key elements of the study setting (e.g., primary care, secondary care, general population) including number and location of centres.	
	5b	D;V	Describe eligibility criteria for participants.	
	5c	D;V	Give details of treatments received, if relevant.	
Outcome	6a	D;V	Clearly define the outcome that is predicted by the prediction model, including how and when assessed.	
	6b	D;V	Report any actions to blind assessment of the outcome to be predicted.	
Predictors	7a	D;V	Clearly define all predictors used in developing or validating the multivariable prediction model, including how and when they were measured.	
	7b	D;V	Report any actions to blind assessment of predictors for the outcome and other predictors.	
Sample size	8	D;V	Explain how the study size was arrived at.	
Missing data	9	D;V	Describe how missing data were handled (e.g., complete-case analysis, single imputation, multiple imputation) with details of any imputation method.	
Statistical analysis methods	10a	D	Describe how predictors were handled in the analyses.	
	10b	D	Specify type of model, all model-building procedures (including any predictor selection), and method for internal validation.	
	10c	V	For validation, describe how the predictions were calculated.	
	10d	D;V	Specify all measures used to assess model performance and, if relevant, to compare multiple models.	
	10e	V	Describe any model updating (e.g., recalibration) arising from the validation, if done.	
Risk groups	11	D;V	Provide details on how risk groups were created, if done.	
Development vs. validation	12	V	For validation, identify any differences from the development data in setting, eligibility criteria, outcome, and predictors.	
Results				
Participants	13a	D;V	Describe the flow of participants through the study, including the number of participants with and without the outcome and, if applicable, a summary of the follow-up time. A diagram may be helpful.	
	13b	D;V	Describe the characteristics of the participants (basic demographics, clinical features, available predictors), including the number of participants with missing data for predictors and outcome.	
	13c	V	For validation, show a comparison with the development data of the distribution of important variables (demographics, predictors and outcome).	
Model development	14a	D	Specify the number of participants and outcome events in each analysis.	
	14b	D	If done, report the unadjusted association between each candidate predictor and outcome.	
Model specification	15a	D	Present the full prediction model to allow predictions for individuals (i.e., all regression coefficients, and model intercept or baseline survival at a given time point).	
	15b	D	Explain how to the use the prediction model.	
Model performance	16	D;V	Report performance measures (with CIs) for the prediction model.	
Model-updating	17	V	If done, report the results from any model updating (i.e., model specification, model performance).	
Discussion				
Limitations	18	D;V	Discuss any limitations of the study (such as nonrepresentative sample, few events per predictor, missing data).	
Interpretation	19a	V	For validation, discuss the results with reference to performance in the development data, and any other validation data.	
	19b	D;V	Give an overall interpretation of the results, considering objectives, limitations, results from similar studies, and other relevant evidence.	
Implications	20	D;V	Discuss the potential clinical use of the model and implications for future research.	
Other information				
Supplementary information	21	D;V	Provide information about the availability of supplementary resources, such as study protocol, Web calculator, and data sets.	
Funding	22	D;V	Give the source of funding and the role of the funders for the present study.	

*Items relevant only to the development of a prediction model are denoted by D, items relating solely to a validation of a prediction model are denoted by V, and items relating to both are denoted D;V. We recommend using the TRIPOD Checklist in conjunction with the TRIPOD Explanation and Elaboration document.

FIGURE 14.4 The TRIPOD checklist for clinical prediction systems. (Reproduced with permission from reference [32].)

how the prediction from the analyzed samples is generalized to the entire population, and indirectly, how the variation of the results could be at different analyzed folds. Such CIs are estimated based on the CI of the proportions [36]. For example, the CI 95% (i.e., $\alpha = 0.05$) of the parameters sensitivity (*Se*), specificity (*Sp*), positive predictive value (*PPV*), negative predictive value (*NPV*), and area under the roc (*AUC*) are estimated using Equations (14.1)–(14.5).

$$SE_{Se} = \sqrt{\frac{Se \times (1 - Se)}{TP + FN}}; Se \pm z_{1-\alpha/2} \times SE_{Se} \qquad (14.1)$$

$$SE_{Sp} = \sqrt{\frac{Sp \times (1 - Sp)}{TN + FP}}; Sp \pm z_{1-\alpha/2} \times SE_{Sp} \qquad (14.2)$$

$$SE_{PPV} = \sqrt{\frac{PPV \times (1 - PPV)}{TP + FP}}; PPV \pm z_{1-\alpha/2} \times SE_{PPV} \qquad (14.3)$$

$$SE_{NPV} = \sqrt{\frac{NPV \times (1 - NPV)}{TN + FN}}; NPV \pm z_{1-\alpha/2} \times SE_{NPV} \qquad (14.4)$$

$$\begin{cases} N1 = TP + FN; N2 = TN + FP; Q_1 = \dfrac{AUC}{2 - AUC}; Q2 = \dfrac{2 \times AUC^2}{1 + AUC} \\ \\ SE_{AUC} = \sqrt{\dfrac{\begin{array}{c} AUC \times (1 - AUC) + (N_1 - 1) \times (Q_1 - AUC^2) \\ + (N_2 - 1) \times (Q_2 - AUC^2) \end{array}}{N_1 \times N_2}}; AUC \pm z_{1-\alpha/2} \times SE_{AUC} \end{cases}$$

$$(14.5)$$

where TP, TN, FP, FN, and SE are true positives, true negatives, false positives, false negatives, and standard error, respectively.

When the CI of a sample statistic is narrower, the estimation of the underlying population parameter is more reliable [37]. In fact, when the performance indices and (or) the sample size decrease, the CI of the performance indices becomes wider (Equations (14.1)–(14.5)), resulting in reduced reliability of the estimated parameters. As an example, suppose that we have $TP = 30$, $TN = 70$, $FP = 4$, and $FN = 7$ (sample size (S) = 111). The performance indices with their CI 95% are as the following: $Se = 0.81$ [CI 95%: 0.68–0.94], $Sp = 0.95$ [CI 95%: 0.89–0.99], $PPV = 0.88$ [CI 95%: 0.77–0.99], $NPV = 0.91$ [CI 95%: 0.84–0.97], and $AUC = 0.88$ [CI 95%: 0.80–0.95]. Increasing the sample size by the factor of two (i.e., $S = 222$), without changing the Type I and II errors, results in narrowed CIs: $Se = 0.81$ [CI 95%: 0.72–0.90], $Sp = 0.95$ [CI 95%: 0.91–0.98], $PPV = 0.88$ [CI 95%: 0.81–0.96], $NPV = 0.91$ [CI 95%: 0.86–0.95], and $AUC = 0.88$ [CI 95%: 0.82–0.93]. Keeping the sample size the same as the first case ($S = 111$), but a slight increase of the Type I and II errors ($TP = 27$, $TN = 67$, $FP = 7$, $FN = 10$), the results are the following: $Se = 0.73$ [CI 95%: 0.59–0.87], $Sp = 0.91$ [CI 95%: 0.84–0.97], $PPV = 0.79$ [CI 95%: 0.66–0.93], $NPV = 0.87$ [CI 95%: 0.80–0.95], and $AUC = 0.82$ [CI 95%: 0.73–0.91]. The lower CI 95% of the performance indices, in this case, would not be acceptable in most applications.

Among COVID-19 diagnosis studies, some studies provided the CI 95% of the performance indices, such as [38–43], while some studies did not [44–50]. As an example, Abraham and Nair [49] designed an automated COVID-19 detection system using X-ray images. Their confusion matrices on two datasets are shown in Figure 14.5. While the sample size of the left dataset is acceptable, the right dataset

Confusion Matrix   Confusion Matrix

		Covid	446	7	0.985

Target Class

Covid	446	7	0.985
Non-Covid	77	420	0.845
	0.853	0.984	0.912
	Covid	Non-Covid	

Output Class

Covid	71	1	0.986
Non-Covid	1	6	0.857
	0.986	0.857	0.974
	Covid	Non-Covid	

Output Class

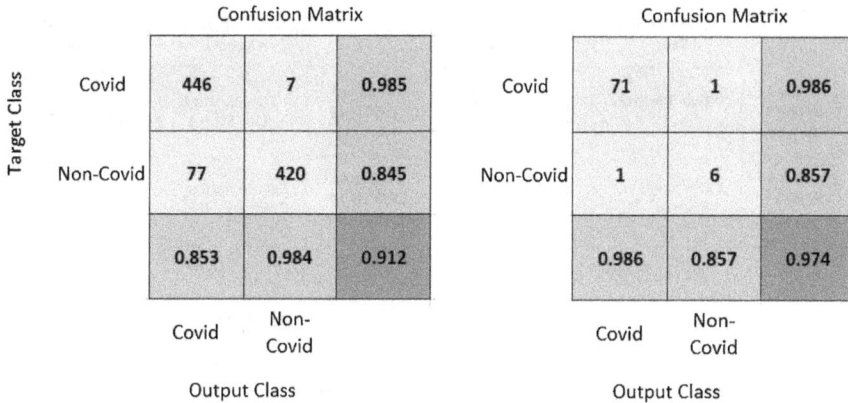

FIGURE 14.5 The confusion matrices of the COVID-19 diagnosis method proposed by Abraham and Nair on two datasets. (Reproduced with permission from reference [49].)

is small, resulting in the wider CIs. For example, $Sp = 0.86$ [CI 95%: 0.60–1.00] for the right dataset, compared with $Sp = 0.85$ [CI 95%: 0.81–0.88] related to the left dataset.

14.2.3 VALIDATION FRAMEWORKS

Some form of internal validation is required for new prediction models to quantify any optimism in predictive performance, such as discrimination and calibration [32]. Bootstrapping or cross-validation are necessary parts of model development [51]. After performing internal validation, to make sure that overfitting does not occur, it is highly recommended to test the performance of the developed model on a different dataset, which is known as external validation [52, 53]. Some primary considerations must be taken into account to guard against testing hypotheses suggested by the data (Type III errors [54]). Hold-out validation is not sufficient, and cross-validation, .632+ bootstrapping, or repeated hold-out must also be used [55]. It was shown in the literature that the hold-out validation could lead to significant overestimation of the performance of the diagnosis system, compared with the cross-validation [48].

In addition to providing the results of the cross-validated confusion matrix (i.e., the analysis of the entire samples used on the test sets), the mean and the standard deviation of the performance indices on the test folds must be reported. These issues are not new. However, if not considered, it leads to misleading results. A good example of providing the variation of the results over the test folds is provided in Figure 14.6. Singh et al. [50] designed an automatic COVID-19 diagnosis system using chest CT images. They performed cross-validation and the variation of the results of different methods over the test folds, as shown in Figure 14.6. However, the CI 95% of the performance indices were not reported. For instance, based on the average confusion matrix values in Figure 14.6(d), some of the performance indices are reported as the following: $Se = 0.92$ [CI 95%: 0.84–0.98], $Sp = 0.89$ [CI 95%: 0.82–0.97], and $PPV = 0.90$ [CI 95%: 0.83–0.97].

	Actual COVID-19 = Yes (+)	Actual COVID-19 = Yes (-)
Predicted COVID-19 = Yes (+)	True Positive TP= 58 ± 5	False Positive FP=9 ±3
Predicted COVID-19 = Yes (-)	False Negative FN =10 ±5	True Negative TN= 55 ±6

a

	Actual COVID-19 = Yes (+)	Actual COVID-19 = Yes (-)
Predicted COVID-19 = Yes (+)	True Positive TP= 59 ± 4	False Positive FP= 10 ±3
Predicted COVID-19 = Yes (-)	False Negative FN =9 ±6	True Negative TN= 54 ±7

b

	Actual COVID-19 = Yes (+)	Actual COVID-19 = Yes (-)
Predicted COVID-19 = Yes (+)	True Positive TP= 60 ± 6	False Positive FP=8 ±3
Predicted COVID-19 = Yes (-)	False Negative FN =9 ±4	True Negative TN= 56 ±6

c

	Actual COVID-19 = Yes (+)	Actual COVID-19 = Yes (-)
Predicted COVID-19 = Yes (+)	True Positive TP= 62±4	False Positive FP= 7 ±3
Predicted COVID-19 = Yes (-)	False Negative FN =6 ±5	True Negative TN= 58 ±6

d

FIGURE 14.6. Confusion matrix of (a) Artificial Neural Network, (b) adaptive neuro-fuzzy inference system, (c) convolutional neural network (CNN), and (d) the proposed multi-objective differential evolution-based CNN model. (Reproduced with permission from reference [50].)

Among the COVID-19 diagnosis methods, the internal validation (cross-validation) was used in some studies, e.g., [40, 46, 48, 50]. However, in some studies such as [38, 56, 57], only hold-out validation was used. Moreover, some studies, including [58–60], used external validation.

14.2.4 PROPER PERFORMANCE INDICES

In epidemiology and medical data mining, each diagnostic accuracy measure has its importance [61]. Each provides essential information, and a missing accuracy index could be confusing to the readers and make it hard for detailed validation [62]. In binary classification problems, the single indices *Se*, *Sp*, *PPV*, *NPV*, and the composite indices F1-score [63], *AUC*, Matthews correlation coefficient (MCC) [64], accuracy, and diagnostic odds ratio (DOR) are usually provided [65]. In multiclass classification problems, on the other hand, proper performance indices such as macro-averaged F-score must be provided, in addition to the overall accuracy [66]. Regardless of the composite indices, Type I and Type II statistical errors must be provided in the diagnosis studies.

Moreover, when the dataset is imbalanced, which is usually the case in COVID-19 studies, the parameter "accuracy" must not be the only composite index that is reported in the paper as it is biased. These basic concepts have been known for years, but they are not considered in some of the COVID-19 studies. For example,

Alakus and Turkoglub [48] compared the performance of deep learning algorithms for COVID-19 diagnosis on 520 non-COVID and 80 COVID patients. The authors performed cross-validation in addition to the hold-out internal validation but did not report the CI 95% of the performance indices. Moreover, in the ten-fold cross-validation results (Table 14.3; [48]), the *Sp* of the best methods is not directly reported in the results; however, it is possible to calculate it based on the other indices, which is 25.58% (Type I error of 0.74). Although the *Se* of the best method is 99.52%, a high false-positive rate is problematic and decreases the diagnosis accuracy and *PPV* value [67], which is surprisingly not the case in this paper.

14.2.5 OVERESTIMATION OF ACCURACY

The parameter *PPV*, the posterior probability that could also be assessed using tree diagrams in screening tests [68], is sensitive to the prevalence of the disease (*P*), based on the Bayes' formula [69, 70]:

$$PPV = \frac{Se \times P}{Se \times P + (1 - Sp) \times (1 - P)} \tag{14.6}$$

The probability that a subject has COVID-19, given that the test result is positive, is the *PPV*. Thus, balancing the test dataset when the prevalence of the disease is low, results in a reduced *PPV* in the population. For example, if a COVID-19 diagnosis test has *Se* = 80% and *Sp* = 95% and the test set is balanced to have the same number of RT-PCR (+) and (–) as the gold standard, the PPV in the test set will be 94%. However, assuming the COVID-19 prevalence is 5% in a population [71], the *PPV* will drop down to 46%. Thus, if the proposed COVID-19 diagnosis test is used, the probability that a subject has COVID-19, given that the test result is positive, is 46%. Imbalanced datasets are usually challenging for classification [72], and balancing the dataset is sometimes used by the authors. For example, Chowdhury et al. [45] designed an automatic COVID-19 diagnosis system using chest X-ray images. Such balancing was used in Table 14.1 for the data without augmentation [45].

The other important issue is using leave-one-subject-out cross-validation when repeated measurements exist for each subject [73]. A COVID-19 diagnosis system must be able to classify a new object as COVID-19 or non-COVID-19. Thus, if we have multiple measurements for each subject, such as in the chest X-ray or CT studies, the entire measurements of a subject must be taken out from the training set, and the performance of the trained system must be reported for the test subject. Otherwise, if other internal validation methods are used, and the training and test set random permutations are performed on the entire measurements, rather than subjects, there is a high probability to have some measurements of a subject in the training set and the rest in the test set. If such repeated measurements are highly correlated, which is the case in image processing problems, the accuracy of the diagnosis system is overstimulated. Using subject-wise cross-validation with a more extensive test sample size is usually preferred to leave-one-subject-out cross-validation as it reduces the estimation variance [74]. For example, Abbasian Ardakani et al. [75] designed a medical diagnosis system to identify COVID-19 based on the deep learning of CT images. The authors used the hold-out internal validation (80% training, 20% test), and the comprehensive list of the performance indices was reported. However, the CI 95% of such indices were not provided. There were 108 COVID-19 and 86 non-COVID-19 patients that participated in the study, and 510 COVID-19 and 510 non-COVID-19 CT image patches were analyzed. Thus, the dataset was balanced, and there were more than one patch for each subject in the dataset. The performance of the proposed system was assessed based on the classified patches (Table 14.3; [75]), resulting in an overestimation of the diagnostic accuracy. Adding subject cross-validation could help the reader rigorously assess the performance of the proposed system. Moreover, considering the low performance of the radiologists in the patch-based visual analysis (e.g., $AUC = 0.60$ [CI 95%: 0.53–0.68], calculated based on the confusion matrix provided by the authors in Table 14.3; [75]), overestimation of the accuracy based on the similarity of the patches of each subject in the training and test sets are more probable.

14.3 COVID-19 PROGNOSIS

COVID-19 prognosis, usually containing survival analysis, is a more difficult task than diagnosis when considering time-varying covariates and multiple outcomes for in-hospital data. Wolkewitz et al. [12] and Wolkewitz and Puljak [11] provided detailed guidelines for proper COVID-19 prognosis and the list of the common pitfalls to avoid as to reduce selection, length, immortal-time, and competing risk bias. For example, cause-specific Cox regression avoids competing risk bias that occurs using standard Cox regression models [76]. Also, to reduce the selection bias, time-to-event models are used instead of standard logistic regression [12]. Moreover, the analysis of time-dependent covariates and competing endpoints require advanced statistical models [77]. In this section, we provide an overview of two advanced complementary methods to analyze COVID-19 data. In the first part, we introduce multi-state models which focus on the time-complex disease progression of COVID-19 including important intermediate events such as admission to intensive care, mechanical ventilation, and terminal events such as death or discharge from the hospital. In the second part, we introduce a risk stratification approach which has been well documented in cardiology to deal with the complexity of several risk factors such as age, gender, co-morbidities, and multiple laboratory-based and non-laboratory covariates.

14.3.1 MULTI-STATE MODELS

14.3.1.1 Valid Parameter Estimates of Disease Progression Are Needed

The general time-dependent disease progression of COVID-19 is displayed in Figure 14.7.

In the COVID-19 pandemic, it is a challenge to identify critical parameters characterizing the disease, including, for example, disease severity, mortality, the need for

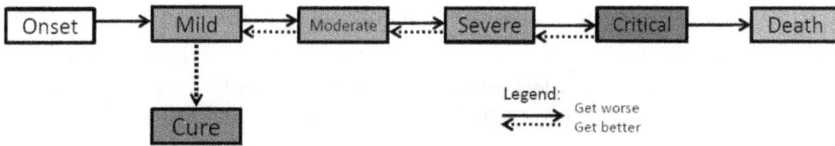

FIGURE 14.7   Disease progression of COVID-19.

hospitalization and intensive care, and the main characteristics of diseased patients. These parameters are essential to model both the future course of the pandemic and the transmission dynamics of the infection. Lack of evidence for quantifying these parameters may have severe consequences on decision making. At an early stage of the pandemic, Ferguson et al. [78] modeled and analyzed possible transmission progressions of COVID-19 under different interventions. Due to limited data, the models were based on major assumptions on disease characteristics partly obtained from personal communication and expert opinions. However, a fundamental pillar of evidence-based medicine is empirical research. Therefore, the parameters must be estimated from data and should not be based on individual experiences. All the available data can be used by estimating the parameters in real-time with frequent updates as data evolves over time.

14.3.1.2 Parameter Estimates Are Often Only Based on Closed COVID-19 Cases

Most of the conditions displayed in Figure 14.7 can be mapped and linked with routinely collected clinical data such as information on stays in normal wards (moderate disease progression), intensive care units (severe disease progression), and need for ventilation (critical disease progression). Data of patients hospitalized with COVID-19 are directly assessable using these routinely collected hospital records. However, due to the time dynamics of the pandemic, most patients are still hospitalized at the time of analysis (active cases). For example, Zhou et al. [79] excluded 613 active cases of a total of 813 active and closed (i.e., discharged or dead) cases. Due to the reported significantly longer hospital stay of survivors (7.5 (5–11) vs 12 (9–15) Zhou et al.), the selected sample of closed cases is likely subject to selection bias.

Similarly, 1032 out of 1099 patients in the descriptive study of Guan et al. [80] were still hospitalized at the time of analysis. Thus, these patients were at different stages in the course of disease than patients discharged or dead by the time of analysis. The same applies to the recent study by Grasseli et al. [81], who analyzed active and closed cases to identify the characteristics of patients admitted to the intensive care unit (ICU) with COVID-19 in the Lombardy region of Italy. Of the 1591 patients 54% were still in the hospital when the data was analyzed. Therefore, to generate fast but unbiased evidence, both closed and active cases must be analyzed relative to the time of infection. For the quantification of hospital capacities, incidences of new cases and hospital admissions must be additionally considered in calendar time.

14.3.1.3   Multi-State Models for Closed and Open Cases

The time-to-event methodology is a powerful tool to study both active and closed cases without introducing selection bias; active cases can be considered as "censored" observations [82]. In the following, we explain how the multi-state methodology can be used to analyze routine data of hospitalized patients with COVID-19 in real-time. For this, we assume that routine data, as displayed in Table 14.1, are available.

Specific administrative dates characterizing a patient's hospital stay are admission and discharge dates to and from the hospital, vital state at discharge, as well as—if applicable—admission and discharge dates to and from the ICU and start–stop dates of mechanical ventilation. By integrating data of case incidence available in calendar time, and individual case data on hospitalization, the requirement of intensive care and mechanical ventilation, essential information for pandemic planning can be generated.

Depending on the availability of the clinical data, patients' progression in the hospital, i.e., transitions between the regular ward and intensive care (with potential ventilation procedures) can be modeled as multi-state models [82]. The progression through the multi-state models is determined by the hazard rates between the corresponding states; see for a general definition in Andersen [82] and a specific COVID-19 example in Hazard et al. [83]. To study the progress of hospitalized patients using this routine data, we propose the multi-state model shown in Figure 14.8.

This model accounts for administrative events, including hospital admission and discharge, ICU admission and discharge, requirement and duration of mechanical ventilation, and mortality. The parameters of interest are estimated for each of the three starting states using the transitions probabilities of the multi-state model.

TABLE 14.1

Routine Data Needed for Real-Time Multi-State Model Analysis of Hospitalized COVID-19 Patients

Routine Data:
- Patient/center identifier
- Age (optional)
- Gender (optional)
- Pre-existing illness (optional)
- Status: active (still in hospital) or closed (discharged or dead)
- Date of the last follow-up (current date if active, hospital discharge date if closed)
- Date of COVID-19 diagnosis
- Date of hospital admission
- Date of admission to intensive care (missing if not required)
- Date of start invasive ventilation (missing if not required)
- Date of stop invasive ventilation (missing if not required)
- Date of discharge from the intensive care unit (missing if not required)
- Date of discharge from hospital (missing if active)
- Vital status at discharge from hospital (missing if active)

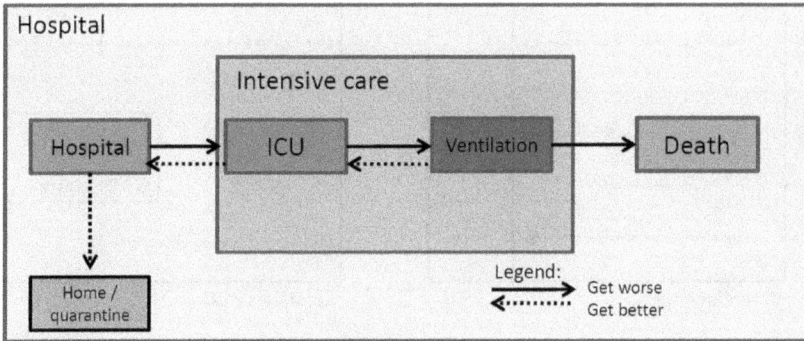

FIGURE 14.8   Multi-state model for clinical COVID-19 data.

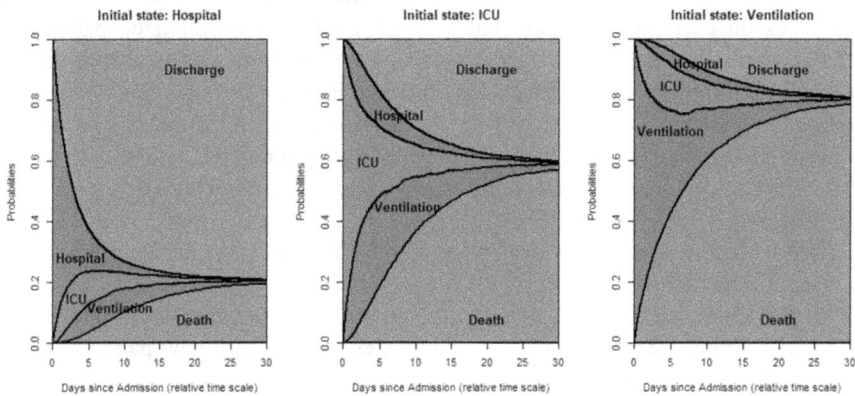

FIGURE 14.9   Stacked probability plot.

These are the length of hospital and ICU stay, duration of mechanical ventilation, and the probabilities of dying or being discharged alive.

14.3.1.4   Stacked Probability Plots

For illustration, we simulated data (in real-time) using time-constant hazard rates for each transition (arrow) for the multi-state model. Figure 14.9 shows the result of stacked probability plots. The plot illustrates the progression of the disease of COVID-19 patients. It includes hospitalization events in the regular ward, in the ICU, under mechanical ventilation, discharge alive, and death.

14.3.1.5   Algorithm for Real-Time Data Analysis

Figure 14.10 illustrates the course of hospital stay in calendar time for five fictional patients. Switching from calendar time to a relative time scale defined as days since hospital admission (see Figure14.10) is an essential step for the estimation of parameters of interest. Patients that are still in the hospital at the time of analysis (active cases) influence estimates as censored observations.

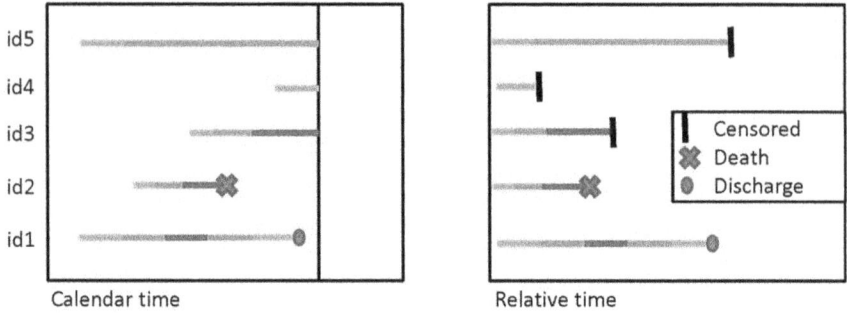

FIGURE 14.10   Illustration of data.

An advantage of analyzing routine data is the possibility to describe hospital capacities on each calendar day. The current number of hospitalized patients in regular wards, in the ICU, and under mechanical ventilation can be displayed and updated each day. At each calendar day, the data used for estimation is updated by adding the patients admitted on that day to the sample. Then, parameters of interest are estimated each day using the updated sample of hospitalized patients.

In order to perform real-time analysis, we suggest the following algorithm:

Step 1: Record routine data from diagnosed COVID-19 cases in calendar time.
Step 2: Transfer to a relative time scale.
Step 3: Calculate the multi-state model and estimate the quantities of interest.
Step 4: Repeat steps 1–3 on every new calendar day with new patients and updated follow-up.

14.3.1.6   Applications, Statistical Code, and Publicly Available COVID-19 Data

Multi-state approaches have already been applied to several COVID-19 datasets. For instance, the data from two studies ($n = 24$ and $n = 53$ patients) are publicly available in Hazard et al. [83]; the paper also contains the statistical code for the analysis using the R-package "mstate." Another COVID-19 example of a more complex multi-state model is given in Rieg et al. [84]. The most advanced multi-state analysis with COVID-19 data has been performed by Roimi et al. [85].

14.3.2   RISK ASSESSMENT

The development of the risk functions has been well documented in cardiology [86–89]. In a cohort, when the time-to-event analysis is performed, it is possible to identify the risk of the subjects having an event of interest by the end of the cohort follow-up. Such a risk assessment procedure usually results in a risk chart that visually identifies the risk of the event based on the selected risk factors in the Cox proportional hazard regression model (Figure 14.11). It could be further divided into laboratory-based and non-laboratory (office-based) categories [90].

Persian Atherosclerotic cardiovascular disease Risk Stratification (PARS)

Women — Low WHR, No CVD family history — Men

	Non-diabetic	Diabetic	Age (years)	Non-diabetic	Diabetic
	non-smoker / smoker	non-smoker / smoker		non-smoker / smoker	non-smoker / smoker

Systolic blood pressure (mm Hg): rows ≥160, 140–159, 120–139, <120 for age groups ≥75, 65–74, 55–64, 45–54, 35–44.

10-year cardiovascular risk (fatal and non-fatal) — Total Cholesterol Categories

Legend: 1% and lower · 2% · 3% – 4% · 5% – 9% · 10% – 14% · 15% and over

Total Cholesterol Categories: 1 2 3 4 5 → 150 200 250 300 mg/dL

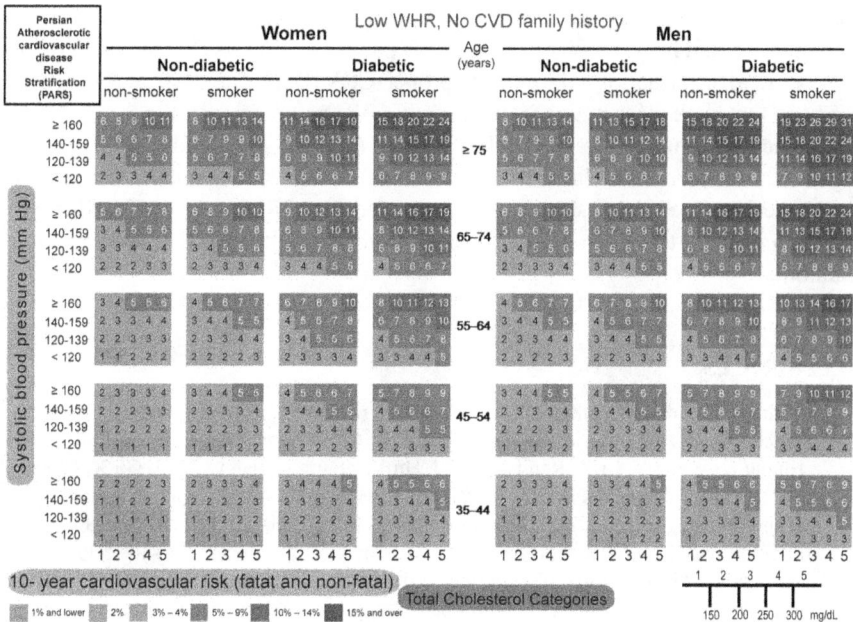

FIGURE 14.11 Persian Atherosclerotic Cardiovascular Disease Risk Stratification (PARS) charts for prediction of ten-year risk of fatal and non-fatal cardiovascular disease in Isfahan Cohort Study population with lower waist to hip ratio (WHR) and no cardiovascular disease family history, 2001–2011. (Reproduced with permission from reference [89].)

Based on the Persian Atherosclerotic Cardiovascular Disease Risk Stratification (PARS) risk chart (Figure 14.11), the risk of ten-year fatal and non-fatal CVD for a diabetic non-smoking woman with low waist to hip ratio (WHR) (WHR < 0.8), without cardiovascular disease family history, who is 50 years old, with an average systolic blood pressure of 150 mmHg, and the total cholesterol of 225 mg/dL is 4%. The same concept could apply in COVID-19 risk assessment when the COVID-19 cohort is performed, and the events of interest are death, discharge, mechanical ventilation, ICU admission, or primary composite endpoint [80]. Note that each aforementioned event has competing events [12], for instance, discharge from hospital or death are competing events for ICU admission [91]. Thus, due to the presence of these competing events, the absolute risk is estimated using specific regression models [89, 92].

Moreover, internal (cross-validation and bootstrapping) and external validation are usually used to avoid overfitting. The goodness-of-fit measures, such as the Nam-D'Agostino chi-square test [93, 94], are used for evaluating time-to-event analysis calibration while C-statistic, and Harrell's C [95], are used for evaluating the discrimination of the model.

14.4 CONCLUSION AND FUTURE SCOPE

COVID-19, a global disaster, has short- and long-term effects on society. It not only affects public health, but it is also highly related to economic and psychiatric consequences. Proper experimental design, statistical analysis, or even peer-review procedures could help fight with the COVID pandemic. In this chapter, we provided some critical information required for proper COVID-19 diagnosis and prognosis. We hope that such information is useful for COVID-19 biomedical research.

Acknowledgments: The research leading to these results has also received funding from the European Union's Horizon 2020 research and innovation program under the Marie Skłodowska-Curie grant agreement No 712949 (TECNIOspring PLUS) and from the Agency for Business Competitiveness of the Government of Catalonia.

REFERENCES

1. Johnson T, Sakya S, Sakya J, Onkendi E and Hallan D. 2020. The top 100 most cited articles on COVID-19. *Southwest Respir. Crit. Care Chronicles* **8**: 42–50.
2. Huang C, Wang Y, Li X, Ren L, Zhao J, Hu Y, Zhang L, Fan G, Xu J, Gu X, Cheng Z, Yu T, Xia J, Wei Y, Wu W, Xie X, Yin W, Li H, Liu M, Xiao Y, Gao H, Guo L, Xie J, Wang G, Jiang R, Gao Z, Jin Q, Wang J and Cao B. 2020. Clinical features of patients infected with 2019 novel coronavirus in Wuhan, China. *Lancet* **395**: 497–506.
3. Sohrabi C, Alsafi Z, O'Neill N, Khan M, Kerwan A, Al-Jabir A, Iosifidis C and Agha R. 2020. World Health Organization declares global emergency: A review of the 2019 novel coronavirus (COVID-19). *Int. J. Surg.* **76**: 71–6.
4. Singhal T. 2020. A review of coronavirus disease-2019 (COVID-19). *Indian J. Pediatr.* **87**: 281–6.
5. Dong E, Du H and Gardner L. 2020. An interactive web-based dashboard to track COVID-19 in real time. *Lancet Infect. Dis.* **20**(5): 533–4.
6. Pak A, Adegboye OA, Adekunle AI, Rahman KM, McBryde ES and Eisen DP. 2020. Economic consequences of the COVID-19 outbreak: the need for epidemic preparedness. *Front Public Health* **8**: 241.
7. Glasziou PP, Sanders S and Hoffmann T. 2020. Waste in covid-19 research. *BMJ* **369**: m1847.
8. Mehra MR, Ruschitzka F and Patel AN. 2020. Retraction-hydroxychloroquine or chloroquine with or without a macrolide for treatment of COVID-19: a multinational registry analysis. *Lancet* **395**: 1820.
9. Yan W. 2020. Coronavirus tests science's need for speed limits. *NY Times* **14**.
10. Callaghan S. 2020. COVID-19 is a data science issue. *Patterns (NY)* **1**: 100022.
11. Wolkewitz M and Puljak L. 2020. Methodological challenges of analysing COVID-19 data during the pandemic. *BMC Med. Res. Methodol.* **20**: 81.
12. Wolkewitz M, Lambert J, von Cube M, Bugiera L, Grodd M, Hazard D, White N, Barnett A and Kaier K. 2020. Statistical analysis of clinical COVID-19 data: A concise overview of lessons learned, common errors and how to avoid them. *CLEP* **12**: 925–8.
13. The Lancet. 2020. COVID-19: fighting panic with information. *Lancet* **395**: 537.
14. Wynants L, Van Calster B, Collins GS, Riley RD, Heinze G, Schuit E, Bonten MMJ, Damen JAA, Debray TPA, De Vos M, Dhiman P, Haller MC, Harhay MO, Henckaerts L, Kreuzberger N, Lohman A, Luijken K, Ma J, Andaur CL, Reitsma JB, Sergeant JC, Shi C, Skoetz N, Smits LJM, Snell KIE, Sperrin M, Spijker R, Steyerberg EW, Takada T, van Kuijk SMJ, van Royen FS, Wallisch C, Hooft L, Moons KGM and van Smeden M. 2020. Prediction models for diagnosis and prognosis of covid-19 infection: systematic review and critical appraisal. *BMJ* **369**: m1328.

15. Lisboa Bastos M, Tavaziva G, Abidi SK, Campbell JR, Haraoui L-P, Johnston JC, Lan Z, Law S, MacLean E, Trajman A, Menzies D, Benedetti A and Ahmad Khan F. 2020. Diagnostic accuracy of serological tests for covid-19: systematic review and meta-analysis. *BMJ* **370**: m2516.

16. Winichakoon P, Chaiwarith R, Liwsrisakun C, Salee P, Goonna A, Limsukon A and Kaewpoowat Q. 2020. Negative nasopharyngeal and oropharyngeal swabs do not rule out COVID-19. *J. Clin. Microbiol.* **58**

17. Chen Z, Li Y, Wu B, Hou Y, Bao J and Deng X. 2020. A patient with COVID-19 presenting a false-negative reverse transcriptase polymerase chain reaction result. *Korean J. Radiol.* **21**: 623–4.

18. Sethuraman N, Jeremiah SS and Ryo A. 2020. Interpreting diagnostic tests for SARS-CoV-2. *JAMA* **323**: 2249–51.

19. American Society For Microbiology ASM Expresses Concern About Coronavirus Test Reagent Shortages. Available at https://asm.org/Articles/Policy/2020/March/ASM-Expresses-Concern-about-Test-Reagent-Shortages (Accessed on Sep 20, 2020)

20. Kubina R and Dziedzic A. 2020. Molecular and serological tests for COVID-19 a comparative review of SARS-CoV-2 coronavirus laboratory and point-of-care diagnostics. *Diagnostics (Basel)* **10**(6): 434.

21. Ghaffari A, Meurant R and Ardakani A. 2020. COVID-19 serological tests: How well do they actually perform? *Diagnostics (Basel)* **10**(7): 453.

22. Ai T, Yang Z, Hou H, Zhan C, Chen C, Lv W, Tao Q, Sun Z and Xia L. 2020. Correlation of chest CT and RT-PCR testing for coronavirus disease 2019 (COVID-19) in China: A report of 1014 cases. *Radiology* **296**: E32–40.

23. Majidi H and Niksolat F. 2020. Chest CT in patients suspected of COVID-19 infection: A reliable alternative for RT-PCR. *Am. J. Emerg. Med.* doi: 10.1016/j.ajem.2020.04.016 [Epub ahead of print]

24. Xie J, Ding C, Li J, Wang Y, Guo H, Lu Z, Wang J, Zheng C, Jin T, Gao Y and He H. 2020. Characteristics of patients with coronavirus disease (COVID-19) confirmed using an IgM-IgG antibody test. *J. Med. Virol.* doi: 10.1002/jmv.25930 [Epub ahead of print]

25. Liu R, Han H, Liu F, Lv Z, Wu K, Liu Y, Feng Y and Zhu C. 2020. Positive rate of RT-PCR detection of SARS-CoV-2 infection in 4880 cases from one hospital in Wuhan, China, from Jan to Feb 2020. *Clin. Chim. Acta* **505**: 172–5.

26. Liu K, Chen Y, Lin R and Han K. 2020. Clinical features of COVID-19 in elderly patients: A comparison with young and middle-aged patients. *J. Infect.* **80**: e14–8.

27. Winter AK and Hegde ST. 2020. The important role of serology for COVID-19 control. *Lancet Infect. Dis.* **20**: 758–9.

28. Wang Y, Hou H, Wang W and Wang W. 2020. Combination of CT and RT-PCR in the screening or diagnosis of COVID-19. *J. Glob. Health* **10**: 010347.

29. Graziadiol S, Hicks T, Joy Allen A, Suklan J, Urwin SG, Winter A, Price DA and Body R. Composite Reference Standard for COVID-19 Diagnostic Accuracy Studies: a roadmap - CEBM Available at https://www.cebm.net/covid-19/a-composite-reference-standard-for-covid-19-diagnostic-accuracy-studies-a-roadmap/ (Accessed on Sep 20, 2020)

30. Cohen J. 1960. A coefficient of agreement for nominal scales. *Educ. Psychol. Meas.* **20**: 37–46.

31. Dramé M, Tabue Teguo M, Proye E, Hequet F, Hentzien M, Kanagaratnam L and Godaert L. 2020. Should RT-PCR be considered a gold standard in the diagnosis of COVID-19? *J. Med. Virol.* doi: 10.1002/jmv.26228 [Epub ahead of print]

32. Collins GS, Reitsma JB, Altman DG and Moons KGM. 2015. Transparent reporting of a multivariable prediction model for individual prognosis or diagnosis (TRIPOD): the TRIPOD Statement. *BMC Med.* **13**: 1.

33. Bossuyt PM, Reitsma JB, Bruns DE, Gatsonis CA, Glasziou PP, Irwig L, Lijmer JG, Moher D, Rennie D, de Vet HCW, Kressel HY, Rifai N, Golub RM, Altman DG, Hooft L, Korevaar DA, Cohen JF and STARD Group. 2015. STARD 2015: an updated list of essential items for reporting diagnostic accuracy studies. *BMJ* **351**: h5527.

34. Collins GS and Moons KGM. 2019. Reporting of artificial intelligence prediction models. *Lancet* **393**: 1577–9.

35. Sounderajah V, Ashrafian H, Aggarwal R, De Fauw J, Denniston AK, Greaves F, Karthikesalingam A, King D, Liu X, Markar SR, McInnes MDF, Panch T, Pearson-Stuttard J, Ting DSW, Golub RM, Moher D, Bossuyt PM and Darzi A. 2020. Developing specific reporting guidelines for diagnostic accuracy studies assessing AI interventions: The STARD-AI Steering Group. *Nat. Med.* **26**: 807–8.

36. Newcombe RG. 1998. Two-sided confidence intervals for the single proportion: comparison of seven methods. *Stat. Med.* **17**: 857–72.

37. Hazra A. 2017. Using the confidence interval confidently. *J. Thorac. Dis.* **9**: 4125–30.

38. Mei X, Lee H-C, Diao K-Y, Huang M, Lin B, Liu C, Xie Z, Ma Y, Robson PM, Chung M, Bernheim A, Mani V, Calcagno C, Li K, Li S, Shan H, Lv J, Zhao T, Xia J, Long Q, Steinberger S, Jacobi A, Deyer T, Luksza M, Liu F, Little BP, Fayad ZA and Yang Y. 2020. Artificial intelligence-enabled rapid diagnosis of patients with COVID-19. *Nat. Med.* **26**: 1224–8.

39. Menni C, Valdes AM, Freidin MB, Sudre CH, Nguyen LH, Drew DA, Ganesh S, Varsavsky T, Cardoso MJ, El-Sayed Moustafa JS, Visconti A, Hysi P, Bowyer RCE, Mangino M, Falchi M, Wolf J, Ourselin S, Chan AT, Steves CJ and Spector TD. 2020. Real-time tracking of self-reported symptoms to predict potential COVID-19. *Nat. Med.* **26**: 1037–40.

40. Sun Y, Koh V, Marimuthu K, Ng OT, Young B, Vasoo S, Chan M, Lee VJM, De PP, Barkham T, Lin RTP, Cook AR, Leo YS and National Centre for Infectious Diseases COVID-19 Outbreak Research Team. 2020. Epidemiological and clinical predictors of COVID-19. *Clin. Infect. Dis.* **71**: 786–92.

41. Lessmann N, Sánchez CI, Beenen L, Boulogne LH, Brink M, Calli E, Charbonnier J-P, Dofferhoff T, van Everdingen WM, Gerke PK, Geurts B, Gietema HA, Groeneveld M, van Harten L, Hendrix N, Hendrix W, Huisman HJ, Išgum I, Jacobs C, Kluge R, Kok M, Krdzalic J, Lassen-Schmidt B, van Leeuwen K, Meakin J, Overkamp M, van Rees Vellinga T, van Rikxoort EM, Samperna R, Schaefer-Prokop C, Schalekamp S, Scholten ET, Sital C, Stöger L, Teuwen J, Vaidhya Venkadesh K, de Vente C, Vermaat M, Xie W, de Wilde B, Prokop M and van Ginneken B. 2020. Automated assessment of CO-RADS and chest CT severity scores in patients with suspected COVID-19 using artificial intelligence. *Radiology RSNA Radiol.* doi: 10.1148/radiol.2020202439 [Epub ahead of print]

42. Bai HX, Wang R, Xiong Z, Hsieh B, Chang K, Halsey K, Tran TML, Choi JW, Wang D-C, Shi L-B, Mei J, Jiang X-L, Pan I, Zeng Q-H, Hu P-F, Li Y-H, Fu F-X, Huang R Y, Sebro R, Yu Q-Z, Atalay MK and Liao W-H. 2020. Artificial intelligence augmentation of radiologist performance in distinguishing COVID-19 from pneumonia of other origin at chest CT. *Radiology* **296**: E156–65.

43. Ni Q, Sun ZY, Qi L, Chen W, Yang Y, Wang L, Zhang X, Yang L, Fang Y, Xing Z, Zhou Z, Yu Y, Lu GM and Zhang LJ. 2020. A deep learning approach to characterize 2019 coronavirus disease (COVID-19) pneumonia in chest CT images. *Eur. Radiol.* doi: 10.1007/s00330-020-07044-9 [Epub ahead of print]

44. Song J, Wang H, Liu Y, Wu W, Dai G, Wu Z, Zhu P, Zhang W, Yeom KW and Deng K. 2020. End-to-end automatic differentiation of the coronavirus disease 2019 (COVID-19) from viral pneumonia based on chest CT. *Eur. J. Nucl. Med. Mol. Imag.* doi: 10.1007/s00259-020-04929-1 [Epub ahead of print]

45. Chowdhury MEH, Rahman T, Khandakar A, Mazhar R, Kadir MA, Mahbub ZB, Islam KR, Khan MS, Iqbal A, Emadi NA, Reaz MBI and Islam MT. 2020. Can AI help in screening viral and COVID-19 pneumonia? *IEEE Access* **8**: 132665–76.
46. Khan AI, Shah JL and Bhat MM. 2020. CoroNet: A deep neural network for detection and diagnosis of COVID-19 from chest x-ray images. *Comput. Methods Programs Biomed.* **196**: 105581.
47. El Asnaoui K and Chawki Y. 2020. Using X-ray images and deep learning for automated detection of coronavirus disease. *J. Biomol. Struct. Dyn.* 1–12. doi: 10.1080/07391102.2020.1767212
48. Alakus TB and Turkoglu I. 2020. Comparison of deep learning approaches to predict COVID-19 infection. *Chaos Solitons Fract.* **140**: 110120.
49. Abraham B and Nair MS. 2020. Computer-aided detection of COVID-19 from X-ray images using multi-CNN and Bayesnet classifier. *Biocybern Biomed Eng.* **40**(4): 1436–45. doi: 10.1016/j.bbe.2020.08.005
50. Singh D, Kumar V, Vaishali and Kaur M. 2020. Classification of COVID-19 patients from chest CT images using multi-objective differential evolution-based convolutional neural networks. *Eur. J. Clin. Microbiol. Infect. Dis.* **39**: 1379–89.
51. Steyerberg EW. 2019. *Clinical Prediction Models: A Practical Approach to Development, Validation, and Updating*, 1st edition. Cham: Springer.
52. Moons KGM, Kengne AP, Grobbee DE, Royston P, Vergouwe Y, Altman DG and Woodward M. 2012. Risk prediction models: II. External validation, model updating, and impact assessment. *Heart* **98**: 691–8.
53. Altman DG, Vergouwe Y, Royston P and Moons KGM. 2009. Prognosis and prognostic research: validating a prognostic model. *BMJ* **338**: b605.
54. Mosteller F. 2006. A k-sample slippage test for an extreme population. IN *Selected Papers of Frederick Mosteller*, edited by SE Fienberg and DC Hoaglin. New York: Springer New York, pp. 101–9.
55. Wah BW. 2007. Pattern recognition. In *Wiley Encyclopedia of Computer Science and Engineering*, vol. EC-14. Hoboken, NJ: John Wiley & Sons, Inc., p. 683.
56. Ahamad MM, Aktar S, Rashed-Al-Mahfuz M, Uddin S, Liò P, Xu H, Summers MA, Quinn JMW and Moni MA. 2020. A machine learning model to identify early stage symptoms of SARS-Cov-2 infected patients. *Expert Syst. Appl.* **160**: 113661.
57. Jaiswal, A., N. Gianchandani, D. Singh, V. Kumar and M. Kaur (2020). Classification of the COVID-19 infected patients using DenseNet201 based deep transfer learning. *Journal of Biomolecular Structure and Dynamics*: 1–8.
58. Wang S, Zha Y, Li W, Wu Q, Li X, Niu M, Wang M, Qiu X, Li H, Yu H, Gong W, Bai Y, Li L, Zhu Y, Wang L and Tian J. 2020. A fully automatic deep learning system for COVID-19 diagnostic and prognostic analysis. *Eur. Respir. J.* **56**.
59. Li Z, Zhong Z, Li Y, Zhang T, Gao L, Jin D, Sun Y, Ye X, Yu L, Hu Z, Xiao J, Huang L and Tang Y. 2020. From community-acquired pneumonia to COVID-19: a deep learning-based method for quantitative analysis of COVID-19 on thick-section CT scans *Eur. Radiol.* doi: 10.1007/s00330-020-07042-x
60. Joshi R P, Pejaver V, Hammarlund NE, Sung H, Lee SK, Furmanchuk A 'ona, Lee H-Y, Scott G, Gombar S, Shah N, Shen S, Nassiri A, Schneider D, Ahmad FS, Liebovitz D, Kho A, Mooney S, Pinsky BA and Banaei N. 2020. A predictive tool for identification of SARS-CoV-2 PCR-negative emergency department patients using routine test results *J. Clin. Virol.* **129**: 104502.
61. Grunau G and Linn S. 2018. Detection and diagnostic overall accuracy measures of medical tests. *Rambam Maimonides Med J.* **9**
62. Marateb HR, Mansourian M and Mañanas MA. 2017. Re: STARD 2015: an updated list of essential items for reporting diagnostic accuracy studies. *BMJ*. Available at https://www.bmj.com/content/351/bmj.h5527/rr-1

63. Sokolova M, Japkowicz N and Szpakowicz S. 2006. Beyond accuracy, F-score and ROC: A family of discriminant measures for performance evaluation. In *AI 2006: Advances in Artificial Intelligence*. Springer Berlin Heidelberg, pp. 1015–21.

64. Boughorbel S, Jarray F and El-Anbari M. 2017. Optimal classifier for imbalanced data using Matthews correlation coefficient metric. *PLoS One* **12**: e0177678.

65. Marateb HR, Mohebian MR, Javanmard SH, Tavallaei AA, Tajadini MH, Heidari-Beni M, Mañanas MA, Motlagh ME, Heshmat R, Mansourian M and Kelishadi R. 2018. Prediction of dyslipidemia using gene mutations, family history of diseases and anthropometric indicators in children and adolescents: The CASPIAN-III study. *Comput. Struct. Biotechnol. J.* **16**: 121–30.

66. Sokolova M and Lapalme G. 2009. A systematic analysis of performance measures for classification tasks. *Inf. Process. Manag.* **45**: 427–37.

67. Kim H, Hong H and Yoon SH. 2020. Diagnostic performance of CT and reverse transcriptase polymerase chain reaction for coronavirus disease 2019: A meta-analysis. *Radiology* **296**: E145–55.

68. Colquhoun D. 2014. An investigation of the false discovery rate and the misinterpretation of p-values. *R. Soc. Open Sci.* **1**: 140216.

69. Ross SM. 2014. *Introduction to Probability and Statistics for Engineers and Scientists*. Elsevier Science, p. 686. doi: 10.1016/C2013-0-19397-X

70. Mohebian MR, Marateb HR, Mansourian M, Mañanas MA and Mokarian F. 2017. A hybrid computer-aided-diagnosis system for prediction of breast cancer recurrence (HPBCR) using optimized ensemble learning. *Comput. Struct. Biotechnol. J.* **15**: 75–85.

71. Pollán M, Pérez-Gómez B, Pastor-Barriuso R, Oteo J, Hernán MA, Pérez-Olmeda M, Sanmartín JL, Fernández-García A, Cruz I, Fernández de Larrea N, Molina M, Rodríguez-Cabrera F, Martín M, Merino-Amador P, León Paniagua J, Muñoz-Montalvo JF, Blanco F, Yotti R and ENE-COVID Study Group. 2020. Prevalence of SARS-CoV-2 in Spain (ENE-COVID): a nationwide, population-based seroepidemiological study. *Lancet* **396**: 535–44.

72. Leevy JL, Khoshgoftaar TM, Bauder RA and Seliya N. 2018. A survey on addressing high-class imbalance in big data. *J. Big Data* **5**: 42.

73. Saeb S, Lonini L, Jayaraman A, Mohr DC and Kording KP. 2017. The need to approximate the use-case in clinical machine learning. *Gigascience* **6**: 1–9.

74. Little MA, Varoquaux G, Saeb S, Lonini L, Jayaraman A, Mohr DC and Kording KP. 2017. Using and understanding cross-validation strategies. Perspectives on Saeb et al. *Gigascience* **6**: 1–6.

75. Ardakani AA, Kanafi AR, Acharya UR, Khadem N and Mohammadi A. 2020. Application of deep learning technique to manage COVID-19 in routine clinical practice using CT images: Results of 10 convolutional neural networks. *Comput. Biol. Med.* **121**: 103795.

76. Latouche A, Allignol A, Beyersmann J, Labopin M and Fine JP. 2013. A competing risks analysis should report results on all cause-specific hazards and cumulative incidence functions. *J. Clin. Epidemiol.* **66**: 648–53.

77. Poguntke I, Schumacher M, Beyersmann J and Wolkewitz M. 2018. Simulation shows undesirable results for competing risks analysis with time-dependent covariates for clinical outcomes. *BMC Med. Res. Methodol.* **18**: 79.

78. Ferguson NM. 2020. Report 9: *Impact of Non-pharmaceutical Interventions (NPIs) to Reduce COVID19 Mortality and Healthcare Demand* (Imperial College London). Available at https://www.imperial.ac.uk/mrc-global-infectious-disease-analysis/covid-19/report-9-impact-of-npis-on-covid-19/ (Accessed on Sep 20, 2020)

79. Zhou F, Yu T, Du R, Fan G, Liu Y, Liu Z, Xiang J, Wang Y, Song B, Gu X, Guan L, Wei Y, Li H, Wu X, Xu J, Tu S, Zhang Y, Chen H and Cao B. 2020. Clinical course and risk

factors for mortality of adult inpatients with COVID-19 in Wuhan, China: a retrospective cohort study. *Lancet* **395**: 1054–62.

80. Guan W-J, Ni Z-Y, Hu Y, Liang W-H, Ou C-Q, He J-X, Liu L, Shan H, Lei C-L, Hui D S C, Du B, Li L-J, Zeng G, Yuen K-Y, Chen R-C, Tang C-L, Wang T, Chen P-Y, Xiang J, Li S-Y, Wang J-L, Liang Z-J, Peng Y-X, Wei L, Liu Y, Hu Y-H, Peng P, Wang J-M, Liu J-Y, Chen Z, Li G, Zheng Z-J, Qiu S-Q, Luo J, Ye C-J, Zhu S-Y, Zhong N-S and China Medical Treatment Expert Group for Covid-19. 2020. Clinical characteristics of coronavirus disease 2019 in China. *N. Engl. J. Med.* **382**: 1708–20.

81. Grasselli G, Zangrillo A, Zanella A, Antonelli M, Cabrini L, Castelli A, Cereda D, Coluccello A, Foti G, Fumagalli R, Iotti G, Latronico N, Lorini L, Merler S, Natalini G, Piatti A, Ranieri MV, Scandroglio AM, Storti E, Cecconi M, Pesenti A and COVID-19 Lombardy ICU Network. 2020. Baseline characteristics and outcomes of 1591 patients infected with SARS-CoV-2 admitted to ICUs of the Lombardy region, Italy. *JAMA* **323**: 1574–81.

82. Andersen PK and Keiding N. 2002. Multi-state models for event history analysis. *Stat. Methods Med. Res.* **11**: 91–115.

83. Hazard D, Kaier K, von Cube M, Grodd M, Bugiera L, Lambert J and Wolkewitz M. 2020. Joint analysis of duration of ventilation, length of intensive care, and mortality of COVID-19 patients: a multi-state approach. *BMC Med. Res. Methodol.* **20**: 206.

84. Rieg S, von Cube M, Kalbhenn J, Utzolino S, Pernice K, Bechet L, Baur J, Lang CN, Wagner D, Wolkewitz M, Kern WV and Biever P. 2020. COVID-19 in-hospital mortality and mode of death in a dynamic and non-restricted tertiary care model in Germany. *medRxiv*. doi: 10.1101/2020.07.22.20160127

85. Roimi M, Gutman R, Somer J, Arie AB, Calman I, Bar-Lavie Y, Ziv A, Eytan D, Gorfine M and Shalit U. 2020. Predicting illness trajectory and hospital resource utilization of COVID-19 hospitalized patients-a nationwide study. *medRxiv*. doi: 10.1101/2020.09.04.20185645

86. D'Agostino RB, Lee ML, Belanger AJ, Cupples LA, Anderson K and Kannel WB. 1990. Relation of pooled logistic regression to time dependent Cox regression analysis: the Framingham heart study. *Stat. Med.* **9**: 1501–15.

87. Schnabel RB, Sullivan LM, Levy D, Pencina MJ, Massaro JM, D'Agostino RB Sr, Newton-Cheh C, Yamamoto JF, Magnani JW, Tadros TM, Kannel WB, Wang TJ, Ellinor PT, Wolf PA, Vasan RS and Benjamin EJ. 2009. Development of a risk score for atrial fibrillation (Framingham Heart Study): a community-based cohort study. *Lancet* **373**: 739–45.

88. Conroy RM, Pyörälä K, Fitzgerald AP, Sans S, Menotti A, De Backer G, De Bacquer D, Ducimetière P, Jousilahti P, Keil U, Njølstad I, Oganov RG, Thomsen T, Tunstall-Pedoe H, Tverdal A, Wedel H, Whincup P, Wilhelmsen L, Graham IM and SCORE project group. 2003. Estimation of ten-year risk of fatal cardiovascular disease in Europe: the SCORE project. *Eur. Heart J.* **24**: 987–1003.

89. Sarrafzadegan N, Hassannejad R, Marateb HR, Talaei M, Sadeghi M, Roohafza HR, Masoudkabir F, Oveisgharan S, Mansourian M, Mohebian MR and Mañanas MA. 2017. PARS risk charts: A 10-year study of risk assessment for cardiovascular diseases in Eastern Mediterranean Region. *PLoS One* **12**: e0189389.

90. Ueda P, Woodward M, Lu Y, Hajifathalian K, Al-Wotayan R, Aguilar-Salinas CA, Ahmadvand A, Azizi F, Bentham J, Cifkova R, Di Cesare M, Eriksen L, Farzadfar F, Ferguson TS, Ikeda N, Khalili D, Khang Y-H, Lanska V, León-Muñoz L, Magliano D J, Margozzini P, Msyamboza K P, Mutungi G, Oh K, Oum S, Rodríguez-Artalejo F, Rojas-Martinez R, Valdivia G, Wilks R, Shaw JE, Stevens GA, Tolstrup JS, Zhou B, Salomon JA, Ezzati M and Danaei G. 2017. Laboratory-based and office-based risk scores and charts to predict 10-year risk of cardiovascular disease in 182 countries: a

pooled analysis of prospective cohorts and health surveys. *Lancet Diabetes Endocrinol.* **5**: 196–213.

91. Nachtigall I, Lenga P, Jóźwiak K, Thürmann P, Meier-Hellmann A, Kuhlen R, Brederlau J, Bauer T, Tebbenjohanns J, Schwegmann K, Hauptmann M and Dengler J.. Clinical course and factors associated with outcomes among 1904 patients hospitalized with COVID-19 in Germany: an observational study. *Clin. Microbiol. Infect.* doi: 10.1016/j.cmi.2020.08.011 [Epub ahead of print]

92. Gerds TA, Scheike TH and Andersen PK. 2012. Absolute risk regression for competing risks: interpretation, link functions, and prediction. *Stat. Med.* **31**: 3921–30.

93. D'Agostino RB and Nam B-H. 2003. Evaluation of the performance of survival analysis models: discrimination and calibration measures. In *Handbook of Statistics*, vol. 23. Elsevier, pp. 1–25.

94. Demler OV, Paynter NP and Cook NR. 2015. Tests of calibration and goodness-of-fit in the survival setting. *Stat. Med.* **34**: 1659–80.

95. Uno H, Cai T, Pencina MJ, D'Agostino RB and Wei LJ. 2011. On the C-statistics for evaluating overall adequacy of risk prediction procedures with censored survival data. *Stat. Med.* **30**: 1105–17.

15 Computer-aided Diagnosis Methods for Non-Invasive Imaging of Sub-Skin Lesions

Ravibabu Mulaveesala, Geetika Dua, and Vanita Arora

CONTENTS

15.1 INTRODUCTION

Skin is an essential organ of the human body. One-third of an average human's body weight is constituted by the skin, and hence it is a major organ [1]. It is, naturally, responsible for many biological and inherent protection functions, such as sensory functions and metabolism [2]. Any imbalance in skin composition could lead to various skin diseases. More importantly, several skin conditions are often symptoms of serious disorders and must be examined on time for early diagnosis. Skin cancer is one of the most prominent cancers in human beings. Every year many cancer cases are reported worldwide. The incidences of skin cancer, and malignant tumors, in particular, have increased substantially over the last few decades [3]. It is a serious skin disorder that develops when errors (mutations) occur in the deoxyribonucleic acid (DNA) of the skin cells. The mutations cause uncontrollable growth of the cell formations, and hence a mass of cancer cells is formed. This uncontrollable growth leads to the formation of malignant tumors [4]. Early detection is essential to improve survival rates of patients [5]. Many new tools to diagnose skin cancer at its initial stage are being developed across the world. Each proposed technique offers some specific advantages and limitations. Therefore, thermal wave imaging (TWI), being a non-invasive, full-field, safe, remote, and economical imaging method, is used for the detection of various cancers in their initial stages, such as breast and skin cancers, as well as thyroid disorders and diabetes [6, 7].

The TWI technique has proven to be a useful method in diagnosing various diseases by measuring the temperature variations over the skin. Recording of the temporal and spatial variations in thermal profiles by detection of the infrared energy emanating from the skin helps in localization of the cancer as well as its lateral dimensions. It is well known that deviation in the normal functioning of the organs is accompanied by changes in the temperature of the body, which further affects the temperature of the skin. Functioning of the diseased tissue differs from that of healthy tissue in terms of heat generation, metabolic rate, blood perfusion, and angiogenesis. Therefore, an accurate temperature map of the skin surface provides information about the causes responsible for heat generation, in particular, the deviation from normal conditions, often caused by the presence of disease. These thermal variations can be imaged by means of thermal wave imaging in clinical diagnostic applications [8–11].

Thermal wave imaging can be either passive or active imaging [12–14]. The preferred implementation is passive, as it can be implemented at ambient conditions without the involvement of any external thermal stimulus on the skin surface. Whereas in a passive approach, the test resolution and sensitivity is not significant to visualize the sub-surface details, so external energy (active thermography) is required to create significant thermal contrast on the skin. This chapter highlights the application of an active approach, in which an external thermal stimulus is imposed on the skin for producing temperature variations. Depending on the applied thermal stimulus, active thermal wave imaging methods can be categorized as pulse-based (pulse and pulse phase) or as modulated (lock-in) thermographic methods [13, 14].

Results obtained with pulse-based thermographic techniques are affected by the emissivity variations on the skin, where illumination variations and intense heat impositions restrict its usage in detecting the presence of malignant tumor or melanoma. On the other hand, the mono frequency thermal excitation used in lock-in thermography (LT) leads to a limited depth resolution in a chosen experimentation cycle due to the single probing wavelength of the induced thermal waves inside the skin. So, in order to detect melanoma at different stages with improved detection sensitivity and resolution using LT, it demands repetitive experimentation at various frequencies which is a time-consuming method. Thus, the requirement is to probe thermal waves within a desired range of frequencies with equal energy inside the object in a test cycle using a comparatively moderate intense heat input. In order to attain the same, an active non-periodic thermographic technique called linear frequency modulated thermal wave imaging (LFMTWI) is presented. The suggested methodology overcomes the demerits associated with traditional pulse-based and lock-in thermal imaging techniques (resolution, limited depth of penetration, and high peak-power requirements).

In LFMTWI, a linear chirp modulated thermal excitation with a suitable band of frequencies having the same energies is applied to the skin. This imposed heat stimulus generates diffusion waves in the skin and causes a similar thermal distribution over it. A tumor inside the skin perturbs the thermal waves, which causes thermal gradients over it. The resulting thermal response is recorded and further processed using frequency and time-domain approaches. Results obtained from phase and pulse compressed images are used to differentiate the regions of healthy and diseased tissues (at different stages) present inside the skin.

15.2 THEORY

In an LFMTWI approach, the test sample is illuminated with modulated heat sources with a predefined range of frequencies for a particular experimental duration. This imposed stimulus generates diffusion waves in the sample, thereby producing temperature gradients over the sample due to the presence of sub-surface anomalies.

The theoretical 1D heat diffusion model for studying the temperature response of the sample is given as:

$$\frac{\partial^2 T(z,t)}{\partial z^2} - \frac{1}{\alpha_s}\frac{\partial T(z,t)}{\partial t} = 0 \tag{15.1}$$

where $T(z, t)$ depicts the thermal gradient over the sample at a location z at a time instant t second and α_s is the thermal diffusivity of the test specimen given by:

$$\alpha_s = \frac{K_s}{\rho_s C_s} \tag{15.2}$$

where K_s is the thermal conductivity (W/m-K), ρ_s is the density (kg/m³), and C_s is the specific heat (J/kg-K). Linear frequency modulated (LFM) incident heat flux, as shown in Figure 15.1, is expressed as:

$$Q(z=0,t) = Qo * \left[1 + \sin\left(2\pi\left((f_a t) + \left(\frac{B}{2T}t^2\right) \right) \right) \right] \tag{15.3}$$

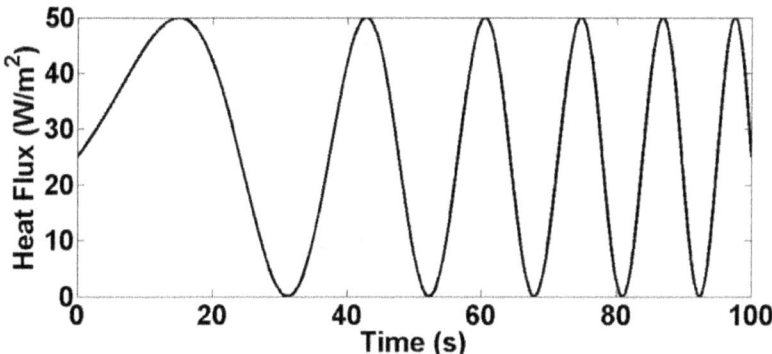

FIGURE 15.1   Incident linear frequency modulated heat flux.

where Q_o is the peak intensity, f_a is the initial frequency, B is the bandwidth, and T is the duration for which the heat flux is incident over the sample. The band of frequencies imposed over the sample is chosen based on its thermo-physical properties and the thickness of the test sample. Figure 15.1 represents the LFM heat flux with a peak intensity of 50 W/m² and a sweeping range of 0.01–0.1 Hz for 100 sec.

For a finite thickness sample, the time-domain boundary value problem is expressed as follows:

$$T_z(z=0,t)=Q(t) \tag{15.4}$$

$$T_z(z=L,t)=0 \tag{15.5}$$

$$T(z,t=0)=To \tag{15.6}$$

where $Q(t)$ is the incident chirp modulated flux, L is the thickness of the test sample, and T_0 is the initial temperature of the test sample.

The thermal diffusion length for LFMTWI can be determined by solving the heat diffusion equation with radiative and conductive boundary conditions. This allows depth scanning in a single test cycle. The test resolution can be further enhanced by increasing the probing bandwidth of the diffusion waves into the sample.

The thermal diffusion length μ_z for LFMTWI is given as:

$$\mu_z = \sqrt{\frac{\alpha}{\pi * \left(f_a + \dfrac{B*t}{T}\right)}} \tag{15.7}$$

where B is the bandwidth given by; $B = f_b - f_a$ and f_a, and f_b are the initial and final frequency of LFM.

The Fourier transform of the heat diffusion Equation (15.1) is represented as:

$$\frac{\partial^2 T(z,\omega)}{\partial z^2} = \frac{i\omega}{\alpha_s} T(z,\omega) \tag{15.8}$$

$$\frac{\partial^2 T(z,\omega)}{\partial z^2} - \sigma^2 T(z,\omega) = 0 \tag{15.9}$$

$$\text{where;} \quad \sigma = \sqrt{\frac{i\omega}{\alpha_s}} = (1+i)\sqrt{\frac{\omega}{2\alpha_s}} \tag{15.10}$$

The frequency-domain boundary value problem can be obtained by applying the infinite integral Fourier transform to the time-domain boundary problem given in Equations (15.4)–(15.6):

Boundary condition at $z = 0$

$$-K_s \frac{\partial T(0,\omega)}{\partial z} = \Im\{Q(t)\} = Q(\omega) \tag{15.11}$$

Boundary condition at $z = L$

$$-K_s \frac{\partial T(L,\omega)}{\partial z} = 0 \tag{15.12}$$

Initial condition at $t = 0$

$$T(z,\omega) = T_0 \tag{15.13}$$

The auxiliary equation can be obtained by putting $\dfrac{\partial}{\partial z} = D$ into Equation (15.9):

$$D^2 T(z,\omega) - \sigma^2 T(z,\omega) = 0 \tag{15.14}$$

$$T(z,\omega)(D^2 - \sigma^2) = 0 \tag{15.15}$$

$$(D^2 - \sigma^2) = 0 \tag{15.16}$$

The roots for Equation (15.16) can be given as follows:

$$D = \pm\sigma \tag{15.17}$$

Therefore, the general solution for the heat diffusion equation can be given as:

$$T(z,\omega) = T_1(\omega,L)e^{\epsilon z} + T_2(\omega,L)e^{-\epsilon z} \tag{15.18}$$

Where $T(\omega, L)$ is the thermal wave in the frequency domain, T_0 is the initial condition, $\epsilon = (1+i)/\mu_z$ is the complex wave number, and $T_1(\omega, L)$ and $T_2(\omega, L)$ are constant values obtained by applying the boundary conditions defined in the Equations (15.11) and (15.12), respectively.

At boundary condition $z = 0$:

$$-K_s \frac{\partial}{\partial z}\left(T_1(\omega,L)e^{+\epsilon z} + T_2(\omega,L)e^{-\epsilon z}\right)\bigg|_{z=0} = Q(\omega) \tag{15.19}$$

$$-K_s\epsilon T_1(\omega,L)e^{\epsilon z} + K_s\epsilon T_2(\omega,L)e^{-\epsilon z}\big|_{z=0} = Q(\omega) \tag{15.20}$$

$$-K_s\epsilon T_1(\omega,L) + K_s\epsilon T_2(\omega,L) = Q(\omega) \tag{15.21}$$

$$-K_s\epsilon\{T_1(\omega,L) - T_2(\omega,L)\} = Q(\omega) \tag{15.22}$$

$$T_1(\omega,L) - T_2(\omega,L) = -\frac{Q(\omega)}{K_s\epsilon} \tag{15.23}$$

Similarly, at boundary condition $z = L$, we get:

$$-K_s \frac{\partial}{\partial z}\left(T_1(\omega,L)e^{+\epsilon z} + T_2(\omega,L)e^{-\epsilon z}\right)\Bigg|_{z=L} = 0 \tag{15.24}$$

$$-K_s\epsilon T_1(\omega,L)e^{\epsilon z} + K_s\epsilon T_2(\omega,L)e^{-\epsilon z}\Big|_{z=L} = 0 \tag{15.25}$$

$$K_s\epsilon T_1(\omega,L)e^{\epsilon z} = K_s\epsilon T_2(\omega,L)e^{-\epsilon z} \tag{15.26}$$

Therefore we get:

$$T_1(\omega,L) = T_2(\omega,L)e^{-2\epsilon z} \tag{15.27}$$

Substituting Equation (15.27) in Equation (15.23) we obtain:

$$T_2(\omega,L)e^{-2\epsilon z} - T_2(\omega,L) = -\frac{Q(\omega)}{K_s\epsilon} \tag{15.28}$$

$$T_2(\omega,L)\left(e^{-2\epsilon z} - 1\right) = -\frac{Q(\omega)}{K_s\epsilon} \tag{15.29}$$

$$T_2(\omega,L) = \frac{Q(\omega)}{K_s\epsilon} \times \frac{1}{1 - e^{-2\epsilon L}} \tag{15.30}$$

and

$$T_1(\omega,L) = \frac{Q(\omega)}{K_s\epsilon} \times \frac{e^{-2\epsilon L}}{1 - e^{-2\epsilon L}} \tag{15.31}$$

The thermal response for a test sample illuminated by a time-dependent heat stimulus can be calculated with the concept of Planck radiation. Therefore, the output detector signal can be calculated in the form:

$$M(t) = \mathfrak{J}^{-1}\left[r\int_0^L \{T(z,\omega) * e^{-zr}\}\,dz\right] \tag{15.32}$$

where $\mathfrak{J}^{-1}$ defines the inverse Fourier transform of the signal and r is the thermal absorption coefficient.

$$M(t) = \mathfrak{J}^{-1}\left[r\int_0^L \{T_1(\omega,L)e^{\epsilon z} + T_2(\omega,L)e^{-\epsilon z}\} * e^{-zr}\,dz\right] \tag{15.33}$$

$$M(t) = \mathfrak{J}^{-1}\left[\left\{rT_1(\omega,L)\int_0^L \left(e^{\epsilon z} * e^{-zr}\right)dz\right\} + \left\{rT_2(\omega,L)\int_0^L \left(e^{-\epsilon z} * e^{-zr}\right)dz\right\}\right] \tag{15.34}$$

$$M(t) = \mathfrak{I}^{-1}\left[\left\{rT_1(\omega,L)\int_0^L\left(e^{(\epsilon-r)z}\right)dz\right\} + \left\{rT_2(\omega,L)\int_0^L\left(e^{-(\epsilon+r)z}\right)dz\right\}\right] \quad (15.35)$$

$$M(t) = \mathfrak{I}^{-1}\left[\left\{T_1(\omega,L)\frac{r}{(\epsilon-r)}\left(e^{(\epsilon-r)L}-1\right)\right\} + \left\{T_2(\omega,L)\frac{-r}{(\epsilon+r)}\left(e^{-(\epsilon+r)L}-1\right)\right\}\right]$$

$$(15.36)$$

$$M(t) = \mathfrak{I}^{-1}\left[\left\{\frac{T_1(\omega,L)}{\frac{\epsilon}{r}-1}\left(e^{(\epsilon-r)L}-1\right)\right\} + \left\{\frac{T_2(\omega,L)}{\frac{\epsilon}{r}+1}\left(1-e^{-(\epsilon+r)L}\right)\right\}\right] \quad (15.37)$$

Using the concept of photo-thermal saturation, this thermal absorption coefficient can be considered infinite for an optically opaque solid. Therefore, the signal obtained of a test sample of thickness L at the infrared detector can be calculated in the form of:

Putting $r\to\infty$, we get

$$M(t) = \mathfrak{I}^{-1}\left[T_1(\omega,L) + T_2(\omega,L)\right] \quad (15.38)$$

15.2.1 MATCHED FILTERING-BASED PULSE COMPRESSION

Matched filtering is a well-known linear filter for maximizing the signal to noise ratio in the presence of random noise. Pulse compression achieved with a correlation-based matched filtering approach will enhance the depth resolution even with the use of low or medium peak-power thermal excitation. This principle allows for the use of low intensity, moderate duration signals during the experimentation, but the depth resolution and sensitivity can be expected to be similar to that of short duration, high peak-power signals, even in the presence of noise to provide a high detection range and better resolution.

In general, convolution-based pulse compression approaches are more familiar, as illustrated in Figure 15.2. Let $M(t)$ be a linear frequency modulated up-chirp (whose frequencies decreases with time) thermal response; then its matched filter will have a linear down-chirp whose frequencies decreases with time $M(-t) = H(t)$. When the signal $M(t)$ passes through a matched filter, the output $G(t)$ obtained will be a narrow duration and high peak compressed pulse (sinc signal), and this can be represented as the convolution of $M(t)$ with $H(t)$ as follows:

$$G(t) = \int_{-\infty}^{+\infty} M(\tau)H(t-\tau)d\tau \quad (15.39)$$

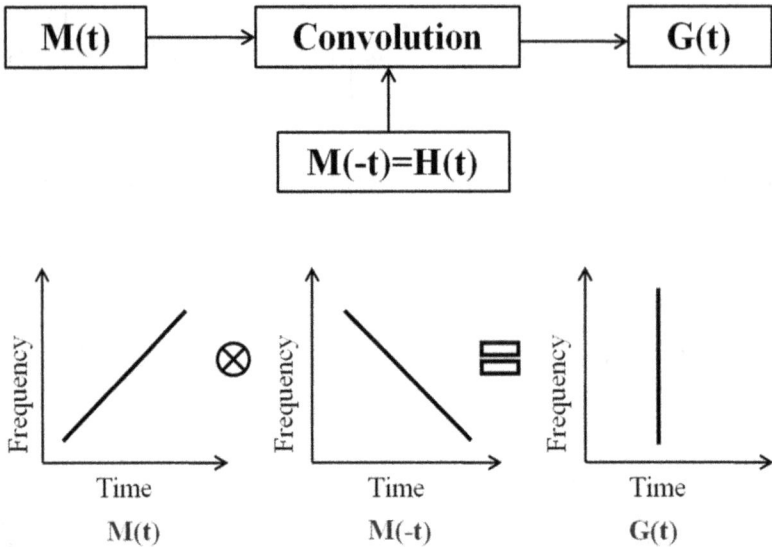

FIGURE 15.2(A) Principle of the convolution-based pulse compression between two signals.

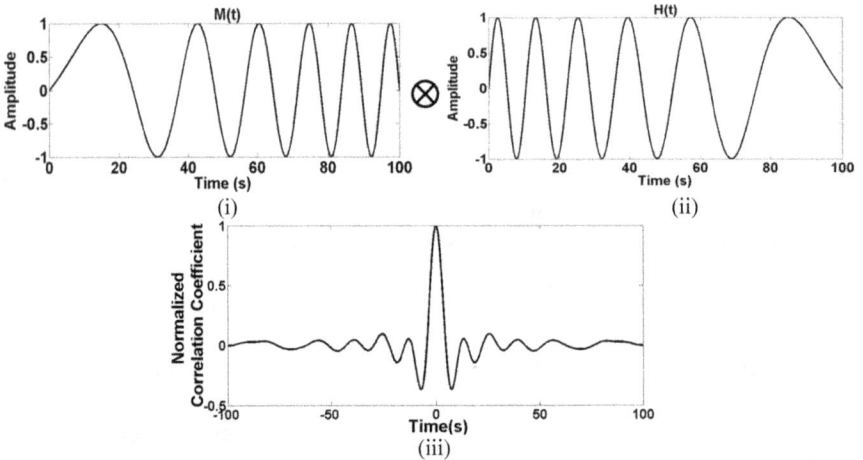

FIGURE 15.2(B) Principle of the convolution-based matched filter for pulse compression. (i) Input linear up-chirp (frequency modulated) signal, (ii) Its matched filter response, (iii) Compressed pulse.

Figure 15.3(a) illustrates the principle of pulse compression using correlation as follows.

In terms of thermal waves, the pulse compression can be performed by correlating the observed thermal response of the abnormal regions with the healthy (normal) region. This facilitates the accumulation of the total supplied energy in the main lobe resulting in a sinc shaped compressed pulse, as shown in Figure 15.3(b).

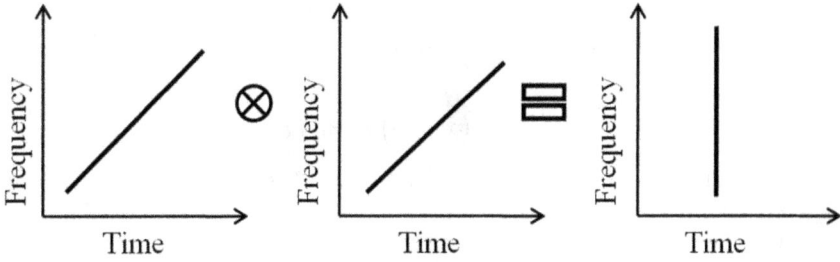

FIGURE 15.3(A) Principle of pulse compression between two signals.

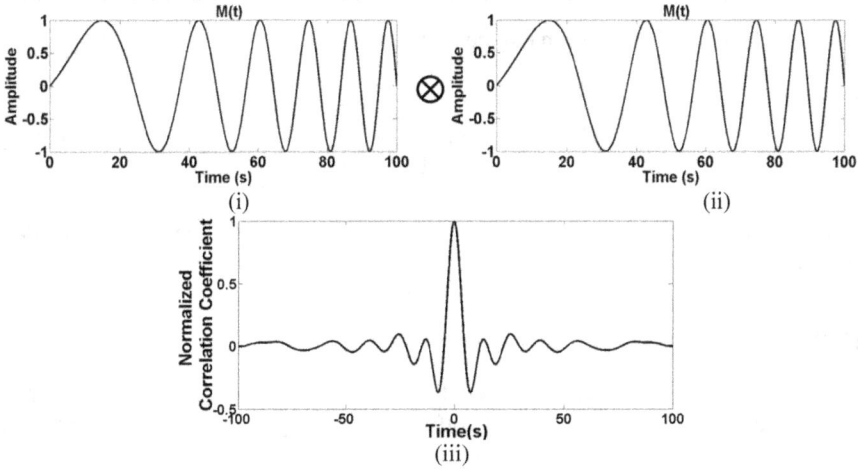

FIGURE 15.3(B). Pulse compression based cross-correlation operation. (i) Input linear frequency modulated signal, (ii) Its matched filter response (without time inversion), (iii) Compressed pulse.

In order to improve the resolution and sensitivity for visualization of sub-surface features, appropriate post-processing on the resultant mean zero thermal distribution data is essential, especially if an anomaly lies deep inside the test specimen with a smaller lateral dimension. Various post-processing schemes have been developed in order to improve detection resolution and sensitivity. In this chapter, phase and correlation coefficient thermograms are formed using frequency and time-domain favorable processing approaches.

15.2.2 FREQUENCY-DOMAIN ANALYSIS APPROACH

In a frequency-domain analysis, the captured thermograms are analyzed using the discrete Fourier transform (DFT) to extract the phase and magnitude from its reconstructed zero mean transformed thermal sequence $(T_{zeromean} (x, y, t))$. DFT is applied to each temporal sequence $T_{zeromean} (x, y, t)$ in the field of view using a well-known formula [13, 14]:

$$T_f(x,y,k) = \sum_{n=0}^{N-1} T_{zeromean}(x,y,t) e^{-\frac{j2\pi kn}{N}}$$

(15.40)

$$= Re\big(T_f(x,y,k)\big) + jImg\big(T_f(x,y,k)\big)$$

where the bin number is k, the number of frames are N, Re and Img are the real and imaginary parts.

Further, the magnitude can be reconstructed as follows:

$$\big|T(x,y,k)\big| = \sqrt{\big(Re\big(T_f(x,y,k)\big)\big)^2 + \big(Img\big(T_f(x,y,k)\big)\big)^2}$$

(15.41)

Whereas the phase information can be retrieved as follows:

$$\angle T_f(x,y,k) = \tan^{-1}\left(\frac{Img\big(T_f(x,y,k)\big)}{Re\big(T_f(x,y,k)\big)}\right)$$

(15.43)

Figure 15.4 illustrates the steps involved to compute the frequency-domain data processing.

15.2.3 Time-Domain Analysis Approach

The time-domain-based correlation coefficient (CCC) image is reconstructed from the circular convolution computed for the selected reference signal and with zero mean thermal data given as follows (Figure 15.5):

$$CCC(\tau) = \Im^{-1}\big\{Ref(x,y,k)^* \cdot T(x,y,k)\big\}$$

(15.44)

where $Ref(x, y, k)$ and $T(x, y, k)$ are the Fourier transforms of the reference thermal response $ref(t)$ and zero mean temperature response $T_{zeromean}(x, y, t)$, respectively. $\Im^{-1}$, and * denote the inverse Fourier transform, multiplication, and complex conjugate operators, respectively. Time-domain analysis helps to compress the imposed thermal energy into a narrow duration correlation peak which enhances the depth resolution of the imaging as shown in Figure 15.4. This not only helps in reducing the width of the main lobe but also enhances the amplitude of the CCC peak. However, a time-domain based phase image is formed from the circular convolution of the in-phase and quadrature-phase as below:

$$\varphi(\tau) = \tan^{-1}\left\{\frac{\Im^{-1}\big\{\big[-isgn(x,y,k)Ref(x,y,k)\big]^* T(x,y,k)\big\}}{\Im^{-1}\big\{Ref(x,y,k)^* T(x,y,k)\big\}}\right\}$$

(15.45)

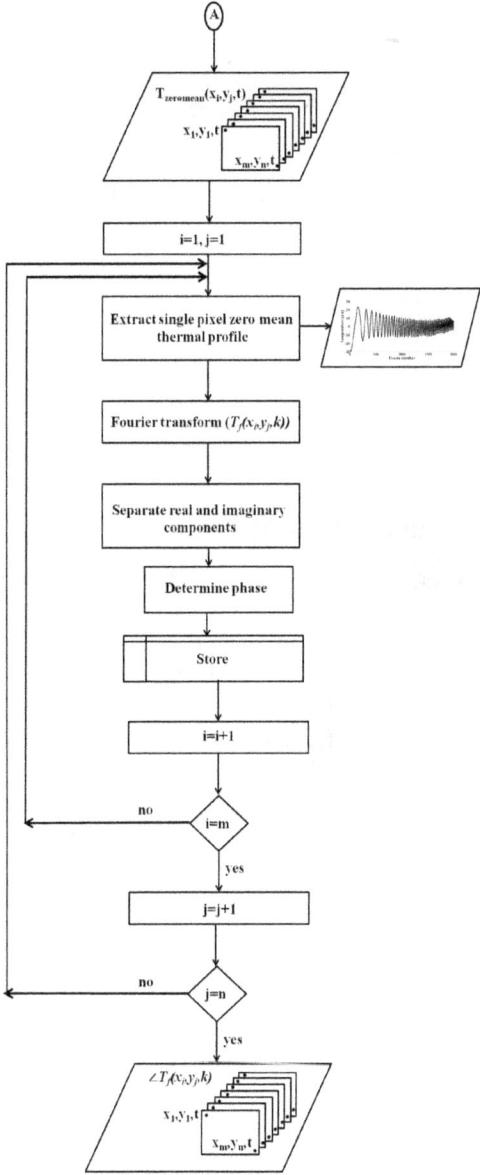

FIGURE 15.4 Flowchart illustrating the frequency-domain analysis approach.

where $sgn(x, y, k)$ is the signum function. The term inside the squared bracket is the frequency response of the quadrature reference signal. The quadrature of the reference temperature response is obtained using the Hilbert transform (HT), and the frequency response is obtained through the discrete Fourier transform.

A flowchart illustrating the steps involved in computing the time-domain analysis is shown in Figure 15.6.

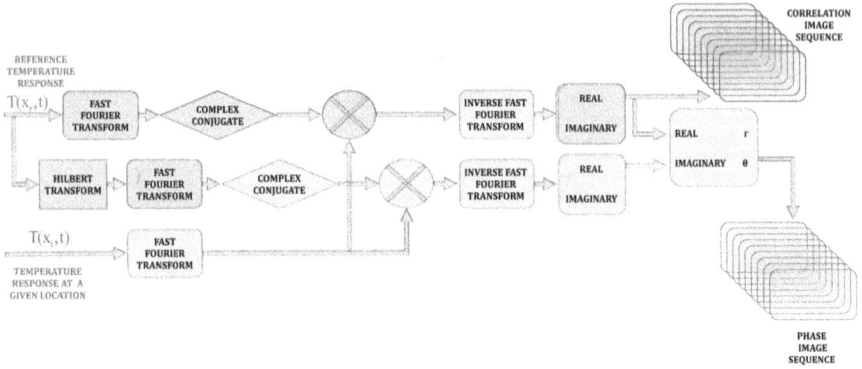

FIGURE 15.5 Illustrating the adopted data processing for the reconstruction of the time-domain CCC and phase images.

15.3 MODELING AND SIMULATION

This present work incorporates numerical modeling of skin tissue using a finite element method (FEM). The layout of the modeled skin tissue sample is shown in Figure 15.7.

A model for the skin tissue sample is considered as a 24 mm x 5 mm x 2 mm cube with 0.06 mm thick epidermis, 0.015 mm thick basal melanin layer, and 1.94 mm thick dermis. Tumors a, b, c, d, e, f, g, and h, underneath the skin surface at a depth of 1.65 mm, are considered as regular ellipsoids with the size of $x = 1$ mm, $y = 0.8$ mm, and z is varying as 0.5 mm, 0.4 mm, 0.3 mm, 0.2 mm, 0.1 mm, 0.05 mm, 0.025 mm, and 0.0125 mm for each tumor, respectively; here, x, y, and z are axial dimensions of the ellipsoid, respectively [10].

Finite element modeling is considered for the skin tissue by imposing a heat flux, as shown in Figure 15.8, with a sweeping range of 0.01 Hz to 0.1 Hz.

The proposed excitation signal $Q(x,t')$ with bandwidth B, is as follows:

$$Q\left(x=0,t'\right)=Q_0\,e^{2\pi j\left(f_0 t'+\frac{Bt'^2}{2\tau}\right)} \tag{15.46}$$

where Q_0 is the amplitude, τ is the duration, f_0 is the initial frequency, and $2\pi j\left(f_0 t' + \dfrac{Bt'^2}{2\tau}\right)$ is the phase.

Pennes bio-heat transfer equation is used to calculate temperature distribution over the skin tissue:

$$\rho \cdot c \frac{\partial T}{\partial t'} = \nabla\left(k'\cdot\nabla T\right)+\omega_b\cdot c_b\cdot\rho_b\left(T_a - T\right)+Q_m \tag{15.47}$$

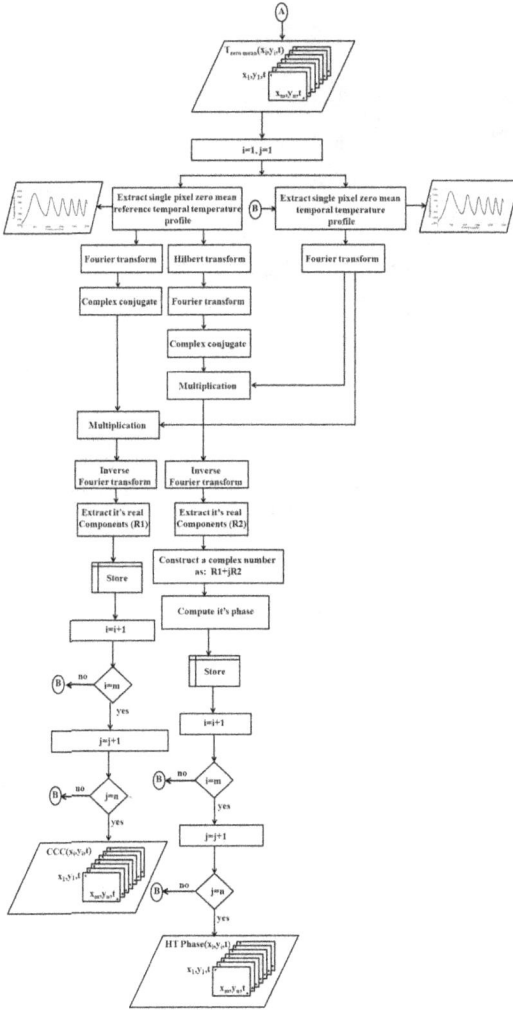

FIGURE 15.6 Flowchart illustrating a time-domain analysis approach.

where ρ_b is the blood density, c_b is the specific heat, ω_b is the blood perfusion rate, T is the temperature of the skin tissue, T_a is the arterial blood temperature, and Q_m is the metabolic heat generation rate, respectively. The thermo-physical properties of the various skin tissue layers and tumors are given in Table 15.1. The density of the blood and specific heat of the blood are taken as 1055 kg/m³ and 3660 J/kg.K [11, 12]. The temperature of the arterial blood is chosen as 310.15 K.

Simulations were carried out under adiabatic boundary conditions, by considering the sample at a temperature of 310.15 K. The corresponding thermal distribution was captured over the sample at 25 Hz frame rate. A signal to noise ratio (SNR) of

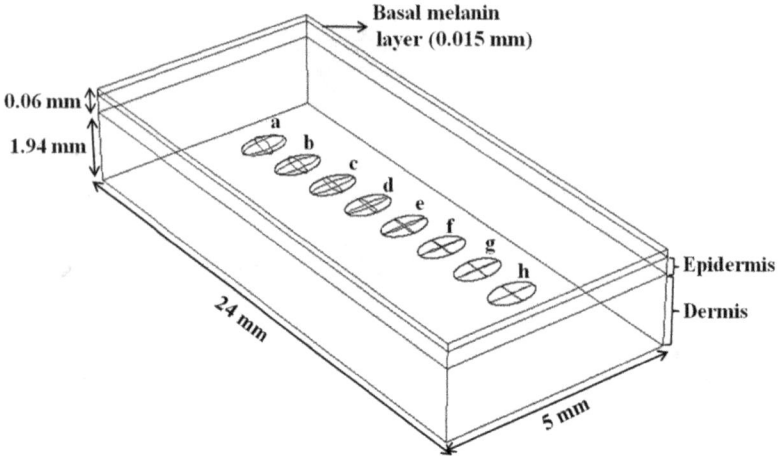

FIGURE 15.7   Layout of the modeled skin sample.

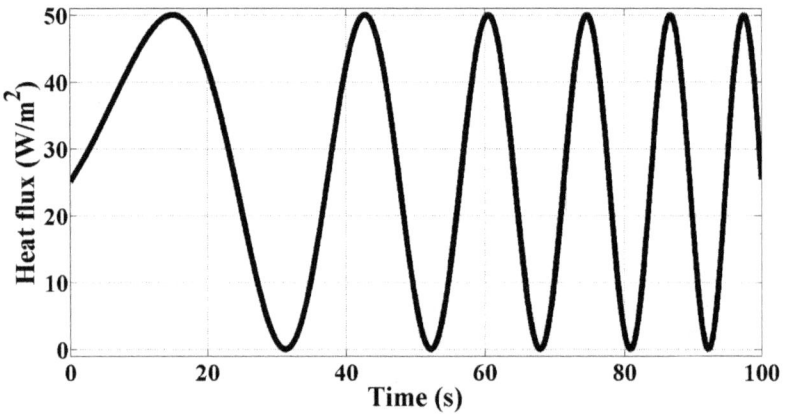

FIGURE 15.8   Illustrates a linear frequency modulated (LFM) heat flux.

50 dB was introduced with the addition of additive white Gaussian noise (AWGN), so that the proposed method could be applied in real-time situations.

15.4   RESULTS AND DISCUSSION

In this section, computational simulation results are presented and discussed. The obtained noisy raw thermogram is shown in Figure 15.9.

The obtained noisy raw thermal data was further pre-processed to reconstruct mean zero thermal sequences. The corresponding mean removed the noisy thermogram and this is as shown in Figure 15.10.

Further, to improve the detection capabilities of the proposed approach, fitted noisy thermal data was processed using frequency- and time-domain approaches.

TABLE 15.1
Properties of the Skin Tissue Layers and Tumor

Properties	Epidermis	Basal melanin layer	Dermis	Tumor (a–h)
Mass density ρ (kg/m³)	1200	1200	1200	1030
Heat capacity c (J/kg °C)	3300	3300	3300	3852
Thermal conductivity k' (W/m °C)	0.445	0.445	0.445	0.558
Blood perfusion rate ω_b (ml/s/ml)	1.25×10^{-3}	1.25×10^{-3}	1.25×10^{-3}	0.0063
Metabolic heat production Q_m (W/m³)	368.1	368.1	368.1	3680

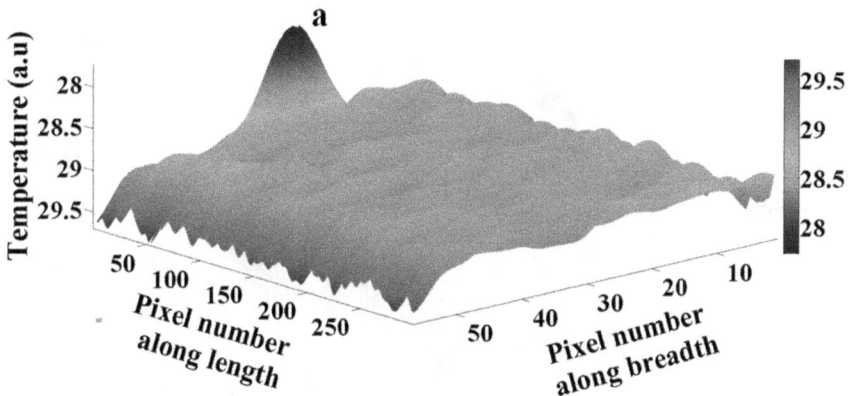

FIGURE 15.9 Noisy raw thermogram.

The resultant images are illustrated in Figure 15.11 and Figure 15.12, respectively. Figure 15.11(a) depicts the resultant magnitude image obtained at 0.03 Hz, and Figure 15.11(b) shows the phasegram at a frequency of 0.04 Hz. Figure 15.12(a) illustrates the time phasegram obtained at 99.28 sec, and Figure 15.12(b) shows the time-domain CCC thermogram at 57.4 sec.

Furthermore, for having a quantitative comparison among the applied post-processing techniques, SNR was considered as a figure of merit. For each tumor, the SNR values computed and compared the frequency and time-domain adopted post-processing approaches. The SNR values are computed using the formula:

$$SNR = 20\log\left(\frac{\text{mean at the centre of tumour region} - \text{mean at the sound skin region}}{\text{standard deviation of sound skin region}}\right)$$

(15.48)

The calculated values are tabulated in Table 15.2.

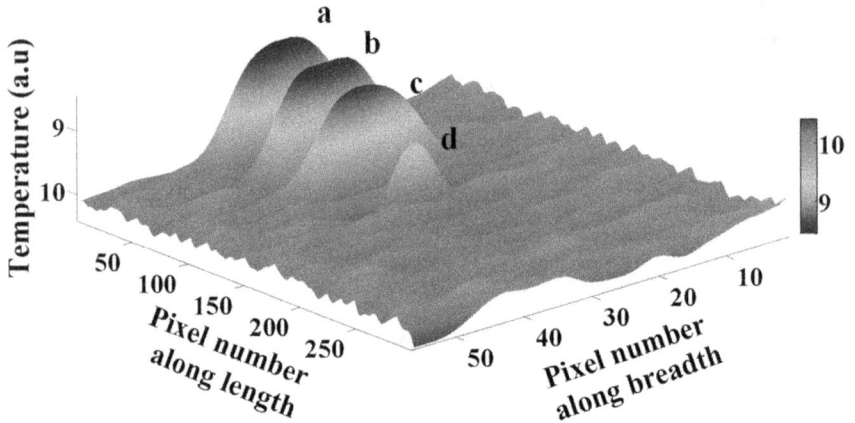

FIGURE 15.10 Fitted noisy thermal image.

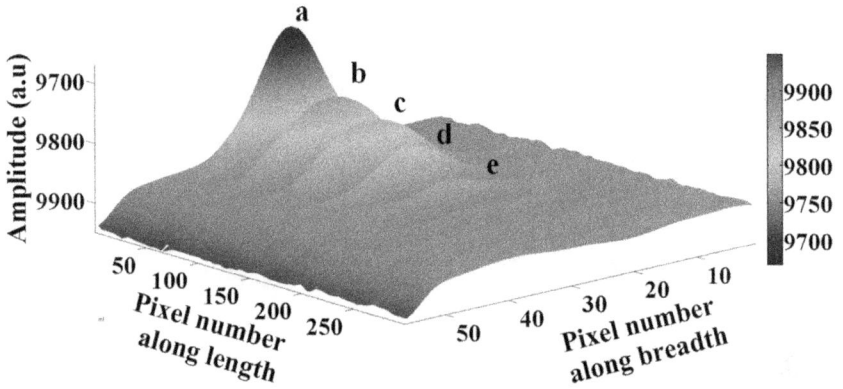

FIGURE 15.11(A) Magnitude thermogram obtained at a frequency of 0.03 Hz.

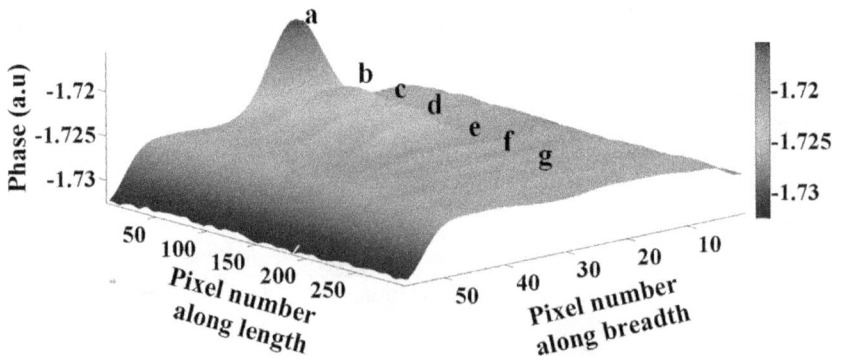

FIGURE 15.11(B) Phase thermogram obtained at a frequency of 0.05 Hz.

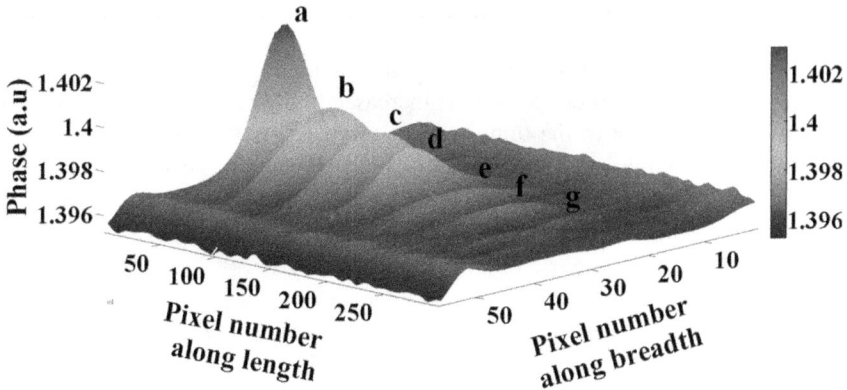

FIGURE 15.12(A) Time phasegram obtained at a time instant of 95.24 s.

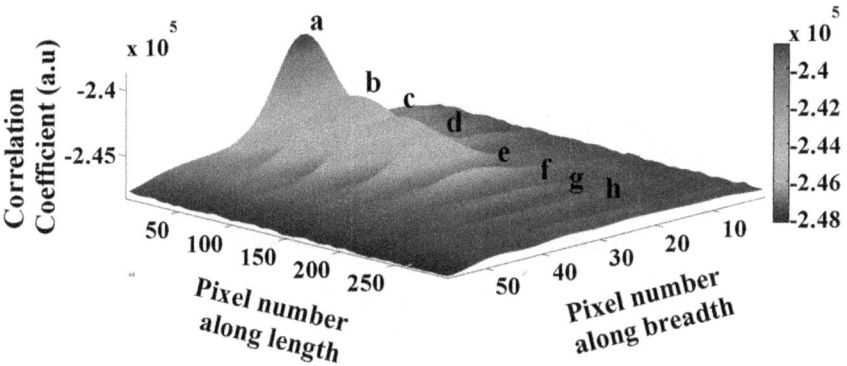

FIGURE 15.12(B) Correlation coefficient thermogram obtained at a time instant of 61.12 s.

TABLE 15.2
Illustrates the SNR (in dB)

Tumor	Fitted noisy	Frequency-domain magnitude	Frequency-domain phase	Time-domain phase (Hilbert transform (HT) phase)	Correlation coefficient (CCC)
a	10.2763	39.5609	39.5121	49.4888	50.5262
b	7.1325	37.6443	35.0892	45.8121	49.2198
c	7.9014	36.1529	32.9035	42.4951	44.9291
d	1.5613	32.6073	31.5959	38.6956	40.7620
e	−1.5401	25.9286	22.4892	28.0518	32.9571
f	−16.0746	17.5448	16.6733	20.5585	25.3470
g	−23.1394	8.2062	15.6407	25.6297	21.9213
h	−24.1125	−0.5134	15.4592	9.4410	15.5490

15.5 CONCLUSION

The results based on the time and frequency domain highlight that a malignant tumor with a higher metabolism rate as well as increased blood perfusion leads to slightly higher thermal contrast in the tumor region compared to the healthy skin region. Further, the detection capabilities of the proposed multi-transform techniques, especially the matched-filter-based pulse compression processing, exhibit superior detection performance compared to that of the traditional frequency-domain-based analysis schemes.

ACKNOWLEDGMENTS

This work was supported financially by the Global Innovation & Technology Alliance (GITA) from the project entitled "The Development of a Portable THERMOgraphy-based Health DeTECTion System (THERMOTECT) in breast cancer screening" with reference number 2016UK0202022 |IN - UK RFP 2016.

REFERENCES

1. Kanitakis, J. 2002. Anatomy, histology and immunohistochemistry of normal human skin. *Eur J Dermatol.* 12(4): 390–399.
2. Nouri, K. 2008. *Skin Cancer.* McGraw Hill, New York.
3. Ruddon, R. W. 2007. *Cancer Biology.* Oxford University Press, Inc., Oxford.
4. Marnett, L. J. and Plastaras, J. P. 2001. Endogenous DNA damage and mutation. *Trends Genet.* 17: 214–221.
5. Wartman, D. and Weinstock, M. 2008. Are we overemphasizing sun avoidance in protection from melanoma? *Cancer Epidemiol Biomarkers Prev.* 17: 469–470.
6. Jones, B. F. 1998. A reappraisal of the use of infrared thermal image analysis in medicine. *IEEE Trans Med Imag.* 17(6): 1019–1027.
7. Jones, B. F. and Plassmann, P. 2002. Digital infrared thermal imaging of human skin. *IEEE Eng Med Biol.* 21: 41–48.
8. Pirtini Cetingul, M. and Herman, C. 2011. Quantification of the thermal signature of a melanoma lesion. *Int J Therm Sci.* 50: 421–431.
9. Pirtini Çetingül, M. and Herman, C. 2010. A heat transfer model of skin tissue for the detection of lesions: Sensitivity analysis. *Phys Med Biol.* 55(19): 5933–5951.
10. Cheong, W. -F., Prahl, S. A. and Welch, A. J. A. 1990. Review of the optical properties of biological tissues. *IEEE J Quantum Electron.* 26(12): 2166–2185.
11. He, B. -H., Wei, H. -J., Chen, X. -M. and Wang, J. 2008. Properties of autofluorescence, absorption coefficient and scattering coefficient spectra for human gastric adenocarcinoma tissues. *World Chin J Digestol.* 16(15): 1692–1695.
12. Vavilov, V., Maldague, X., Dufort, B., Robitaille, F. and Picard, J. 1993. Thermal non-destructive testing of carbon epoxy composites: detailed analysis and data processing. *NDT E Int.* 26(2): 85–95.
13. Dua, G. and Mulaveesala, R. 2017. Infrared thermography for detection and evaluation of bone density variations by non-stationary thermal wave imaging. *Biomed Phys Eng Express* 3(1): 017006.
14. Mulaveesala, R. and Dua, G. 2016. Non-invasive and non-ionizing depth resolved infrared imaging for detection and evaluation of breast cancer: A numerical study. *Biomed Phys Eng Express* 2(5): 055004.

Index

For Product Safety Concerns and Information please contact our EU
representative GPSR@taylorandfrancis.com
Taylor & Francis Verlag GmbH, Kaufingerstraße 24, 80331 München, Germany

www.ingramcontent.com/pod-product-compliance
Lightning Source LLC
Chambersburg PA
CBHW060755220326
41598CB00022B/2437

* 9 7 8 0 3 6 7 6 3 8 8 4 9 *